The Science of Ethanol

The Science of Ethanol

Walter E. Goldstein

CRC Press
Taylor & Francis Group
Boca Raton London New York

CRC Press is an imprint of the
Taylor & Francis Group, an **informa** business

CRC Press
Taylor & Francis Group
6000 Broken Sound Parkway NW, Suite 300
Boca Raton, FL 33487-2742

Printed on acid-free paper
Version Date: 20161024

International Standard Book Number-13: 978-1-4987-2615-3 (Hardback)

Library of Congress Cataloging-in-Publication Data

Names: Goldstein, Walter E., author.
Title: The science of ethanol / Walter E. Goldstein.
Description: Boca Raton : CRC Press, 2017.
Identifiers: LCCN 2016028051| ISBN 9781498726153 (hardback : acid-free paper) | ISBN 9781498726160
Subjects: LCSH: Ethanol as fuel. | Ethanol.
Classification: LCC TP339 .G656 2017 | DDC 662.6/692--dc23
LC record available at https://lccn.loc.gov/2016028051

Visit the Taylor & Francis Web site at
http://www.taylorandfrancis.com

and the CRC Press Web site at
http://www.crcpress.com

This book is dedicated to those who persevere in developing valuable energy sources undeterred by opposition and challenges.

To my wife, Paula, for her love, support, and editing suggestions.

To our wonderful children, Susan and Marc, and grandchildren, Alex, Noah, Reena, and Daniel.

Contents

Preface

The Science of Ethanol is directed toward helping the ethanol industry to be successful. Success means the industry will have greater sales and profits allowing it to flourish. Success also means that in achieving business goals, the companies in this industry will benefit the common good. The companies in the industry (and their associations) have to help make the energy supply more plentiful and less costly. The industry must be a force to improve the environment and also contribute to improving health and well-being. These objectives are very noble, worthwhile, and make good sense. The objectives can be addressed in very specific and practical ways that are a consequence of direct and indirect actions of the ethanol industry.

Conversely, the industry and its companies can take other paths that seem convenient, logical, and maybe save money in the short term. However, some of these paths are fraught with negative consequences that can lead to the demise of the industry.

With this in mind, this book addresses the subject by focusing on aspects that are believed to be most important. This book emphasizes the biological approach to producing ethanol. The following benefits of ethanol are noted:

- Ethanol produced by biological means is commercially viable and profitable (or expected to be profitable) for the parties involved (including farmers and producers)
- Biologically formed ethanol is another valued entity in the mix of viable energy alternatives, any of which can apply depending on specific circumstances
- Ethanol therefore protects our nation (and any nation) by offering an alternative that is based on technology development for situations where foreign influences threaten our automobile energy supply as they did once before (in the mid-1970s).
- The ethanol industry is *de facto* a valued partner of the petroleum industry by helping to preserve our energy supply that provides fuel for automobiles.
- The ethanol industry can be a primary pathfinder, seeking ways via research to show how fuel blends can be more efficient.
- The industry can serve as a primary resource to lead research to improve the environment based on its actions and the products it can develop and refine.
- The ethanol industry can provide ways to introduce health benefits to the animal feed industry (and perhaps indirectly to the human food supply) by improving an important by-product of manufacture (Distillers Dry Grains).

Chapter 1 introduces the science of ethanol and objectives in writing this book.

Chapter 2 focuses on best ways to develop, understand, and use yeast to manufacture bulk ethanol. Yeast is emphasized since there are not sufficient advantages presently to consider other organisms that may be replacement microbes in a commercial setting. This can, of course, change since research often changes things dramatically.

The organism is the key aspect to developing a process that can be successfully scaled up and employed. Currently, and for some time to come, this involves an ever-improving yeast organism. Laboratory and pilot plant research, and the practical process development work that is so important in research, pilot plants, and manufacturing, will continue to improve yeast and the ethanol process since there are variable nuances specific to each facility site.

Chapter 3 focuses on use of the ethanol/gasoline fuel blend in internal combustion engines. This topic emphasizes research and development to make ethanol a better and more efficient fuel. Ethanol's energy density per kilogram is less than that of gasoline. Therefore, it is necessary to show how ethanol use is beneficial to counter this disadvantage. This chapter suggests ways to make ethanol beneficial in a number of ways. One way is to effect friction reduction in the automobile power plant, either directly or indirectly as a consequence of ethanol usage in the blend with gasoline. Such work will be preferably co-sponsored by the ethanol industry and its partners.

Chapter 4 focuses on computer-based process control (and process control specifically) as a means to continually tune processes to produce ethanol in a superior way in manufacturing. This begins in research, is tested and verified in pilot plant operations, and then proven in manufacturing. The end result is scale up of modifications that significantly reduce operating costs and the capital cost of facility modifications, or the next new ethanol plant. Recommendations are based on personal experiences that involve developing biocatalysts, yeasts, and enzymes, and using them in processes to produce ethanol.

Chapter 5 focuses on feedstock, considering agricultural resources available throughout the world. The information in this chapter demonstrates that ethanol facilities can be located in any country and have spinoff benefits to the locale, whether in developing countries or more established nations. The database developed in this section should be useful to interested parties. Arguments made in this section help one select their preferred feedstock source to produce ethanol.

Chapter 6 deals with helping to understand the biocatalytic properties of yeast and how to optimize its performance. This is facilitated through mathematical modeling. This model is useful in its own right. It is also beneficial to use as a basis to adapt the mathematical formulation to a specific facility for captive benefit. The application of this work is expected to improve yields, production rates, and cost reduction in a number of ways. The model can certainly be adapted in a practical way to specific operating facilities.

Chapter 7 provides an examination of cost parameters and patterns that emerge in the ethanol industry from available data. This information (and other data like it that one may have in their own facility) can be used to minimize costs in operation and in capital investment. This information can be used to examine a specific facility and adapt the methods to ones needs and database. A key aspect is to judge your specific feedstock with what others are using. This chapter provides an example comparison for ethanol production from feedstocks provided by corn wet milling, corn dry milling, and lignocellulose sources.

Chapter 8 deals with pollution aspects. Ethanol produces pollution (mainly carbon dioxide) as a natural consequence of being manufactured or used in combustion of the ethanol/gasoline blend. There are existing uses for carbon dioxide. Development of new uses will require research. Work in this area will be quite valuable to the company involved (and to society as a whole), since action is being taken to use the waste by-product in a productive/profitable way that lessens the pollution load (at least in principle).

As discussed in this chapter, such research work can have greater impact than one might initially anticipate. It hoped that carbon dioxide accumulation does not have the dire global impact as proposed by some. All data are not in on scenarios that involve global warming due to carbon dioxide. Taking action in a positive manner through research is the best course of action allowing there is some cause and effect of carbon dioxide accumulation on weather and other environmental areas. Action should be taken by conducting speculative research that could result in discoveries that have great benefit.

Chapter 9 points out that the ethanol processes manufacture a key product and many by-products. These can be exploited commercially as opportunities. By-products, such as Distillers Dry Grains used for animal feed, impact significantly on facility economics. Such impact will apply to ethanol produced anywhere by the selected feedstock. The by-products from these worldwide feedstocks need to be developed in a manner that has been accomplished by the U.S. corn refining industry. However, it is expected that experts in the animal feed industry will agree that their work is far from complete. Suggestions to improve such products and also potentially benefit animal and human health are covered in this chapter.

Chapter 10 notes that the main alternative route to ethanol (through ethylene) is possibly appropriate under specific circumstances. The path emphasized in this book is based on the use of an agriculturally sourced feedstock.

Chapter 11 provides conclusions, recommendations, and research directions for what can (and should) be done to protect the industry and help it to prosper.

Parties that should benefit from reading this book include

- Those who are designing and operating ethanol facilities and trying to improve yield, productivity, and reduce costs.
- Parties in industry organizations who wish to promote ethanol (since this book tries to do that very thing).
- Those in the agricultural/farming community who are involved in ethanol or wish to participate in what can be a lucrative enterprise.
- Those who are in government and are trying to help developing countries. Specific suggestions are made in this book to help them.
- Those in all segments of the animal feed industry (since specific suggestions should be of interest to them in regard to nutrition and safety).
- Those parties in the human food industry interested in new product opportunities and ways to improve food safety.
- Parties in government concerned with regulating the ethanol industry since suggestions made herein may be helpful to allow the industry to successfully develop.
- Those in the environmental areas who are assessing impact of pollutants. As noted in this book, answers to important questions affecting our well-being may be addressed through measurements of constituents in air layers

and specific experiments that can be conducted to simulate effects beyond *in silico* work.

- Attorneys who are responsible for assuring that their company complies with regulations and wish to monitor, predict, and plan for legislation and laws that may be enacted.
- Those designing catalytic converters and who are responsible for research on the many aspects of automobiles, as they may find important partners in the ethanol industry (if parties accept recommendations for research and development such as those made in this book).
- Those in the general public who wish to learn more about the importance of ethanol and how it can affect their lives in a positive manner.

Acknowledgments

To my former educators and mentors who helped me develop skills so that I could have the opportunity to contribute professionally and reciprocate by helping others.

To the publisher, Taylor & Francis, for originally suggesting that I undertake this project and for issuing this book. In particular, the guidance of Joseph Clements, acquisitions editor, is very much appreciated.

Contributions of colleagues at Taylor & Francis (USA and globally) and Nova Techset Private Limited (India and UK) are appreciated.

Author

Dr. Walter E. Goldstein is president of Goldstein Consulting Company, offering services in chemical engineering and biotechnology (see http://www.goldconsul.com/). He is called upon to improve processes and products in health care, consumer products, food, pharmaceuticals, chemicals, and biofuels. He provides expertise to analyze product and process components, material defects, construction practice defects, and analyses of accidents causing injury.

Many of his projects involve biotechnology processes that include mammalian cell, bacterial, and fungal/yeast propagation. He has extensive background in fermentation technology and development of products and processes in this field. He is also president and co-founder of a research and development firm with an active patent. Its objective is to produce universal blood from stem cells in a bioreactor process thereby avoiding the use of donor blood for transfusion.

He has been involved in bioprocess energy-related projects throughout his career. He was instrumental in leading research and development resulting in a commercial facility to produce ethanol (New Energy of Indiana, $180 million capital investment, 100 million gallon per year facility). His involvement in this book stems from his desire to impact on the alternative energy field in a manner that is synergistic with protecting the environment and helping businesses.

Dr. Goldstein was vice president for Biotechnology Research for Miles, Inc., a former division of Bayer, Inc. from 1982 to 1987. He was also vice president and director of research for ESCAgenetics Corporation, a plant sciences biotechnology company from 1988 to 1994. He founded Goldstein Consulting Company in 1994 and has been engaged in several entrepreneurial enterprises since that time. He founded and developed a forensic sciences DNA profiling training laboratory at the University of Nevada Las Vegas from 2003 to 2008. His past publications for Taylor & Francis are *Sick Building Syndrome and Related Illness: Prevention and Remediation of Mold Contamination* (2011) and *Pharmaceutical Accumulation in the Environment: Prevention, Control, Health Effects, and Economic Impact* (2014).

He obtained his MBA from Michigan State University in 1968, and is a registered professional engineer. Dr. Goldstein earned his doctorate in chemical engineering from the University of Notre Dame in 1973. He is a member of the American Academy of Forensic Sciences, the American Institute of Chemical Engineers, and Sigma Xi.

He and his wife, Paula, reside in Las Vegas, Nevada. They have two children and four grandchildren.

1 Introduction

Ethanol is an important component of the energy supply. This book demonstrates the importance of ethanol in the energy supply by showing how scientific/engineering innovation and technology development can enhance the value of ethanol. Therefore, this book will describe how ethanol can (1) be a better fuel, (2) maximally protect the environment, (3) reach cost minima, (4) attain its best commercial value, and (5) help protect our security.

The aim is to improve or enhance the biological processes for manufacturing ethanol. Biological processes are dominant in the production of ethanol for use as a fuel, and this is expected to continue. Chemical processes also provide a means to produce ethanol. However, they are best used to help find ways to manufacture valuable products from ethanol and its by-products. Study of chemical processes to produce ethanol from substances such as ethylene (derived from petroleum) can be a means to identify ways to improve the biological routes.

Catalysis (using an entity and conditions to allow chemical and biochemical reactions to be feasible and to go faster) is a critical step for producing ethanol, as well as to burn it in combustion (e.g., in automobiles). In the biological processes that produce ethanol, catalysis is part of the internal biological (enzyme) reactions in an organism or the catalysis is carried out externally or outside the organism by enzymes present in a liquid or perhaps even a solid phase. Improvement of both internal and external biocatalytic reactions is essential for the further development of ethanol technology.

In the case of biological production, many raw materials are used. Advantages of using one or another raw material as a resource will depend on the availability, locale, cost, pollution considerations, safety, and other factors. If many raw material resources are available in a particular location, then one can switch from one to another depending on the supply and cost. The sources can be corn, corn waste such as stover (leaves, straw, and a lignocellulose source), special crops such as amaranth, waste, and refuse. If a particular source is favored in a locale, then a means to convert that source to ethanol must be available commercially (i.e., at a competitive cost). The biological catalyst sources (organisms and/or enzymes) must be suitably developed so that they are appropriate to raw material(s) used. If the organism or enzyme is not available, then it is necessary to make sure that the catalyst is made available in the future. The raw materials, organisms, and enzymes to use (or to potentially use) are covered in this book.

Furthermore, since all processes produce waste, utilization of that waste to create additional economic return through added by-products (or recoverable added energy) is important due to the impact of such return on economics. Utilization of the waste (e.g., lignin, corn stover, waste biomass) is dependent on the form in which the waste is generated. For example, an intact form of lignin may allow a superior secondary product application without incurring the cost of breaking the lignin to convert it to something else.

This form of the waste is dictated by raw materials, the process, and the biocatalysts used. It will be shown that it is critical to enhance the commercial value of by-products of ethanol manufacture (e.g., waste corn proteins or leaves from the corn plant in the case of corn sources, lignin in the case of lignocellulose sources) to attain the best economic value for ethanol manufacture. This economic value comes in the form of a by-product credit or outright sales. Raw material selection is influenced then by the by-product credit or sales that will be available.

Raw materials, organisms, and biocatalysts are interdependent in many ways, which complicates the subject matter but also makes the subject very interesting. This is interesting because it leads to the optimization of use of raw materials, catalysts, and processes in unexpected ways. The interaction of raw materials, catalysts, processes, and their optimization is an important subject in this book since this partly defines possibilities for the enhancement of ethanol as a product designated to make a significant impact on its use in the provision of energy and possibly other areas.

Models are applied in chemical engineering to optimize processes. Mathematical models can be used in early stages in research and applied to processes to produce ethanol, either biologically or chemically. These processes are controlled, and there are appropriate models applied for such control. These models can be directly applied to production processes to minimize cost and maximize quality. The models provide a useful way to evaluate how raw materials, catalysts, and processes can be best conducted and applied to produce ethanol.

The use of models is important no matter the value of the entity involved. In the case of ethanol, the cost per unit entity produced is relatively small compared to other nonethanol process examples (such as pharmaceutical production). The costs are magnified for ethanol since the volume of production is so large compared to that for a pharmaceutical where the cost per unit volume is high but the volume is relatively low.

Therefore, optimizing each step in a process and then for the process as a whole can enable significant differences in the choice of raw material, biocatalyst, process steps, ethanol yield, and ethanol quality. The objective is minimal cost and best quality attributes for use in a fuel blend. Use of ethanol in fuel is the most significant ethanol product example in the present or foreseeable future. Such costing and quality attributes are covered in this book extensively. Optimization of ethanol manufacture and its use may be critical to the enhancement of ethanol as a major product in energy fuel blends.

The computer-based cost and raw material utilization models for a process facility should also extend to the development and acquisition of raw materials (e.g., corn, lignocellulose, wastes of these, generalized waste that can be converted biologically or chemically to ethanol). This complicates the optimization process. However, this also makes the optimization process more realistic and useful since raw material sources and costs play an important role in the preparation of ethanol. The use of raw material resources for ethanol production can affect the availability of resources for other means, such as allocation of land for growing food crops as opposed to crops to produce ethanol for fuel blends. The contention between use of a raw material to form ethanol or in food processing is addressed in this book.

The scale of the operation is an important consideration since it can affect cost as well as quality. It can also affect whether or not ethanol is appropriate for a locale. For example, in some locales not all factors allow for the construction of large facilities to take advantage of economies of scale. Some locales may only allow for relatively smaller ethanol production facilities. This may be due, for example, to the availability of raw material supply, energy to power the facility, labor to operate the facility, or availability of capital to construct the facility. Therefore, the optimization has to allow for the application to small plants since factors in these plants may not be appropriate for larger facilities. Consideration of smaller plants allows the application to areas that have lesser resources for the production and use of ethanol. This can be significant for new businesses, for example, in the developing areas of the world.

The most significant present application is ethanol's use in fuel blends with gasoline. Ways to enhance this application can have a dramatic effect on the use of ethanol. The process of combustion may be modified and controlled by catalytic means as we look at ways to improve how ethanol is burned to power a vehicle.

The influence of ethanol in gasoline blends is changing, of course, due to the use of gasoline–battery hybrids and the possibility of total battery-powered cars, solar-powered cars, or cars powered by fuel cells fed by hydrogen and air. Given these examples, automobiles powered by gasoline–ethanol blends will still continue to grow in a significant mode for some time to come.

Several factors dictate the future of powering of automobiles and trucks by fuel blends containing ethanol (which can vary by ethanol content relative to gasoline-content). Some of these relate to tax incentives and other kinds of influences due to commercial and political interests, and market manipulations, which have short-term and often disruptive effects. However, the sustainable utilization of ethanol depends on science and technology.

Therefore, part of this book deals with developing the scientific basis to use ethanol. This deals with thermodynamics, kinetics, and factors related to internal combustion engines (how they work and how they can be made to work better). This involves many factors in engine design and operation, usage by consumers, and factors that may make ethanol a preferred combustible. This subject will be dealt with as a special extension of chemical engineering applied to "The Science of Ethanol." Truly, the real opportunity for ethanol in applications is to find best ways for its use in combustion to power vehicles and to optimize advantages of fuel blends to improve engine efficiency, lower fuel cost, and minimize the pollution of the environment.

Atmospheric pollution is a very real problem caused by many sources. Certainly, effluent from internal combustion is a contributing source as are many other sources, such as animal waste and effluent from power plants. Some argue that this is inevitable and not important, whereas others argue that we are polluting the atmosphere and must eliminate all sources at all costs. These are two extremes of an argument that may go on forever. It is important that we find ways to control the negative consequences of such pollution since that will be beneficial as a contributing factor to enhancing and promoting the value of ethanol as will be discussed in this book.

The extremist arguments are (in a way) irrelevant to analysis of ethanol/gasoline mixtures as fuel since one seeks compromises that are (in combination) best for our society.

The problem of pollution is very real as pollution from internal combustion does contribute to fouling of the atmosphere. Steps to lower pollution are important. For example, pollution should be reduced in large cities with automobile traffic or in areas where pollutants concentrate, even where few automobiles are present. This should be done in a manner where business and commerce are protected.

There are differences in the production of pollutants from oil refineries that produce gasoline and other hydrocarbons from those facilities that produce ethanol, biologically or otherwise. There are differences in pollution effluent from burning ethanol, gasoline, or combinations of these. Some of these pollutants can be minimized or controlled by using catalytic converters (devices that break down organic compounds in internal combustion effluent). The completely reacted fuel (if the fuel were ideally a mixture of ethanol and gasoline, which it is not) would be carbon dioxide and water. This ideal conversion is not possible since the feed has impurities and the reactions do not go to such completion. Also, carbon dioxide is attributed to affecting the upper layers of the earth's atmosphere leading to retention of heat and warming of the planet. It stands to reason that carbon dioxide should be removed from automobile effluent in innovative and best economical ways that are possible just in case assertions of extreme negative consequences of carbon dioxide accumulation are correct. The important subject of carbon dioxide accumulation is also discussed in this book.

The subject of pollution is complicated and argumentative. However, it is clear that for health reasons, at least, controlling and reducing organic and inorganic content in the effluent from internal combustion is desirable. This book then addresses how this may be best accomplished for ethanol and in a manner so that ethanol's use as a product is enhanced.

Following Chapter 1, Chapter 2 deals with yeast biology since that is the most essential component to producing ethanol. Chapter 3 deals with enhancing the value of ethanol in internal combustion in motor vehicles since combustion is the most important application for ethanol. Such an enhancement is essential for ethanol to have a sustained value compared to competition.

Chapter 4 deals with computer-based optimization of the ethanol process, since the author believes that this leads to best process control and business (essential for cost reduction and sustained sales). Chapter 5 deals with raw material resources since raw material selection and use is the dominant cost factor and a key aspect to parties becoming involved with this subject of producing ethanol by biological means. This chapter includes raw materials for use in conversion to ethanol and those entities produced that are a by-product of the ethanol process.

Chapter 6 deals with the next critical subject covering enzymes, application of yeast, and chemical engineering process modeling to seek ways to improve the process from a technological standpoint and have the best impact on cost reduction. Chapter 7 examines some aspects in this subject, the content of previous chapters, and the best way to control and reduce costs to secure profits and the potential of profits.

Chapter 8 deals with the subject of how to minimize negative aspects of harming the environment and health since this can derail the success for ethanol. More importantly, this is the right subject to consider to protect the health and well-being of society.

Chapter 9 deals with maximizing the value of entities produced besides ethanol and the value of products that are derived from ethanol and its by-products. This is essentially examining the amplifications of ethanol sales to new sales and profits.

Chapter 10 deals with petrochemicals as sources of ethanol to effect comparison to biological routes. That route is mentioned. However, the real interest and driver for the industry is the biological route and its scale up to commercial levels.

Chapter 11 deals specifically with key points made in this book, which substantiate ethanol as a valued entity. The points made signify what must be done to progress to improve its competitive status and to ensure that it fits into commerce and improves its applications based on justifiable research.

2 Yeast Biology and Bulk Manufacture of Ethanol

2.1 PROCESSES TO PREPARE ETHANOL USING BIOTECHNOLOGY

There are two central topics for examining the "Science of Ethanol." The first is the selected process for making ethanol (principally by biotechnological means). The second is the main application for using ethanol, which is the combustion of ethanol as part of a fuel system to power a vehicle. Economics and determination of value revolve around these two topics. It is good to be sure that basic economics of the development and the use of ethanol is soundly considered and implemented. However, there is widespread disagreement on this matter in many sectors. When there are problems in justifying work on this subject, it is up to research, technology development, and manufacturing to resolve such problems and show what can be done. This subject has many facets to it beyond the two principle questions of making ethanol and burning it as a fuel.

Figure 2.1 shows the basic biochemical process used to prepare ethanol. A raw material is provided to a system to convert it to ethanol (and by-products plus waste). The raw material is physically treated, converting it to a form that is best for ethanol production later in an adjunct facility (or elsewhere in the process) plus with the attribute of also maximizing the possibility for valuable by-products and/or energy recovery.

Then the next step is presently breakdown or conversion of the raw material (biochemically or chemically) to simpler substances so that it is easier for the key organism such as yeast to convert the raw material to ethanol. It is possible that the breakdown or conversion may be postponed until the actual bioconversion to ethanol if the process and the resulting product attributes allow this. One example of a conversion process is to use enzymes that treat ground and finely milled whole corn to convert the corn directly to simple carbohydrates and a valuable waste by-product. Another example would be an enzyme complex to treat lignocellulose for conversion also to simple carbohydrates and valuable waste by-products or energy sources, similar and analogous to those obtained after digesting the corn.

Lignocellulose may assume a variety of forms such as waste materials from a corn plant, waste paper or wood, or a plant not normally used as a food source such as a weed. Another source may be trash that is largely carbon-based amenable to conversion if the waste can be broken down to a form where enzymes can digest it or where the enzymes themselves or special microbes can directly perform such a breakdown. It is clear that the possibilities even within this short description for variation and optimization are considerable. These different options are considered within this book to allow optimization by raw material availability and region of the world.

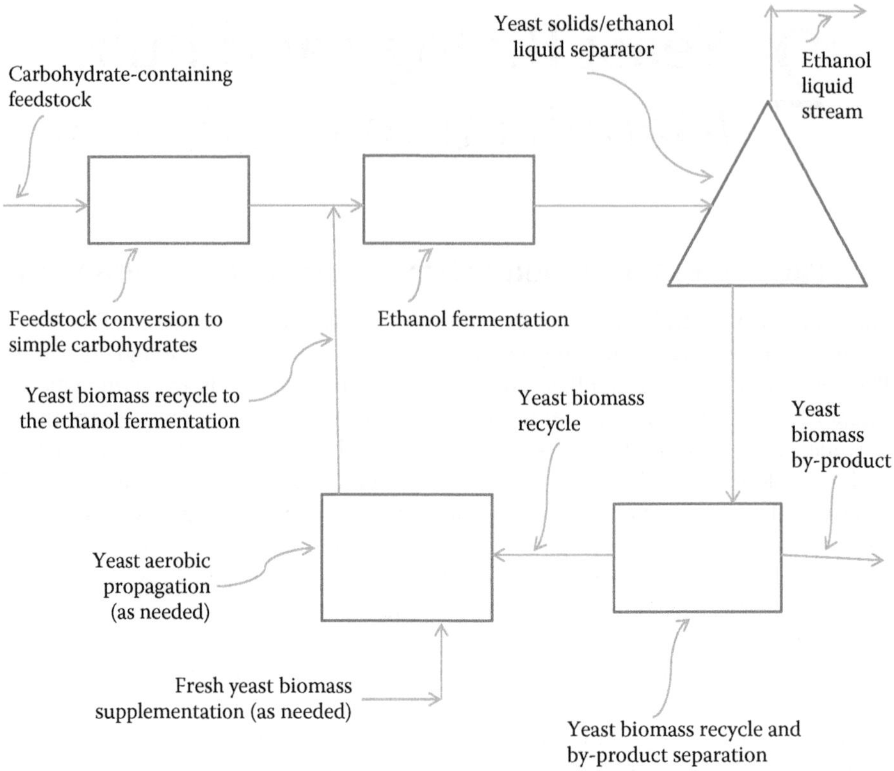

FIGURE 2.1 Ethanol process flow chart.

The bioreactors that can be used may be batch reactors. They may be fed batchwise, which means ingredients to fuel the fermentation are provided as a function of time. Preferably they are continuous reactors with biomass recycle. This is represented in Figure 2.1.

2.2 YEAST BIOLOGY DEVELOPMENT

The yeast most commonly applied to ethanol is *Saccharomyces cerevisiae*, a eukaryotic microbe. It has a nucleus containing reproductive DNA just like in humans. *Saccharomyces* is the genus and *cerevisiae* is the species. More specifically (see Figure 2.2), it is a globular-shaped, yellow-green yeast belonging to the kingdom Fungi, which includes multicellular organisms such as mushrooms and molds (Blaylock, B. 2016; MicrobeWiki, 2016). *S. cerevisiae* is a unicellular yeast even though it is in the class of fungi that usually consist of multiple cells connected together in strands called *hyphae*. This yeast reproduces typically by budding in the small entities on the surface of the yeast cell as shown in Figure 2.2 (Jay, J.M. 2000). These buds fall off to form new yeast cells. The yeast, *S. cerevisiae,* is a diploid organism with two sets of eight chromosomes, or a diploid set of 16, though

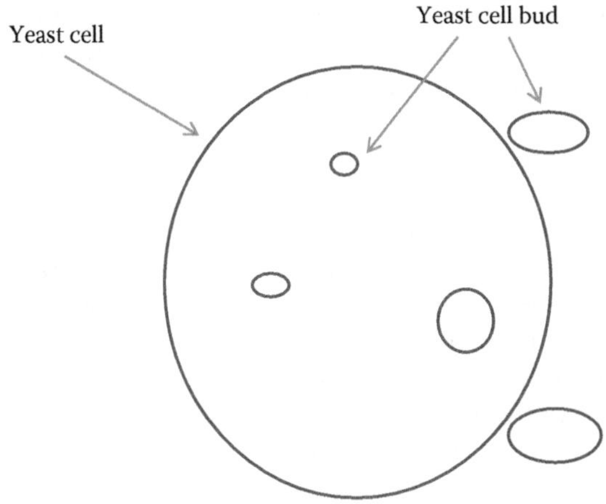

FIGURE 2.2 Yeast cell sketch.

this can vary (Gordon, J.L. et al. 2011). Chromosomes are the reproductive strands containing DNA. In comparison, humans have 23 diploid sets of chromosomes (Butler, J.M. 2005).

Natural strains of the yeast have been found on the surfaces of plants, the gastrointestinal tracts and body surfaces of insects and warm-blooded animals, soils from all regions of the world and even in aquatic environments. Most often it is found in areas where fermentation can occur, such as on the surface of fruit, in storage cellars, and on the equipment used during the fermentation process (Bekatorou, A. et al. 2006).

S. cerevisiae is famously known for its role in food production. It is the main component in the fermentation process that converts sugar into alcohol—an ingredient shared in beer, wine, and distilled beverages. It is also used in the baking process as a leavening agent; yeast releasing gas into the environment results in the spongy-like texture of breads and cakes. Because of its role in fermentation, humans have known and used *S. cerevisiae* for thousands of years. Archaeologists have found evidence of a fermented beverage in a pot in China as early as 7000 BC (Legras, J. et al. 2007). Molecular evidence of yeast being used in fermentation was found in a wine jar dating back to 3150 BC (Landry, C.R. et al. 2006).

An advantage of *S. cerevisiae* is its fast growth rate. On a normal yeast medium, it takes 90 minutes for the yeast population to double (Sherman, F. 2002). Colonies are usually visible 2–3 days after placing them on the fresh medium.

S. cerevisiae can exist in two different forms: haploid or diploid. It is usually found in the diploid form (Landry, C.R. et al. 2006). The diploid form is ellipsoid shaped with a diameter of 5–6 μm, while the haploid form is more spherical with a diameter of 4 μm. In the exponential phase, haploid cells reproduce more than diploid cells. Haploid and diploid cells can reproduce asexually in a process called budding, where the daughter cell protrudes off a parent cell. The buds of haploid cells are adjacent to each other, while the buds of diploid cells are located in opposite poles. Additionally, diploid cells can exhibit pseudohyphal growth if they are grown on a poor carbon source and exposed to heat or high osmolarity.

In addition to budding, diploid cells can undergo a meiotic process called sporulation to produce four haploid spores. Sporulation can be induced if the yeast is exposed to either a poor carbon or nitrogen source or lack of a nitrogen source. Spores also have a higher tolerance to conditions such as high temperature (Landry, C.R. et al. 2006; Spore-Biology_online dictionary. 2016).

As a eukaryote, *S. cerevisiae* contains membrane-bound organelles. Its chromosomes are located in the nucleus and it uses mitochondria to conduct cellular respiration. Like all other fungi, the cell's shape is based on its cell wall. The cell wall protects the cell from its environment as well as from any changes in the osmotic pressure. The inner cell wall has a high concentration of β-glucans, while the outer cell wall has a high concentration of mannoprotein. Chitin is usually located in the septum (Cabib, E. et al. 1991).

Yeast is typically used to produce bulk ethanol. Other organisms may also be used. However, the history of the use of yeast is quite extensive, and it is a very appropriate organism for this application. Yeast is extremely robust. It can be handled and separated typically without damage since it is not sensitive to shear compared to other organisms. Yeast is facultative. This means it can grow and reproduce under aerobic (oxygen) conditions and form products under anaerobic (nonoxygen) conditions. Conditions can be changed conveniently so that aerobic condition to fuel yeast reproduction and growth can be changed to an anaerobic environment, where the oxygen content of the broth is lower to allow ethanol to form. Aerobic and anaerobic conditions may be established in separate vessels for convenience. However, the yeast can manage this change quite well and readily adapt to the broth condition.

When a process to use yeast is being developed, it is most logical and practical to begin that development at a small scale in the laboratory and gradually increase equipment size, applying chemical engineering principles, so that it is easy to get an idea of what may happen in the scale up. Such a scale up can be modeled to help the process be efficiently developed and avoid wasting money on a larger scale device that does not work. However, sometimes models only partially help and are inaccurate to a great degree, so a greater amount of experimental testing is required. Such testing can still be conducted at a small scale, developing scientific and engineering data and understanding. This requires effective cooperation of scientists and engineers involved in such work. Better industries understand this quite well, since not understanding this leads to expensive failures and perhaps lost jobs.

For yeast biology, or any microbial biology, this starts at the bench scale, trying to understand the organism and how it grows, under what conditions, and how it can be made to make product efficiently. In the modern era, understanding of the genetics of yeast has advanced considerably to help understand how the organism copies DNA (deoxyribonucleic acid) to make copies of itself so that it can reproduce, how it translates the DNA into RNA (ribonucleic acid), and then how the RNA is used as a template to synthesize proteins at the organelle termed the ribosome in the cell (Butler, J.M. 2005). The proteins (and other molecules) are part of the fabric for cell propagation and maintenance and also for how the cell produces molecules for different purposes such as surviving. Understanding this and how these proteins work together with other molecules has (and will) made a big difference in advancing this field. This extends to the better understanding of the genetics of plants producing feedstock for the ethanol fermentation, and also understanding of how ethanol and other fuels affect our environment and our health and well-being. Basic reviews of yeast genetics have been completed (e.g., see Botstein, D. and Fink, G.R. 2011).

2.3 DEVELOPING YEAST STRAINS TO PRODUCE ETHANOL

Yeast strain selection is based on its ability to propagate in a consistent fashion, be stable, and produce product economically. At the bench scale, this starts with understanding parameters important to the strain. It is necessary to understand levels of oxygen ranges needed to sustain growth and levels of oxygen absence to help cause ethanol to be synthesized and excreted into the fermentation broth. Furthermore, the organism has the ability to ferment the available feedstock (e.g., corn or cassava starch converted into glucose, sugar, and feedstock conversions from lignocellulose sources). The organism will be dependent on major nutrients such as nitrogen sources and specific ions. Its productivity (amount of ethanol produced per unit time) may be inhibited or enhanced by certain ions or their concentration. The uptake of entities into the yeast cell needs to be determined to see how this affects cell growth, maintenance, and ethanol production rate. Ethanol is an inhibitor so the ability of the yeast cell to overcome such inhibition must be determined (Jay, J.M. 2000).

In the laboratory, research is conducted on various yeast strains to find out which may work best for the feedstock available. We are interested in the nutrients to be supplied and the ranges for them to maximize stable yeast growth and maintenance and stable production of ethanol. One way of doing this is to set small bioreactors for the observation of cell growth and yeast productivity. There is already evidence in the literature for being able to run these as chemostats to establish steady-state populations dependent on a particular nutrient, and then perhaps on combinations of continuous stirred tank reactors that are used for production at much larger scales. With the right experimentation, it is possible to examine the uptake and results and correlate these to genetic changes and find out what factors help keep these genetic changes from drifting or changing. Some work on yeast mutants provides an example of a portion of what might be done with the right experimentation (Gresham, D. et al. 2008).

2.4 NUTRIENTS AND CONDITIONS

An initial step in studying and evaluating yeast strains is to examine nutritional requirements. If the objective is to study intracellular aspects of the strains at fixed concentrations of nutrients, then such nutrients are fixed at a particular concentration to find out the intracellular conditions and component concentrations that may affect metabolic pathways, growth, capacity to produce ethanol, DNA and RNA content, and so on (Saldanha, A.J. et al. 2004). It is possible to make cell growth and function dependent on a limiting nutrient; that is, the nutrient that is held at a particular value by feeding in a continuous fermentation by running a chemostat experiment in continuous fermentation (Gresham, D. et al. 2008). The fermentation yeast growth rate becomes equal to the rate of the supply of the limiting nutrient. The mathematical formulation justifying this approach is covered in the literature (Blanch, H.W. and Clark, D.S. 1997).

Studies of nutrients and conditions for best yeast growth, stability, and ethanol production may be made in situations where several variables need to be looked at using a mathematical process model or an experimental design. In this case, the aim is to seek a combination that will maximize yeast growth and the ethanol productivity of the yeast. To try and reduce costs at the same time requires a different and more complicated approach to resolve the optimization problem.

In some cases, if the entities that will inhibit the yeast growth and the ethanol production are ignored (the known susceptibility of the yeast to ethanol inhibition as covered in part of Chapter 6 in this book), then the aim is to make concentrations of nutrients and growth as high as necessary while avoiding limiting the growth of the yeast or the production of ethanol. This is an optimum to be found for the selected yeast in a particular production situation (Howell, G. 2011).

Table 2.1 is a collage of examples of data produced by different experimenters with different objectives to make the point about perhaps studying the organism under limiting nutrient conditions. In this case, it is necessary to control a particular entity to find out what its limitation might do to intracellular entities and metabolic pathway characteristics. As indicated in some cases, glucose or phosphate or sulfate may be limiting (Gresham, D. et al. 2008). In other cases, it is possible to experiment on various conditions of several nutrients in combination to find out the best conditions of not using any more nutrients than necessary, which result in the maximum ethanol production in the least amount of time. In Table 2.1, for example, the entries at carbohydrate feedstock concentration near or above 200 g/L show the results of the experiments that are trying to maximize ethanol production to the best extent possible (Martınez-Moreno, R. et al. 2012; Lam, F.H. et al. 2014).

In this case, it is necessary to consider all the variables since changing one variable may affect another. For example, if glucose concentration is raised, it might be necessary to raise the concentration of nitrogen. Too little nitrogen can put pressure on the yeast and cause it to sporulate, which may be undesirable if the purpose is to control yeast growth by means of "budding" (Hanson, P.K. 2008; Lallemand 2002). Often, nutrients may best be programmed and changed during the course of the fermentation involving the growth of yeast and then the formation of alcohol. This may apply to the dynamic fermentation system best managed using a model that involves using a recipe and precise computer control for additions. There are a number of other factors such as the type of nitrogen sources, such as particular amino acids or simply a source of ammonium as indicated in Table 2.1. The micronutrients shown at the bottom of Table 2.1 may have critical levels of importance in a particular yeast and its process situation (Stephanopoulos, G.R. 2014). For reasons of economics, it may be wise to use an inexpensive source such as corn steep liquor in part, yeast extract in part, or soy extract, to find something that works consistently and is inexpensive to be able to put this matter aside. The minimum quantity needed would be used to achieve results.

In the case where models are used for programming nutrients, an extension of the Monod model is of use (Goldstein, W.E. and Schreuder, W.A. 2011).

TABLE 2.1
Yeast Nutrients and Process Factors

Process Factor	Measurement Method		References
Feedstock flow rate	Flow meter		
Feedstock density	Densitometer		
Oxygen concentration in biomass propagation	Sterilizable electrode		
Oxygen concentration in ethanol fermentor	Sterilizable electrode		
Oxygen concentration in biomass propagation	Sterilizable electrode		
Oxygen concentration in ethanol fermentor	Sterilizable electrode		
Carbon dioxide concentration in off-gas	Mass spectrometer	Manual method lab and pilot plant only	
Oxygen concentration in off-gas	Mass spectrometer	Manual method lab and pilot plant only	
Carbon dioxide concentration in off-gas from biomass propagator	Mass spectrometer	Manual method lab and pilot plant only	
Oxygen concentration in off-gas from biomass propagator	Mass spectrometer	Manual method lab and pilot plant only	
Carbon dioxide concentration in off-gas from ethanol fermentor	Mass spectrometer	Manual method lab and pilot plant only	
Oxygen concentration in off-gas from ethanol fermentor	Mass spectrometer	Manual method lab and pilot plant only	
Nitrogen concentration in off-gas from biomass propagator	Mass spectrometer	Manual method lab and pilot plant only	
Nitrogen concentration in off-gas from ethanol fermentor	Mass spectrometer	Manual method lab and pilot plant only	

(Continued)

TABLE 2.1 (*Continued*)
Yeast Nutrients and Process Factors

Process Factor		Measurement Method				References
Other gasses in off-gas from propagator		Mass spectrometer	Manual method lab and pilot plant only			
Other gasses in off-gas from ethanol fermentor		Mass spectrometer	Manual method lab and pilot plant only			
Yeast settling, floc size, diameter, visual images		Grad samples by laboratory	Lab and pilot plant special meter			
Ethanol flow rate in streams and effluent		Meter	Meter in pilot plant			
Ethanol concentration and other alcohols		Online HPLC	Manual lab HPLC			
Ingredient	**Source**	**Source Concentration (mg/L)**	**Source Concentration (g/L)**	**Measurement Method**		
Glucose			300	Online HPLC	Manual lab HPLC	Trafton, A. (2014)
Glucose			10	Online HPLC	Manual lab HPLC	Gresham, D. et al. (2008)
Maltose				Online HPLC	Manual lab HPLC	
Maltotriose				Online HPLC	Manual lab HPLC	
Maltotetraose				Online HPLC	Manual lab HPLC	
Sucrose			250	Online HPLC	Manual lab HPLC	Ortiz-Julien, A. et al. (2002)
Sucrose			150	Online HPLC	Manual lab HPLC	
Sucrose hydrol			280	Online HPLC	Manual lab HPLC	Martínez-Moreno, R. et al. (2012)

(*Continued*)

TABLE 2.1 (*Continued*)
Yeast Nutrients and Process Factors

Ingredient	Source	Source Concentration (mg/L)	Source Concentration (g/L)	Measurement Method			References
Nitrogen	Urea	300–500		Online nitrogen indicator	Nitrogen indicator	Nitrogen indicator	Heist, P. (2008)
Nitrogen	Variety aa	5–200 N		Online nitrogen indicator	Nitrogen indicator	Amino acid analysis	Martínez-Moreno, R. et al. (2012)
Nitrogen	DAP	10–60 N		Online nitrogen indicator	Nitrogen indicator	By outside lab	Ortiz-Julien, A. et al. (2002)
Nitrogen	NH_4Cl	500		Online nitrogen indicator	Nitrogen indicator	Total nitrogen	Vieira, E.D. (2014)
Phosphate	KH_2PO_4	13.3		Online phosphate indicator	Manual phosphate	By outside lab	Saldanha, A.J. et al. (2004)
Phosphate	KH_2PO_4	10			Anion exchange chrom		Gresham, D. et al. (2008)
Phosphate	KH_2PO_4	5			Manual phosphate		Vieira, E.D. (2014)
Sulfate	$(NH_4)_2SO_4$	3		Online sulfate indicator	Anion exchange chrom		Saldanha, A.J. et al. (2004)
Sulfate	$(NH_4)_2SO_4$		5				Gresham, D. et al. (2008)
	$MgSO_4$		1				Vieira, E.D. (2014)
	$MgSO_4$		0.5				Gresham, D. et al. (2008)
Uracil							Hanson, P.K. (2008)
Adenine							Hanson, P.K. (2008)

(Continued)

TABLE 2.1 (*Continued*)
Yeast Nutrients and Process Factors

Ingredient	Source	Source Concentration (mg/L)	Source Concentration (g/L)	Measurement Method		References
Potassium	KOH		Elevated	Online potassium indicator	Manual potassium	Trafton, A. (2014)
Potassium	KCl	100			Cation exchange chrom	Vieira, E.D. (2014)
Vitamins and Supplements						
Yeast extract	Vitamin, nitrogen, factors		6	None	Certificate of analysis	Vieira, E.D. (2014)
Yeast extract	Vitamin, nitrogen, factors		3–5		Outside lab and supplier	Ortiz-Julien, A. et al. (2002) Acumedia (2011)
Calcium chloride		100		None	Manual lab cation exchange	Gresham, D. et al. (2008)
Sodium chloride		100		None	Manual lab cation exchange	Gresham, D. et al. (2008)
Boric acid		0.5		None	Micronutrient certificate of analysis by supplier	Gresham, D. et al. (2008)

(*Continued*)

TABLE 2.1 (*Continued*)
Yeast Nutrients and Process Factors

Vitamins and Supplements	Source	Source Concentration (mg/L)	Source Concentration (g/L)	Measurement Method	References
Copper sulfate		0.04			Gresham, D. et al. (2008)
Potassium iodide		0.1			Gresham, D. et al. (2008)
Ferric chloride		0.2			Gresham, D. et al. (2008)
Manganese sulfate		0.4			Gresham, D. et al. (2008)
Sodium molybdate		0.2			Gresham, D. et al. (2008)
Zinc sulfate		0.4			Gresham, D. et al. (2008)
Biotin		0.001			Gresham, D. et al. (2008)
Calcium pantothenate		0.2			Gresham, D. et al. (2008)
Folic acid		0.001			Gresham, D. et al. (2008)
Inositol		1			Gresham, D. et al. (2008)
Niacin		0.2			Gresham, D. et al. (2008)
p-Aminobenzoic acid		0.1			Gresham, D. et al. (2008)
Pyridoxine		0.2			Gresham, D. et al. (2008)
Riboflavin		0.1			Gresham, D. et al. (2008)
Thiamine		0.2			Gresham, D. et al. (2008)

For example,

$$\mu_v = \frac{\mu_{v0}}{1 + k_{v1}/N_{v1} + k_{v2}/N_{v2} + \cdots} \quad (2.1)$$

where N_{vi} are nutrients and the set is $\sum_i k_{vi}/N_{vi}$. The k's are operating parameters analogous to K_m and K_i Michaelis–Menten constants (Michaelis, M. and Menten, M. 1913). In Equation 2.1, μ_v is the growth rate of the yeast, and μ_{v0} is the maximum growth rate of the yeast (units of s^{-1}).

It seems desirable to look at ratios of nutrients to quantity of biomass formed and maintained, since that is the most logical from an experimental standpoint. It is possible that some of these get depleted to too low a level during the fermentation and stress the organism. Perhaps, it is best to start with the key nutrients given in Table 2.1 and evaluate them in the laboratory at bench scale followed by pilot plant development if nutrient programming is necessary. It may be that introducing and developing this aspect could optimize the fermentation and also optimize ways to lower costs. This development was practiced in a system that was developed at the Biotechnology Group Miles Laboratories in the 1980s (Schlager, S.T. et al. 1978–1982).

REFERENCES

Bekatorou, A., Psarianos, C., and Koutinas, A.A. 2006. Production of food grade yeasts. *Food Technology Biotechnology*. Vol. 44. p. 407–415.

Blanch, H.W. and Clark, D.S. 1997. *Biochemical Engineering*. Marcel Dekker, Inc., New York. p. 282.

Blaylock. B. 2016. *Saccharomyces cerevisiae* photograph. https://microbewiki.kenyon.edu/index.php/UserBlaylock.

Botstein, D. and Fink, G.R. 2011. Yeast: An experimental organism for twenty-first century biology. *Genetics*. Vol. 189, p. 695–704.

Butler, J.M. 2005. *Forensic DNA Typing: Biology, Technology, and Genetics of STR Markers*. 2nd Ed. Chapter 2. Elsevier Academic Press, London, UK.

Cabib, E., Silverman, S.J., Shaw, A., Das Gupta, S., Park, H., Mullins, J.T., Mol, P.C., and Bowers, B. 1991. Carbohydrates as structural constituents of yeast cell wall, and septum. *Pure and Applied Chemistry*. Vol. 63. p. 483–489.

Goldstein, W.E. and Schreuder, W.A. 2011. Mathematical model of mold propagation and product formation in building materials, inherent transport phenomena and applications, in *Sick Building Syndrome and Related Illness: Prevention and Remediation of Mold Contamination*, W.E. Goldstein, Ed., Taylor & Francis Group, CRC Press, Boca Raton, FL.

Gordon, J.L., Byrne, K.P., and Wolfe, K.H. 2011. Mechanisms of chromosome number evolution in yeast. *PLoS Genet* Vol. 7(7). p. e1002190. doi: 10.1371/journal.pgen.1002190.

Gresham, D., Desai, M.M., Tucker, C.M. et al. 2008. The repertoire and dynamics of evolutionary adaptations to controlled nutrient-limited environments in yeast. *PLoS Genet* Vol. 4(12). p. e1000303. doi: 10.1371/journal.pgen.1000303.

Hanson, P.K. 2008. Yeast Genetics I: Analysis of gene complementation and mating types. Birmingham-Southern College. http://faculty.bsc.edu/phanson/yen/BSC_Yeast_Genetics_1.pdf.

Howell, G. 2011. Yeast nutrition and successful fermentations. Australian and New Zealand Grapegrower and Winemaker. Issue 573. Copyright 2013 Vintessential Laboratories.

Jay, J.M. 2000. *Modern Food Microbiology*. 6th Ed. Aspen Publishers, Inc., Gaithersburg, MD.

Lallemand. 2002. Nutrients for alcoholic fermentations. http://www.lallemandwine.com/.

Lam, F.H., Ghaderi, A., Fink, G.R., and Stephanopoulos, G. 2014. Engineering alcohol tolerance in yeast. *Science*. Vol. 346(6205). p. 71–75.

Landry, C.R., Townsend, J.P., Hartl, D.L., and Cavalieri, D. 2006. Ecological and evolutionary genomics of *Saccharomyces cerevisiae*. *Molecular Ecology*. Vol. 15. p. 575–591.

Legras, J., Merdinoglu, D., Cornuet, J. and Karst, F. 2007. Bread, beer and wine: *Saccharomyces cerevisiae* diversity reflects human history. *Molecular Ecology*. Vol. 16. p. 2091–2102.

Martınez-Moreno, R., Morales, P., Gonzalez1, R., Mas, A., and Beltran, G. April 10, 2012. Biomass production and alcoholic fermentation performance of *Saccharomyces cerevisiae* as a function of nitrogen source. Federation of European Microbiology Societies. *Yeast Research*. Vol. 12. p. 477–485.

Michaelis, M. and Menten, M. 1913. Die Kinetik der Invertinwirkung. *Biochemistry Zeitschrift*. Vol. 49. p. 333–339.

MicrobeWiki 2016. *Saccharomyces cerevisiae*. https://microbewiki.kenyon.edu/index.php//saccharomyces_cerevisiae.

Saldanha, A.J., Brauer, M.J., and Botstein, D. 2004. Nutritional homeostasis in batch and steady-state culture of yeast. *Molecular Biology of the Cell*. Vol. 15. p. 4089–4104.

Schlager, S.T., Schwartz, A., Goldstein, W.E. et al. 1978–1982. Computer Process Control for Fermentation Improvement, Biotechnology Group, Division of Miles Laboratories, Division of Bayer, A.G.

Sherman, F. 2002. Getting started with yeast. *Methods Enzymology*. Vol. 350. p. 3–41.

Spore-Biology_online dictionary. 2016. http://www.biology-online.org/dictionary/Spore.

Stephanopoulos, G.R. 2014. Engineering alcohol tolerance in yeast. *Science*. Vol. 346. (6205). p. 71–75.

3 Optimization of Ethanol Use As a Fuel

3.1 INTRODUCTION TO THE BASIS OF FUEL USE IN THE INTERNAL COMBUSTION ENGINE

The ethanol/gasoline fuel mixture passes into the carburetor to be mixed with air and then ignited in the car's engine cylinders to propel the piston to drive the engine's mechanical drive train. The overall energy balance dealing with this involves examining inputs and outputs of potential energy, kinetic energy, pressure–volume (pv) energy, internal energy, heat, and work. The system may lose or gain heat reversibly (in small steps that can be reversed) or in large irreversible steps such as when fuel and oxygen react.

For internal combustion, this presupposes that no other source of energy to provide work output is inherent, only pv work is involved. Other forms of work sources could be, for example, surface-related energy, gravity for space flight, or electrical as from fuel cells, and so on. The energy input to run the engine and power the cylinders is provided by the reaction of the fuel and oxygen in the air. This raises the internal energy since it raises the temperature. It also raises the pv energy since pressure increases (to push the piston in the cylinder to allow the volume to expand). Since the volume expands, moving a piston with a surface area a given distance, this force by distance is carrying out work by definition (Johnston, R.M. et al. 1992).

In the automobile engine, an ignited fuel–air mixture enters a defined small volumetric space. This gas in this space is compressed somewhat to aid the chemical reaction to occur, this space then being the clearance of the cylinder (Johnston, R.M. et al. 1992). The compression ratio is the clearance expanded under the effects of the reaction and resulting pressure to reach the expanded (displaced) volume. The ignition of the fuel–air mixture rapidly raises the temperature due to combustion, raising the pressure further as a consequence, and resulting in the cylinder traveling to expand the volume. The expansion of the gas propels the cylinders to expand with mechanical linkages then turning the drive shaft. Hot gases are then exhausted and the cylinder is returned to its initial position for another cycle beginning with introduction of the gas mixture. This is a four-cycle process (the Otto cycle) that is the basis of present-day automobiles powered by the combustion of ethanol–gasoline fuels and oxygen (Johnston, R.M. et al. 1992).

There are many causes of energy losses in this process (Johnston, R.M. et al. 1992). Burning of the gas is not ideally optimal, heat is lost (though minimally) from the automobile engine cylinders, and certainly in the exit gasses. The power train of the automobile has friction losses every place power is transferred from one mechanical device to another. There is friction loss in the flow involving cylinders where ignition occurs and most particularly in the area where piston rings seal the piston from the cylinder wall. The piston rings act as a moving seal against the cylinder wall in order to separate the gas mixture from the other oil-laden side that lubricates the piston-cylinders and the engine components. Also, perhaps the heat of combustion is suboptimal, since the air-combusted fuel mixture is heated beyond that sufficient to raise the pressure in the cylinder ahead of the piston since hot combustion gasses subsequently exit. The extra heat provides energy in part to overcome friction losses. If we can somehow help counter these losses using ethanol as a base (irrespective of the energy density of ethanol being fundamentally less than that of gasoline), we can then make the case for ethanol as being more effective as an energy source if the ethanol/gasoline stream is "engineered" to be optimal for internal combustion (Cesur, I. et al. 2014). The challenge is then how to do that and thus find ways to save energy in the use of ethanol.

Where can this energy savings due to ethanol be effected? First, it appears that since the octane number of ethanol is relatively higher, the combustion can be run at a higher compression ratio since it burns hotter without the tendency to "auto-ignite" as much resulting in "knocking" (By, A. et al. 1981). This inherently increases work output and more energy derived from the fuel. This may be an advantage restricted to new designs since it cannot necessarily affect current vehicles on the road. A better advantage is to somehow find ways to reduce the friction between the piston and cylinder wall at the piston ring surface. The literature points out that approximately 10% of the losses in friction in the automobile occur during the piston ring operation (Smedley, G. 2004). Perhaps, friction loss at this point is higher (24%) as noted in the literature (Brooke, L. 2015). If the range is 10%–24%, this is a significant target to reduce ethanol–gasoline fuel use.

Work is in progress to improve this area. One interesting article analyzes the unsteady state nature of the piston–cylinder process since this is a way to see what can be done about friction from a dynamic standpoint (Flagan, R.C. and Seinfeld, J. 1988). Another examines thermodynamics and kinetics (Nbu-Nada, E. et al. 2008). This makes utmost sense since changes to friction may be best addressed and analyzed from a kinetic standpoint. This will ultimately come down to piston ring sealing material and structure benefited synergistically by association with an optimal ethanol–gasoline mixture, with possibly additives to the mixture to help this happen. The use of optimal ethanol–gasoline mixtures has already been studied to an extent in the literature (Cesur, I. et al. 2014).

Another opportunity lies in the use of ethanol and other alcohols produced say by fermentation with perhaps their modification through the use of additives more environmentally friendly than methyl tertiary butyl ether (MTBE) or embracing aspects of the brand STP (Ahmed, N.S. and Nassar, A.M. 2011). Ethanol is a more positive oxygen containing and octane-enhancing fuel without the pollution and toxicity of MTBE. An additive that embodies what STP is supposed to provide may be beneficial if it can reduce friction at the piston ring without worsening leakage of oil or combusted ethanol–gasoline past the piston ring seal.

Numerous authors are seeking ways to reduce friction at this point. This includes new designs for piston rings (Gangopadhyay, A. et al. 2012). There are also steps that propose unusual coatings (Kennedy, M. et al. 2012). Specific measurements of this region have been taken (Wei, N. et al. 2014). An opportunity lies in an approach that examines friction from a fundamental basis. Some interesting analogies could be derived from examining the basic mechanics of skiing on snow and ice (Karloef, L. et al. 2005). Also, some analogies may be derived from the area of hydroplaning that is so important in the design and analysis of tire performance (Sinnamon, J.F. and Tielking, J.T. 1974). What might be desired is to apply a fluid and/or coating that would create a hydroplaning layer of fluid sealing that would be impenetrable to hot fuel gas from combustion or lubricating oil from the other direction. Based on some reading, it may be possible to create bubbles in the lubricant that act to reduce friction (Ferrante, A. and Elghobashi, S. 2005). A useful literature survey reference should be valuable in helping to analyze possibilities (Andersson, P. et al. 2002).

3.2 LINKAGE OF FRICTION REDUCTION TO ETHANOL FUEL USE

Friction reduction linked to ethanol means better mileage per gallon thus countering the point that the energy density of ethanol is less than that of gasoline since ethanol is already partially oxidized. If it comes down to nothing special from ethanol (which the author doubts will occur, since research, if done well, always results in something of value), then the parties producing and selling ethanol should invest in research in the area of friction reduction and acquire patents that allow for horizontal integration of their businesses. There may be a relationship of friction reduction to ethanol that will be discovered. If not that, then intellectual property may result that is valuable for licensing as a consequence of the investment in R&D. There are reported studies where researchers are examining ethanol and similar alternatives or substances complementary to gasoline in order to see if there are effects on friction and performance (Knothe, G. 2012; Cesur, I. et al. 2014; Wei, N. 2014).

A key aspect to reducing friction is to examine the relationship between viscosity and interfacial tension at the cylinder liner wall and the sealing rings wall. The viscosity is that of the lubricant in between the piston and the cylinder wall (Smedley, G. 2004). Momentum transfer through layers of fluid to the wall is due to viscous flow between layers. This passes through the lubricant oil layer which can vary between the cylinder (the cylinder lining) and ring which are typically sealing rings made of various metals, sometimes coated with chromium. At the wall, the force (derived from the momentum transfer) that causes friction loss is related to the velocity gradient with gap distance at the wall which is governed by the viscosity and the velocity of the fluid rubbing the wall. The gap can vary. It is undesired for the metals to rub (this is a microscopic point contact) or for the lubricant layer to be very thick since this will build up the momentum transfer to the surface which will raise the pressure.

The oil is modified with additives from varied research. In reality, what is being done is intended to change the friction at the surface. This rubbing action is governed by surface tension. Surface tension is small if the surface is akin to a slippery Teflon-like surface (Andersson, P. 2002; Ahmed, N.S. and Nassar, A.M. 2011). The right lubricant will have this property as the excellent lubricating oil mixtures do. What we would like to do is to develop improvements by effecting a material that coats the walls to make a step change in slipperiness (this is tied to surface tension control). Perhaps additives based on ethanol or alcohols related to ethanol can be made from special esters formed from reacting alcohol with an anhydride like substance.

Perhaps this can be a special kind of polymer. There have been many compounds and in testing in this regard (Ahmed, N.S. and Nassar, A.M. 2011). This may be a key to the ethanol program in different ways (use research conducted by ethanol parties to develop new additives, and incorporate them with an inherent connection to ethanol).

For some reason (to be discovered) coatings such as the carboglide may be synergistic with ethanol from an operating standpoint or ethanol will be used as an additive that imparts a slippery surface to cut the friction. The low friction exhibited already represents a very slippery surface. But even small changes can result in enormous savings in fuel usage. Biodiesels are burned as fuels and they are basically esters (Knothe, G. 2012). So, maybe ethanol can consist of modified esters in part as a fuel additive and bring friction relief from another direction besides the engine lubricant.

It is desirable to conserve fuel without sacrificing performance—that is, how much mileage can one get out of a tank of gasoline? If this can be stretched though octane changes as-well-as reducing friction losses, then that compensates for the relatively lower energy density of ethanol. If somehow, as noted and suggested, ethanol, perhaps including other performance chemicals like other alcohols, can be made to reduce friction in the piston/cylinder, helping to ease the friction of the piston ring, that can save energy, and lower fuel costs, making ethanol more attractive as a fuel. If friction loss can be reduced at the piston rings through tactics noted, then this at least partially compensates for the difference in energy density between ethanol. If the friction loss at the piston rings is 10% of the total energy expended to power the automobile, then if part of this is covered through means noted, that at least partially compensates for the difference in energy density. R&D has to be expended to accomplish such objectives. Such R&D is already underway.

In summary then, as a preamble to thermodynamic and reaction analyses, using a sample calculation, in the cylinder gap (the piston leaves a small space in the cylinder for injection of gasses to be burned—the gasses are compressed slightly before being ignited), the combustion reaction takes place causing a spike of temperature which can be calculated as we will show by example. This temperature spike raises the pressure in the small volume quickly and also adds to the internal energy of the ethanol–gasoline mixture that is being ignited and converted into products of combustion. The cylinder then expands increasing the volume in the cylinder, changing the internal energy as noted, and cooling the gas due to expansion. Since volume is increased as a consequence of this expansion, useful work is done (powering the drive train).

This theoretical work is partially lost in the piston ring, and the rest delivered to turn the crank shaft in a timing sequence with other cylinders. More power must be supplied and fuel used to compensate for this loss. The power train losses are inherent to the vehicle and unless some special optimization is apparent, these losses are not affected by ethanol. Therefore, we cannot do anything about that loss through the use of ethanol unless for some reason, the burning of fuel at a particular state using the elevated compression ratio and friction reduction can result in less friction losses elsewhere in the drive train. It would seem that controlling the boundary layer formed of gas in an optimal ethanol–gasoline mixture to cause minimal friction or perhaps using alcohol mixtures could be a route to reducing friction, or else, an additive related to ethanol and produced from it can help reduce friction. It would make sense for ethanol producers to arrange for work to be done in the area of friction reduction related to the use of ethanol as an extension of application of their product.

3.3 THERMODYNAMICS OF FRICTION DEPENDENCE ON SYSTEM VARIABLES

A calculation to illustrate this process follows, starting with a simple steady state, non-flow analysis based on thermodynamics, referenced in the literature, including a good book authored by instructors from the United States Naval Academy (Johnston, R.M. et al. 1992). This will proceed to looking at an unsteady state analysis involving mass flow in the literature where we can better examine friction reduction from a dynamic aspect.

Consider the first law of thermodynamics as applied to a gas cylinder in an automobile. The change in internal energy in the gas (the total energy possessed by the gasoline–ethanol–air mixture) is given by the heat added and lost and the work done by the system. For steady-state thermodynamics, putting the reaction heat added aside for the moment,

$$dU = T\,dS - P\,dV \tag{3.1}$$

Here, the first law is expressed in terms of differentials from calculus (tiny changes in entities). The term dU is the change in internal energy (all energy possessed by the gas from all sources). T is temperature and dS is the change in entropy. Entropy is a measure of disorder and the T dS term reflects the increase in that disorder, and the reversible loss of energy as heat during the process (Callen, H.B. 1960). Reversible heat losses result from small and slow steps and should have sufficient time to occur. In the case of the gas cylinder, the steps are fast and not slow. However, the time is so short that heat loss this way does not have time to occur, so the TdS terms is relatively small enough to be ignored. In effect, certain steps in the process, aside from combustion, are effectively isentropic (entropy does not change enough to influence the process). This helps us derive expressions to examine system thermodynamics.

P is pressure and dV is incremental volume. The term P dV is the pressure–volume work done by the process (in this case, driving the piston). There can be other kinds of energy sources such as electrical or even interfacial. However, these are not present in the case of the internal combustion engine used in automobiles in this analysis.

The gas mixture can be assumed to be ideal. That means it follows the Perfect Gas Law (Callen, H.B. 1960, p. 51). This assumption helps us derive an expression relating temperature, volume, and pressure so we can begin to analyze this process from a thermodynamic aspect.

The Perfect Gas Law is given by

$$PV = nRT \tag{3.2}$$

Since the gas in the cylinder is reasonably assumed to behave close to a perfect gas, this equation be used as an equation of state (Callen, H.B. 1960, p. 33). Here n is the quantity of material introduced, and R is the gas constant (Johnston, R.M. et al. 1992, p. 63). The change in internal energy for an ideal gas is only a function of temperature so that

$$dU = nC_v dT \tag{3.3}$$

where C_v is defined as the $(\partial U/\partial T)_v$, meaning the change in internal energy with temperature at constant volume (the specific heat at constant volume), and n is the quantity of material introduced into the system to be combusted to provide energy.

If we differentiate the perfect gas equation, we obtain $PdV + VdP = nRdT$ for the case where n (the amount of material in the system) is constant. Basically, in a cycle involving one stroke of the piston, an aliquot of material is introduced into the cylinder to be converted in a reaction following spark ignition. Therefore, the differential quantity of work is given by

$$PdV = -VdP + nRdT \tag{3.4}$$

Note also that V (the volume) in Equation 3.4 can be replaced by nRT/P which is desired in order to relate temperature to pressure. Temperature change or rise is derived from the combustion and that results in an elevated pressure that will drive the piston in the cylinder.

Making the substitution for V by nRT/P as noted, and substituting Equations 3.3 and 3.4 into Equation 3.1, and letting TdS be relatively small as described, results in

$$nC_v\, dT = nRT\frac{dP}{P} - nR\, dT \tag{3.5}$$

The term n can be cancelled from Equation 3.5. Collecting terms and dividing Equation 3.5 through by T results in

$$C_v\frac{dT}{T} = R\frac{dP}{P} - R\frac{dT}{T} \tag{3.6}$$

Collecting terms,

$$\left(C_v + R\right)\frac{dT}{T} = R\frac{dP}{P} \tag{3.7}$$

Since $C_p - C_v = R$ for an ideal gas, Equation 3.7 can be expressed as Equation 3.8 (Callen, H.B. 1960, p. 56)

$$\left(C_p\right)\frac{dT}{T} = R\frac{dP}{P} \tag{3.8}$$

Integrating this expression, where it is assumed that C_p is constant with temperature (strictly true only over short temperature changes), results in

$$C_p \ln T = R \ln P + \text{constant} \tag{3.9}$$

Before compression prior to combustion, the temperature is T_1 and the pressure is P_1. Therefore, setting these values in Equation 3.9 allows the constant to be expressed as

$$C_p \ln T_1 - R \ln P_1 = \text{Constant} \tag{3.10}$$

Substituting for the constant then in Equation 3.9 results in

$$C_p \ln T = R \ln P + C_p \ln T_1 - R \ln P_1 \tag{3.11}$$

Equation 3.11 is rearranged so that the specific heat and gas constant terms are shown as exponents so that system pressure can be related to the change in system temperature, or

$$\frac{P}{P_1} = \left(\frac{T}{T_1}\right)^{Cp/R} \tag{3.12}$$

Equation 3.12 can then provide the pressure attained due to temperature increase for an ideal gas.

Since pressure is used to drive a piston in the cylinder to do work, and that work is measured by the change in volume (plus losses due to friction and what may escape as energy not used), it is useful to express the pressure change in terms of volume change so that the cumulative value of the pressure volume work ($\int P\,dV$), the integral of the pressure volume change, can be estimated.

Starting with Equation 3.12, we substitute for T from the perfect gas equation 3.2, and similarly for T_1, and obtain

$$\frac{P}{P_1} = \frac{(PV/nR)^{Cp/R}}{(P_1V_1/nR)^{Cp/R}} \tag{3.13}$$

Noting that nR cancels out from Equation 3.13, and rearranging this expression to express pressure in terms of volume change, and noting that C_p–R is equal to C_v (Equations 3.14 and 3.15 are obtained (see Appendix 3A for the derivation)

$$\frac{P}{P_1} = \left(\frac{V_1}{V}\right)^{Cp/Cv} \tag{3.14}$$

Equations 3.14 and 3.15 show the volume change associated with a pressure change. In the case of the piston–cylinder system, this equation applies to what can be assumed to be a near isentropic change to compress the gas mixture prior to combustion. This equation also applies to the near isentropic expansion of the gas after combustion due to expansion from piston movement (note that the temperature increase from the combustion reaction increases the temperature to then raise the pressure further after prior compression to drive the piston). The pressure increase from the temperature causes the piston to move and expand the volume to then reduce the pressure. So when the piston moves to start the expansion, it is starting from the elevated pressure produced by elevated temperature in the reaction.

For the case of the expansion, let the starting pressure be P_3, and the starting volume V_3. The starting volume V_3 is the initial volume per mass or mole entity of gas before expansion following combustion.

The pressure volume work ($\int P\,dV$) involves integrating $P\,dV$ from V_3 the starting volume per unit mass or mole to the final volume per unit mass or mole (V_4).

In terms of P_3 and V_3, Equations 3.14 and 3.15 are expressed as

$$\frac{P}{P_3} = \left(\frac{V_3}{V}\right)^{Cp/Cv} \tag{3.15}$$

Substituting for P, using Equations 3.14 and 3.15, and substituting into the equation for work ($\int P\,dV$), Equation 3.16 is obtained, which is the expression for the integral in terms of volume only.

$$\text{Work} = W_{cylf} = \int P\,dV = \int P_3\left(\frac{V_3}{V}\right)^{Cp/Cv} dV \quad \text{with limits from } V_3 \text{ to } V_4 \tag{3.16}$$

Carrying out this integration results in the equation for PV work by the cylinder or

$$\begin{aligned} W_{cylf} &= \text{Frictionless Work on cylinder} \\ &= P_3 V_3^{Cp/Cv}/(1 - C_p/C_v)\,[V_4^{1-Cp/Cv} - V_3^{1-Cp/Cv}] \end{aligned} \tag{3.17}$$

The required additional work to compensate for the friction loss at the piston rings per cylinder is ΔW_{cylf}.

The total work required per cylinder is

$$W_{cyl} = W_{cylf} + \Delta W_{cylf} \tag{3.18}$$

This work is required to contribute to turning the crankshaft as pictorially indicated in Figure 3.1 as the expansion portion. Figure 3.2 shows the four stroke or cycle process showing fuel/air intake, compression, ignition, and expansion to do the work, and the return stroke to exhaust the gasses. This chart is similar to that provided in the literature reference already cited (Johnston, R.M. et al. 1992, p. 224). The gasses leaving are hot so energy is wasted in this exhaust (an area to

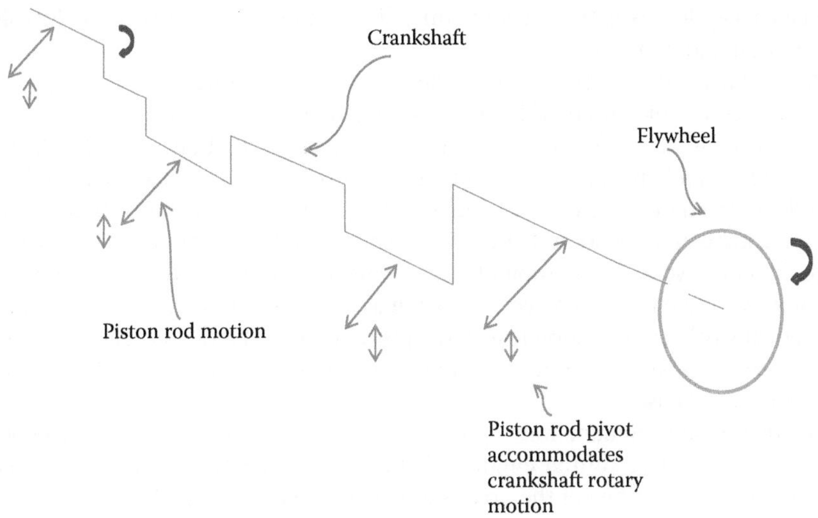

FIGURE 3.1 Crankshaft and piston rod motion.

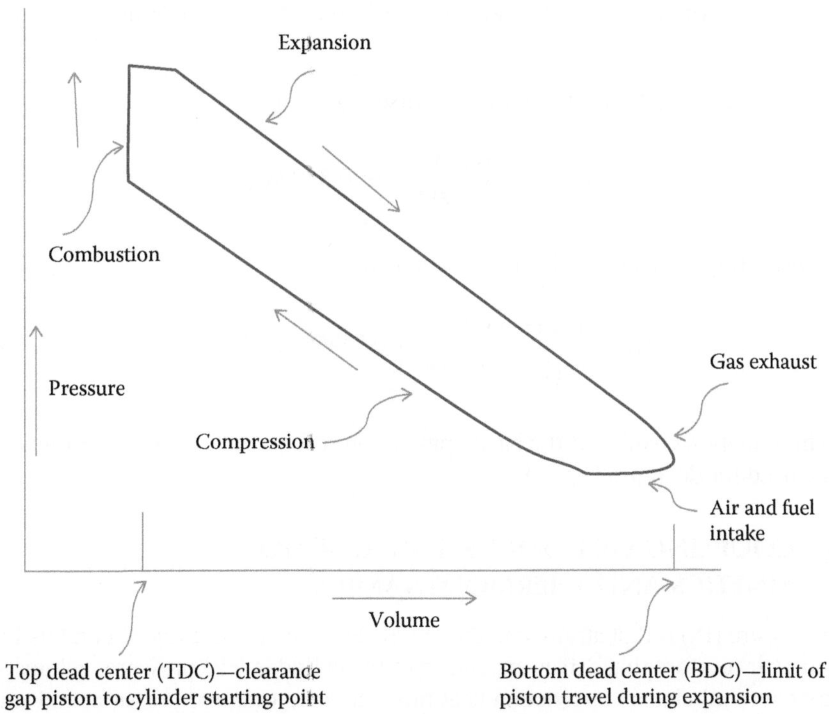

FIGURE 3.2 Piston–cylinder expansion, compression, and combustion cycle.

reduce energy losses is to minimize this). The simplified chart (Figure 3.2) shows the ignition and expansion.

As noted, friction losses occur at the crankshaft and further in the engine, and so on, to operate the automobile. It is assumed that this is the same no matter what fuel is used. However, it is possible that changes to the fuel and cylinder operation can change what happens further on in terms of friction loss. That is a complication considered in automotive engineering. For our purposes, the focus is on the cylinder and the combustion reaction that causes the piston rod to move in the cylinder.

In the Otto cycle, the ignition of the fuel/air mixture (e.g., ethanol, gasoline, and air) provides the energy to power the piston as discussed. Equation 3.17 provides the theoretical work (sans friction loss in the piston-cylinder) and it is desired to convert this to power (or work per unit time) to be able to examine some details in regard to energy input and use.

A 2007 Honda Sedan, four cylinder automobile indicates 166 horsepower at 5800 rpm. For this or similar engines, there is one stroke of each of the four pistons for every two revolutions of the drive shaft (Honda Motor Sales 2007).

Starting with Equation 3.17, the horsepower required (HP_i) (with zero friction consideration) is

$HP_i = W_{cylf}$ (work in ft-lb$_f$ per stroke per cylinder) times (4 cylinders) times (5800 rpm) times (one stroke/two revolutions of the crankshaft)

$$\text{or } HP_i = W_{cylf} 4\ (5800) \left(\frac{1}{2}\right) = 11{,}600\ W_{cylf} \text{ft} - \text{lb}_f/\text{min} \tag{3.19}$$

Since there are 33,000 ft-lb$_f$/min per horsepower, then

$$HP_i = \frac{11{,}600\ W_{cylf}}{33{,}000} \text{ or } 0.352\ W_{cylf} \tag{3.20}$$

Substituting Equation 3.17 into Equation 3.20

$$HP_i = \frac{0.352\ P_3 V_3^{C_p/C_v}}{\left(1 - C_p/C_v\right)} \left[V_4^{1-C_p/C_v} - V_3^{1-C_p/C_v}\right] \tag{3.21}$$

This equation changes if the basis rpm or power required is less, for example, due to the need for decreased speed.

3.4 COUPLING OF COMBUSTION REACTION KINETICS AND THERMODYNAMICS

The pressure (P_3) is that attained in the initial clearance volume due to combustion of the fuel/air mixture that drives expansion of the cylinder. The pressure, volume, and temperature all change as reaction heat raises temperature and pressure, and causes cylinder expansion. This is a dynamic process that is a function of time within a stroke. The pressure will go through a peak and then decline as the piston travels to create expansion in the cylinder (Johnston, R.M. et al. 1992). This is best analyzed by a force balance that varies with time since this may best account for static and dynamic friction effects as the piston rings are bathed by lubricating oil in proximity of the cylinder liner at the inside of the cylinder wall. However, we can estimate the pressure attained by estimating the temperature and then pressure given the fuel gas mixture approximates an ideal gas (Johnston, R.M. et al. 1992).

Initially, assume that the inlet pressure of fuel and air entering the system is at entry pressure P_1 and entry temperature T_1. This automatically defines the volume per amount of gas since the perfect gas law applies. At these conditions, following the example in the literature, the specific volume per unit mass is V_1 (Johnston, R.M. et al. 1992). Following initial compression, conditions are designated P_2, T_2, and V_2. These are the conditions prior to ignition. The ignition changes pressure to P_3, temperature to T_3, and specific volume per unit mass to V_3.

The heat of reaction of a mixture of 15% ethanol and 85% gasoline by volume is 18,600 BTU/lb, or 10,312 kcal/kg (see Appendix 3B for derivation of this quantity). A reasonable figure for the amount of fuel/air mixture delivered per cycle to drive a piston and return it to its initial position is 0.019 lbs or 8.6 g. The amount of air provided is about 14.7 times the amount of fuel given in weight units (Toyota Motor Sales U.S.A., Inc. 2014). This is approximately the amount required by stoichiometry for heptane. At 18,600 BTU/lb, this is 350 BTU or 88.2 kcal (Johnston, R.M. et al. 1992, p. 238).

The fuel is subject to combustion following ignition to raise the temperature based on the heat supplied by the reaction (the 350 BTU above) in an amount equal to this heat quantity divided by C_v, the heat capacity at constant volume (Equation 3.3). This temperature rise increases the pressure and drives to the piston to expand the gas and carry out work for a cycle. The heat from the combustion is assumed to increase the pressure instantaneously, keeping the total volume constant for that instant at the value of clearance volume. The pressure then propels the piston causing expansion of the air and moving the drive shaft. This expansion causes the pressure to fall. The air and product gasses are exhausted removing remaining heat content. This remaining heat is of course wasted and is an area to improve energy efficiency if this heat could be reused somehow.

Of course, part of this inputted energy goes into elevating the pressure to overcome friction losses that occur. For example, such losses, as commented on previously, occur at the piston ring system as it rubs against the cylinder wall liner lubricated by oil bathing the piston but isolated from the combustion side.

3.4.1 Equations Showing the Dependence of Friction Loss on System Variables

The equations to be shown include those for combustion, those for pressure rise, and those including volume expansion. The first step is to derive the expression for compression prior to ignition. Next, expressions are derived for the combustion phase. Note that we wish to separate the reaction and temperature increase in the clearance volume from the ensuing piston expansion, incorporating terms for friction contribution during the expansion. This assumption of discrete separation of combustion and expansion is allowed if the time for combustion, temperature, and pressure buildup is much less than that for expansion of the piston in the cylinder. Finally, expressions are shown for expansion in terms of thermodynamics and in terms of forces for piston motion. The expressions are useful to help identify and comment on the key variables linked to friction and its potential reduction (the main objective).

3.4.2 Combustion and Temperature/Pressure Rise in the Clearance Volume

The combustion is complex in terms of number of reactions and species formed (Martinez, I. 1995). However, it can basically be viewed as a second-order reaction involving the fuel and oxygen (Johnston, R.M. et al. 1992, p. 507). In the case of the E15 fuel (15% ethanol and 85% gasoline by volume), the reactions are:

Ethanol + Oxygen→ Carbon Dioxide + Water
Gasoline + Oxygen→ Carbon Dioxide + Water

Ethanol is a pure compound, C_2H_5OH, while gasoline is a mixture of hydrocarbons, specifically alkanes, represented by the general formula C_NH_{2N+2}, for example, heptane, C_7H_{16}, or octane, C_8H_{18} (Curtis, S. et al. 2008; Noller, C. 1958).

Assume the combustion can be represented on a first basis as a second-order reaction (Levenspiel, O. 1999)

$$\frac{dC_{fuel}}{d\theta} = -k\, C_{fuel} C_{oxy} \tag{3.22}$$

where C_{fuel} is the concentration (grams or moles per unit volume) of fuel fed into the cylinder clearance volume, θ is the time, and d is the differential operator from calculus, so that $dC_{fuel}/d\theta$ is the rate of change of the concentration of the fuel with time. The second-order rate constant k has units of reciprocal time multiplied by reciprocal concentration (e.g., s^{-1} grams/unit volume^{-1}). The term C_{oxy} is the concentration of oxygen in the clearance which we assume is a batch reactor space. Since air is fed and not oxygen, the concentration is the oxygen contained in the air (about 21 volume or mole %).

The amount of air fed in by the carburetor control mechanism is proportional to the fuel rate required so the air is not independent and is related to the concentration of the fuel. Typically, the mass amount of air fed is between 10 and 15 times the mass amount of fuel (Flagan, R. and Seinfeld, J. 1988, Section 2.3, p. 66). This is close to the stoichiometric amount required for burning of gasoline, considering gasoline is represented by heptane or octane. Therefore, we can assume as a practical matter that Equation 3.23 applies.

$$C_{oxy} = \alpha\, C_{fuel} \tag{3.23}$$

where α is constant reflecting the mass amount of oxygen per mass amount of fuel. In the literature in one example, α is 14.7 in terms of weight of air supply (Toyota Motor Sales U.S.A. Inc. 2014).

Therefore, substituting for C_{oxy} in Equation 3.22, we obtain

$$\frac{dC_{fuel}}{d\theta} = -\alpha k C_{fuel}^2 \tag{3.24}$$

Integrating Equation 3.24 (the derivation is provided in Appendix 3C), we obtain

$$C_{fuel} = \frac{C_{fuel0}}{(1+\alpha\ k\ C_{fuel0}\theta)} \tag{3.25}$$

From this expression, we can see that the fuel is dependent (and decreases) with time, as we would intuitively expect since we are burning it up. As time (θ) gets large, the fuel content C_{fuel} will tend to zero. This is also as one would expect.

C_{fuel0} is the initial fuel concentration in the clearance volume at the onset of the combustion reaction (when time θ is zero).

Substituting Equation 3.25 back into Equation 3.24 results in the revised expression for the rate of reaction (or rate of combustion) as

$$\frac{dC_{fuel}}{d\theta} = -\alpha k \left[\frac{C_{fuel0}}{(1+\alpha\ k\ C_{fuel0}\theta)}\right]^2 \tag{3.26}$$

The reaction of this fuel with oxygen in the air (in the short time interval in the clearance volume to drive the piston) generates heat. The heat generated raises temperature and pressure. This process is really continuous since the cycle repeats itself after the piston has returned from travel by the action of the drive shaft cam mechanism on the piston rod. This means that the cyclic process really proceeds in a sinusoidal form over time. Analysis of the sinusoidal steady-state process is interesting. However, it does not add new information to analyze it this way unless the sinusoidal pulsing somehow alters friction losses, for example, at the piston surface as it proceeds along the cylinder wall. If the pulsing results in a minor vibration so that the "microgap" between piston and cylinder varies, this variation in the "microgap" can thereby alter the rubbing action that can affect starting and rolling friction between the piston and the cylinder wall. Under this premise, controlling that frequency could help reduce friction and losses of power potentially. If ethanol can help this by interacting somehow at the piston–cylinder interface then this could result in advantages for ethanol. Control of surface tension at the surface due to ethanol effects might be a relevant topic. In this case, analysis of this continuum as a sinusoidal process can be useful.

Consider this reaction for the simpler case as a batch process per cycle, the change in concentration due to combustion of fuels results in emission of heat that varies with time during the cycle.

The heat generated per cycle due to combustion is given by

$$Q_{comb}(\theta) = V_{clear}\Delta H_{comb}\frac{dC_{fuel}}{d\theta} \tag{3.27}$$

where $Q_{comb}(\theta)$ is the heat generated per unit time in combustion in clearance volume V_{clear} and where ΔH_{comb} is the heat of combustion or heating value of the fuel (in units, e.g., BTU/lb or kcal/kg).

Substituting Equation 3.26 into Equation 3.27 (replacing $dC_{fuel}/d\theta$) results in

$$Q_{comb}(\theta) = V_{clear}\Delta H_{comb}\alpha\ k\left[\frac{C_{fuel0}}{(1+\alpha\ k\ C_{fuel0}\theta)}\right]^2 \quad (3.28)$$

The heat of combustion raises the temperature nearly instantaneously in the clearance volume. Letting this volume be fixed for the short time instantaneous reaction, then $Q_{comb}(\theta)$ also equals the expression for time-dependent temperature change given as

$$Q_{comb}(\theta) = V_{clear}(C_{fuel0} + \alpha\ C_{fuel0})\ \frac{C_v dT}{d\theta} \quad (3.29)$$

where the term V_{clear} $(C_{fuel0} + \alpha\ C_{fuel0})$ expresses the total mass of fuel and air injected into the clearance volume, and C_v is the specific heat for the temporarily constant volume. Most of this volume is air where approximate figures for the specific heat values are 1.0048 kj/kg K or 0.240 BTU/lb R for C_p and 0.7159 kj/kg K or 0.171 BTU/lb R for C_v (Johnston, R.M. et al. 1992, p. 70).

Using Equation 3.29, substituting for $Q_{comb}(\theta)$ into Equation 3.28, and rearranging to solve for $dT/d\theta$, Equation 3.30 is obtained. Note that V_{clear} cancels from both sides of the equation.

$$\frac{dT}{d\theta} = \frac{[\Delta H_{comb}\alpha\ k\ C_{fuel0}]}{\{[(1+\alpha\ k\ C_{fuel0}\theta)^2][(1+\alpha)\ C_v]\}} \quad (3.30)$$

From, Appendix 3D, Equation 3.30 is integrated to obtain

$$T = T_i + \left(\frac{\Delta H_{comb}}{[(1+\alpha)\ C_v]}\right)\left\{1 - \left[\frac{1}{(1+\alpha\ k\ C_{fuel0}\theta)}\right]\right\} \quad (3.31)$$

As time gets large, the temperature approaches steady state, given by

$$T = T_i + \left(\frac{\Delta H_{comb}}{[(1+\alpha)\ C_v]}\right) \quad (3.32)$$

The region of combustion occurs at constant volume under the assumption that the reaction is fast compared to piston movement (which seems quite reasonable). Also, the combustion reaction is an irreversible process and generates internal heat so it is non-adiabatic in that sense and not isentropic since the entropy should increase in this process as disorder occurs. Therefore, Equations 3.14 and 3.15 where pressure is related to volume as indicated do not apply to the combustion phase. However, the perfect gas law does apply, and also the volume is fixed until the piston begins to move as a consequence of the pressure generated by the heat of combustion.

Given this, applying the perfect gas law, Equation 3.2 applies

$$PV = nRT \tag{3.2}$$

In this case, volume is fixed at a specific volume per unit mass. During the compression, the volume per unit mass was changed from V_1, the initial volume per unit mass of gas entering to the value V_2. The pressure and temperature of the gas mixture at inlet are fixed, so the volume is fixed and known from the perfect gas law. The volume change in compression prior to ignition is also known, since it is the initial specific volume times the ratio of the piston clearance volume to the sum of the displacement volume and the clearance volume. So the specific volume of the gas after compression is given by

$$\frac{V_2}{V_1} = V_1\left[\frac{V_{clear}}{(V_{clear} + V_{disp})}\right] \tag{3.33}$$

Since V_2, the specific volume after compression is known or estimated, then P_2 and T_2 can be calculated from the perfect gas equation (Equation 3.2) and from the equation for isentropic compression (Equations 3.14 and 3.15). The specific volume after combustion, before the cylinder moved, will be V_3 (the same specific volume as V_2). Again, this assumes that the moles before and after combustion are the same. This is not strictly true, though the assumption is reasonable for our purposes (to identify variables deemed important to helping establish the value of ethanol and point to how to evaluate the importance of such variables).

The mass of gas in the system is the total supplying the reaction or $V_{clear}\ C_{fuel0}$ $(1 + \alpha)$. Note the moles in the system is represented by mass divided by the molecular weight. The molecular weight is essentially that of air for our purposes. For purposes of computing pressure, we can, however, use Equation 3.2.

$$\text{Therefore, } P_3 = \frac{RT_3}{V_3} \tag{3.34}$$

Substituting into Equation 3.28a, we obtain

$$P_3 = \frac{R}{V_3}\left\{\left[T_i + \left(\frac{\Delta H_{comb}}{[(1+\alpha)\ C_v]}\right)\right]\left(1 - \left[\frac{1}{(1+\ \alpha\ kC_{fuel0}\theta)}\right]\right)\right\} \tag{3.35}$$

Equation 3.35 provides the pressure that exists due to the combustion reaction initiated in the clearance volume. As time θ gets large so that $\alpha\ k\ C_{fuel0}\theta$ is much greater than 1, Equation 3.35 reduces to the steady-state form given by

$$P_3 = \frac{R}{V_3}\left\{T_i + \left(\frac{\Delta H_{comb}}{[(1+\alpha)C_v]}\right)\right\} \tag{3.36}$$

This pressure is the starting point for driving the piston in the cylinder and is the starting point for the next expression to be derived dealing with piston motion in cylinder retarded by friction. What is of interest with regard to Equations 3.35 and 3.36 is to find the way to obtain the minimum pressure resulting from combustion that minimizes energy in order to drive the cylinder piston in the manner necessary. It may be that the value of C_v that is dependent on temperature is important so there may be an optimal temperature to attain. Since this expression assumes the perfect gas law is valid, deviations from this may dictate a particular temperature range during combustion.

The magnitude of ΔH_{comb} is important as this is the heating value of the fuel, decidedly different for ethanol and gasoline as discussed. However, the model in this chapter is the E15 (15% ethanol and 85% gasoline by weight). It is entirely possible that exact expressions for this complex system may dictate that there is an optimal blend to minimize energy usage (Curtis, S. et al. 2008). Certainly, the quantity of fuel per cycle will be important as well as the exact kinetics of how that fuel is burned. Apparently, from Equations 3.35 and 3.36, the temperature and also the pressure at the onset of combustion are important. Further the rates of reaction and fuel/air ratio come into play if times to steady state are relatively very long.

3.5 SAMPLE CALCULATION

The following sample calculation (using Equations 3.31 through 3.36) facilitates describing this subject:

The sample calculation will cover the combustion area following compression of the introduction of the gas mixture following a piston stroke and the expansion following the combustion since this is the region where the fuel selection may have the most impact. These are the areas that are expected to impact on the value of ethanol as a fuel component.

3.5.1 Sample Calculation for Adiabatic Isentropic Compression Prior to Ignition

The incoming gas mixture is assumed to be at a pressure of 15 lbs-force/sq.in., equivalent to 1034 dynes/sq. cm. Assume a temperature of 75°F, which is 24°C or 297 K for this gas. Assume the gas is compressed before the combustion reaction. Figures for appropriate conditions are provided in a cited reference example (Johnston, R.M. et al. 1992, p. 238).

V_{clear} is the clearance volume. For the example of the Honda Sedan XE, the compression ratio is 9.7:1 (Honda Motor Sales 2007). The compression ratio is defined as

$$\frac{V_{disp} + V_{clear}}{V_{clear}} = \text{Compression Ratio} \tag{3.37}$$

Therefore, the clearance volume is

$$V_{clear} = \frac{V_{disp}}{\text{Compression Ratio} - 1} \tag{3.38}$$

Since for this model automobile, $V_{disp} = 2354$ cc, and the compression ratio is 9.7, the clearance volume is calculated to be 271 cc from Equation 3.38.

The frontal area of a piston in the four cylinder 2007 Honda Accord is computed from the bore of 8.7 cm as 59 cm^2.

Note, it may be that the choice of the piston/cylinder size and the number of cylinders can impact on the efficiency of fuel usage, and perhaps the use of four cylinders is optimal. This may be something to examine further in engine design. However, obviously this does not matter for automobiles already on the road. One must optimize for existing vehicles and their design. This then falls upon fuel design and usage as well as lubricant design and usage. This area has been (and is) receiving considerable research interest (West, B.H. et al. 2007; Carden, P. 2010).

The incoming fuel mixture is subjected to compression before ignition and combustion. In this case, the initial compression involves going from the displacement volume of 2354 cc to a volume of 271 cc, a compression ratio of 9.7 as noted. The pressure attained is based on isentropic compression and follows Equations 3.14 and 3.15 $P_2/P_1 = (V_1/V_2)^{Cp/Cv}$. The specific heat ratio assumed ($k = C_p/C_v$) is 1.4 (Johnston, R.M. et al. 1992, p. 238). The absorption of heat is critical to understanding optimal use of energy. This argues for calorimetric experiments to see if there is an optimal compression ratio for absorption of heat with the equations as a guide since that may help determine best conditions for engine operation for new designs and existing automobiles that may be tuned to such conditions if that is not done so already (Balich, G.W. and Ashenbach, C.R. 2004). Again, the exact dependence of k on temperature and conditions may be important in order to determine where to operate in conducting such compression prior to ignition. In this case, according to Equations 3.14 and 3.15, the compression results in a pressure of 309 psia, or 21,325 dynes/cm^2. Since hot gasses are exhausted, the initial temperature is based on that of the new cycle of incoming air fuel mixture or 297 K. For this ideal gas, by an adiabatic process and assuming a perfect gas, the temperature resulting after adiabatic (isentropic) compression is given by Equation 3.12.

$$\left(\frac{P_2}{P_1}\right) = \left(\frac{T_2}{T_1}\right)^{Cp/R}$$

Rearranging this equation results in

$$T_2 = T_1\left(\frac{P_2}{P_1}\right)^{R/Cp} \tag{3.39}$$

Inserting values (assuming the gas is air), where $R/C_p = 0.0689$ kcal/(kg K)/0.2395 kcal/(kg K) = 0.288 (Johnston, R.M. et al. 1992, p. 70).

$T_2 = 297$ K (21,325 dynes/cm^2)/(1034 dynes/cm^2)$^{0.288}$ or
$T_2 = 710$ K

Therefore, the conditions following adiabatic or isentropic compression (assumed for our purposes here) are a pressure of 21,325 dynes/cm^2 or 309 psia, and a temperature of 710 K in the clearance volume (271 cc). This is the condition prior to ignition of the fuel/air mixture already injected in the clearance volume.

3.5.2 Sample Calculation for the Ignition and Combustion Phase

In a previous cycle, the gas mixture has been drawn into the clearance volume. The calculation in the previous section describes a sample calculation for the compression of that gas prior to ignition. From before, the equations for combustion kinetics in relation to pressure achieved are shown below.

Therefore, Equations 3.40 and 3.41 show the elevation in temperature in the clearance volume dependent on the heating value and the reaction kinetics.

$$T_3 = T_2 + \left(\frac{\Delta H_{comb}}{[(1+\alpha)\ C_v]}\right)\left\{1 - \left[\frac{1}{(1+\ \alpha\ k\ C_{fuel0}\theta)}\right]\right\} \tag{3.40}$$

$$T_3 = T_2 + \left(\frac{\Delta H_{comb}}{[(1+\alpha)C_v]}\right) \tag{3.41}$$

As time θ gets large so that α k $C_{fuel0}\theta$ is much greater than 1, Equation 3.40 reduces to the steady-state form given by Equation 3.41. Note that the temperatures are noted as being T_2 for the end of the compression cycle and T_3 for the end of the combustion cycle.

Continuing with the sample calculation, P_2 is equal to 21,325 dynes/cm^2 or 309 psia and T_2 is equal to 710 K. The volumetric specific heat $C_v = 0.1706$ cal/g K.

The grams in the system (g) have to be expressed in gram-moles (g-moles). This means the molecular weight has to be estimated.

With the basis being a gallon of fuel, 0.15 gal is ethanol and 0.85 gal is gasoline.

Since the density of ethanol is 6.836 lbs/gal, then this involves 1.0254 lbs of ethanol or 466 g. Similarly, the density of gasoline is 6.152 lbs/gal, so 0.85 gals of gasoline represents 5.2292 lbs of gasoline or 2377 g (Appendix 3B).

Since the molecular weight of ethanol is 46, then 466 g of ethanol represents 466 g/46 g per gram-mole or 10.13 gram-moles of ethanol The average molecular weight of gasoline approximates that of heptane so use 105 g per gram-mole. Therefore, the gram-moles of gasoline in the fuel mixture is 2377 g/105 g/gram-moles or 22.64. Therefore, adding the gram-moles of ethanol to the gram-moles of gasoline in a gallon of fuel (10.13 gram-moles + 22.64 gram-moles) results in 32.77 gram-moles total. The total gram amount in the gallon of fuel is 466 g of ethanol and that of gasoline is 2377 g. Adding these together gives the total weight of the gallon of fuel or 2843 g. Dividing the total grams by the total gram-moles gives the average molecular weight (2843 g/32.77 gram-moles) of 86.8 g/gram-mole for the fuel mixture, excluding the air involved.

The specific heats used represent the specific heats of the air and gasoline/ethanol mixture. The gas mixture entering the carburetor (by our example in this chapter) contains 14.7 g of air for each gram of fuel. The molecular weight of air is 29 g/gram-mole. On a mole or volumetric basis, air is 21% oxygen. Therefore, about

3.4 g of oxygen is provided for each gram of fuel. This is the figure that is relevant for combustion (the amount of oxygen is approximately that required by stoichiometry to burn heptane). Basically, the specific heat of the fuel/air mixture that can be used in calculations is approximately that for air.

The heat of combustion is for the mixture of ethanol and gasoline, being 15% ethanol by volume. From before, this is calculated to be 18,600 BTU/lb or 10,312 cal/g. At 2 g/s with 0.01 s/stroke (3000 rpm crankshaft at two strokes per revolution yields 6000 strokes per minute, and at 1 minute per 60 seconds the result is 0.01 seconds per stroke), 0.02 g of fuel per stroke is consumed. The fuel concentration is 0.02 g/271 cm^3, where 271 cc is the clearance volume, or a concentration of 7.4×10^{-5} g/cm^3 for the combustion reaction per stroke.

From before, the heating value or heat of combustion is 18,600 BTU/lb or 10,312 cal/g fuel. We will substitute numbers into Equation 3.42, the steady-state solution

$$T_3 = T_2 + \left(\frac{\Delta H_{comb}}{[(1+\alpha)C_v]} \right) \tag{3.42}$$

The term $1 + \alpha$ is a parameter representative of air-fuel ratio. If $1 + \alpha$ is 15.7, then

$T_3 = 710$ K + (10,312 cal/g/[(15.7) (.171)] or
$T_3 = 710$ K + 3841 K = 4551 K

This number is too high, since if real it would damage the piston and cylinders. The correction to this (in order to use the model in this chapter and the kinetics assumptions, prior to confirming matters by experiment) is to have the system not reach steady state but effectively cease at an earlier time point. Then the temperature and pressure would not get that high. This is plausible since the system could reach a point where the piston started moving before steady state is reached in temperature and pressure after combustion. The temperature attained should be about half of the final above figure (4551 K) according to the literature (Johnston, R.M. 1992, p. 238).

In Equation 3.40, given by

$$T = T_i + \left(\frac{\Delta H_{comb}}{[(1+\ \alpha)\ C_v]} \right) \left(1 - \left[\frac{1}{(1+\ \alpha\ k\ C_{fuel0}\theta)} \right] \right)$$

let $\alpha\, k\, C_{fuel0}\theta = 0.5$ to achieve this 50% reduction.

With $C_{fuel0} = 7.4 \times 10^{-5}$ g/cm^3 per stroke, and $\alpha = 14.7$, then k θ becomes 540. If $\theta = 0.001$ s then k is 5.4×10^5 (g/cm^3-s)$^{-1}$. This is not unreasonable for such extremely fast reactions (Heghes, C. I. 2006; Sivaramakrishnan, R. et al. 2010; Westley, F. 1980). In any case making this assumption, then

$T_3 = 710$ K + 3841(0.35) K = 2054 K. This is a more realistic figure (Johnston, R.M. et al. 1992). If the concentration is 10 times higher (7.4×10^{-4} g/cm^3 per stroke), then k is 10 times lower or 5.4×10^{-4} (g/cm^3-s)$^{-1}$

Since we are assuming that the system still obeys the Perfect Gas Law, as the temperature and pressure conditions allow this, we can compute the pressure using this assumption. The system is mostly air and largely nitrogen gas since air is 79% nitrogen by volume and mole content. If one examines the van der Waals assumption as an equation of state rather than the perfect gas law, some error is introduced (see Appendix 3E). Since we are dealing with small changes that can impact on energy usage, it will be useful to obtain actual measurements for purposes of examining the proper equation of state for the combustion situation as a separate project (Reid, R.C. et al. 1987; Johnston, R.M. 1992). This is relevant to more accurately predicting the pressure resulting to drive the piston. Note there is interest in obtaining measurements of pressure, temperature, and specific volume to check theory versus actual data (Kuo, P.S. 1996; Ratiu, S. et al. 2009; Johnston, R.M. 1992, p. 70).

As was noted earlier, the volume per mole is fixed at V_{clear} (at the resulting pressure and temperature) prior to the piston volume motion up until the point where the piston moves. The number of moles n is based on the total of that for air and fuel. The average molecular weight for the fuel as determined previously is 86.8 g per mole. On the basis of 1 g of fuel and 14.7 g of air, the moles of air is 14.7/29 (= 0.507 moles) and that of fuel is 1/29 (= 0.0345 moles), so the average molecular weight of the mixture is 15.7/0.542 or 28.97 or still 29. The total amount per unit volume in the clearance volume from this example is $(1 + \alpha)\ C_{fuel0}$ or $(1 + 14.7)\ 7.4 \times 10^{-4}$ g/cm^3 or 1.16×10^{-2} g/cm^3. At 29 g/g-mole, this is 4.01×10^{-4} moles/cm^3. This calculation allows to get a feeling for the approximate amount of the air and fuel combination involved in the combustion reaction.

We can compute the pressure resulting after combustion using the perfect gas law since the volume and total moles are fixed. The piston is assumed to be fixed in position during the reaction, or at least well into the reaction.

$$\text{So, } \frac{P_3}{P_2} = \frac{T_3}{T_2} \tag{3.43}$$

Therefore, $P_3 = P_2\ (T_3/T_2)$ or 21 atm (2054 K/710 K) = 60.8 atm or 893 psia.

Therefore, at this point the piston is postulated to begin to move and carry out expansion while the combustion finishes. Since expansion is occurring, the temperature and pressure rise will begin to abate and then decline in the expansion phase. This expansion is assumed to be adiabatic and isentropic since little heat can be transferred in the time allotted. Again, we assume that the friction effect is such that we can still assume isentropic behavior though of course this is not strictly true.

3.5.3 Sample Calculation for Adiabatic Isentropic Expansion Following Combustion

The equation for the isentropic expansion is given (again)

$$P_4 = \frac{P_3}{(V_4/V_3)^{C_p/C_v}} \tag{3.44}$$

Here as noted, the conditions P_4 and V_4 represent those after expansion to power the engine. The ratio of V_4/V_3 is the compression ratio. In our example, this is 9.7. Again, Cp/Cv is 1.4. P_3 is 60.8 atm or 893 psia. Therefore, P_4 is calculated to be 2.53 atm or 37.1 psia.

From Equation 3.49. $T_4 = T_3\ (P_4/P_3)^{R/Cp}$

From before R/Cp = 0.288 (for air).

From the above, $P_4/P_3 = 2.53$ atm/60.8 atm or 0.0416

From the above example after combustion, $T_3 = 2054$ K

Therefore, $T_4 = 2054\ \text{K}\ (0.0416)^{0.288} = 2054\ \text{K}\ (0.408) = 822$ K.

This hot gas mixture is subsequently exhausted to begin the next stroke cycle where again all this heat is wasted. It should be reclaimed in improved designs as noted unless it can somehow be reclaimed and used for other purposes.

3.6 CONSIDERATIONS OF PISTON EXPANSION WITH REGARD TO FORCES AND FRICTION

The next step in the analysis is to see how the pressure attained, as given in Equation 3.44 impacts on equations related to piston flow. This should be a guide for experiments to understand piston ring and piston expansion so as to aid in experiments that seek to reduce energy usage when ethanol is used. To begin with, examining Figure 3.2, the force across the surface of the piston divided by the piston cross-sectional area yields the driving pressure (P in Equation 3.44). Expressing this mathematically, applying Newton's First law of motion, in the presence of resisting friction, Equation 3.45 is assumed to apply as a first basis approach.

$$P(\theta)A_1 = M\,a_1(\theta) + \tau_{11} + \tau_{01}e^{-\beta(\theta)} + \text{resisting forces due the linkage to the crankshaft} \qquad (3.45)$$

where $P(\theta)$ depicts pressure varying with time, A_1 is the frontal area of the piston, and M is the mass of the piston. τ_{11} is part of the initial shear stress (friction) to overcome at the cylinder wall in order to move the piston that is continually in contact with the cylinder wall through a lubricant layer during motion (it may be proportional to the driving pressure with the proportionality constant being a coefficient of moving friction (Andersson, P. et al. 2002; Shenoy, P.S. and Fatemi, A. 2006; Postma, R.W. 2004). $a_1(\theta)$ is the acceleration of the piston as it begins expansion from rest, $\tau_{01}\ e^{-\beta(\theta)}$ is part of the shear stress (friction-causing) during motion that has to be overcome initially to move the cylinder under pressure which decreases with time quite rapidly.

The resisting forces on the piston rod are complex to analyze. For our purposes, focusing on the piston and the cylinder, these forces are taken to be against the force on the piston and perhaps proportional to the applied force $P(\theta)A_1$ during the expansion. They may be considered to be contributors (in a sense) to the shear forces $\tau_1 + \tau_{01}e^{-\beta(\theta)}$ to focus on factors that may relate to ethanol usage. If the resisting forces due to the linkage to the crankshaft mechanism are taken to be proportional to the driving pressure only during the period of piston expansion, then the proportionality factor (designated k_1) can be combined into the remaining factors on the right-hand-side of Equation 3.45. Introducing this proportionality factor, Equation 3.45 is changed to

$$P(\theta)A_1 = M\ a_1(\theta) + \tau_{11} + \tau_{01}e^{-\beta(\theta)} + k_1P(\theta)A_1 \tag{3.46}$$

Subtracting k_1 $P(\theta)A_1$ from both sides of Equation 3.46 and separating out the factor $1 - k_1$ (k_1 is assumed to have a range of $0 \leq k_1 \leq 1$) results in Equation 3.47

$$(1-k_1)\ P(\theta)A_1 = M\ a_1(\theta) + \tau_{11} + \tau_{01}e^{-\beta(\theta)} \tag{3.47}$$

Dividing through by $1 - k_1$ results in Equation 3.48

$$P(\theta)A_1 = M\ a_1(\theta)/(1-k_1) + \tau_{11}/(1-k_1) + \tau_{01}/(1-k_1)\ [e^{-\beta(\theta)}] \tag{3.48}$$

For simplicity, we designate

$$a(\theta) = \frac{a_1(\theta)}{(1-k_1)} \tag{3.49}$$

$$\tau_1 = \frac{\tau_{11}}{(1-k_1)} \tag{3.50}$$

and

$$\tau_0 = \frac{\tau_{01}}{(1-k_1)} \tag{3.51}$$

Substituting Equations 3.49 through 3.51 into Equation 3.48 results in

$$P(\theta)A = M\ a(\theta) + \tau_1 + \tau_0e^{-\beta(\theta)} \tag{3.52}$$

This definition of friction allows the resisting friction force due to piston/cylinder interaction to be $\tau_1 + \tau_0$ initially before the piston is moving. At large times, resistant frictional shear reduces to τ_1 (where β is an adjustable parameter that can be determined experimentally).

The point at which the piston can move is when the pressure overcomes resistant shear friction pressure equal to the $\tau_1 + \tau_0$. This point is a point in time is close to when the combustion reaction ceases and may be just prior to that point as indicated in the previous example. In a well-running engine, tuning will preferably be timed for optimal transition from the period of combustion reaction (where pressure is produced) to the period where piston expansion occurs (due to the driving pressure).

Note the piston in actuality moves (in theory) when the combustion reaction is still proceeding resulting in a situation where pressure is increasing due to combustion and diminishing due to expansion. It may be (as expected) that the combustion and expansion states are quite separate so that the two effects can be analyzed separately. If the two areas will interact, then this causes a peak in the pressure from the combustion phase to the expansion phase (see the graph in Johnston, R.M. et al. 1992, p. 224). Presumably, the two phases will interact minimally in a well-timed engine control system.

It may be that factors such as τ_1 and τ_0 can be experimentally measured when examining materials in construction of the cylinder, cylinder wall, and piston plus penetrants that can aid in reducing friction (Ahmed, N.S. and Nassar, A.M. 2011; Brooke, L. 2015; Gangopadhyay, A. et al. 2012; Kennedy, M. et al. 2012).

The pressure given in Equation 3.52 decreases due to expansion, starting from the pressure attained at the end of combustion. As in the case for compression, the expansion can be assumed to occur isentropically since the friction is relatively small compared to total force required to move the piston, and since essentially no heat is lost from the system except when expansion is over and the gas is discharged. The gas gets cool due to expansion.

Assuming a perfect gas as the piston expands following combustion, and that this expansion can be assumed to be isentropic, then Equation 3.53 applies for the expansion phase. The initial pressure is that at the end of combustion (P_3) and the initial volume is V_3. In Equation 3.53, P_4 and V_4 are the pressure and volume, respectively, that result during the expansion of the piston in the cylinder.

$$P_3/P_4 = \left(V_4/V_3\right)^{C_p/C_v} \tag{3.53}$$

At the end of piston travel, the volume (V_4) is that of the cylinder displacement plus the clearance volume, or $V_{disp} + V_{clear}$. The cylinder displacement (V_{disp}) is known and the total distance of travel times the cylinder frontal area is V_{disp}. The pressure (P_4) is the final pressure following expansion when the volume is $V_{disp+}V_{clear}$. In Equation 3.52, pressure P(θ), is P_3. Therefore, rewriting Equation 3.53, expressing it in terms of P_3, results in

$$P_3 = P_4\left(\frac{V_4}{V_3}\right)^{Cp/Cv} \tag{3.54}$$

Substituting Equation 3.54 into Equation 3.52 results in

$$P_4\left(\frac{V_4}{V_3}\right)^{Cp/Cv} = M\,a(\theta) + \tau_1 + \tau_0 e^{-\beta(\theta)} \tag{3.55}$$

The volume V_4 is given by piston movement and piston area. The acceleration a(θ) may be experimentally and theoretically determined by measurements of piston position and piston velocity resulting in expressions as a function of time according to Newton's Law of motion (Nolan, P.J. 1993).

In Equation 3.55, P_4 and V_4 are the final pressure and total volume, respectively, after piston expansion (displacement). The term V_3 is the volume change as a function of time as the piston is moving so this is $V_3(\theta)$. This volume, $V_3(\theta)$, can be written as axial length or position (x) of expansion (that changes with time) times the front area of the piston (A_p). Similarly, V_4 can be written as the final axial length after expansion is completed (x_f), times the frontal area (A_p). Substitution into Equation 3.55 (cancelling the A_p term which appears in both the numerator and denominator on the left hand side of the equation) results in

$$P_4\left(\frac{x_f}{x}\right)^{Cp/Cv} = M\ a(\theta) + \tau_1 + \tau_0 e^{-\beta(\theta)} \tag{3.56}$$

The acceleration of the piston [$a(\theta)$] can be expressed as the second derivative of the piston position with time or $d^2x/d\theta^2$. Substituting the expression for this second derivative into Equation 3.56 results in

$$P_4(x_f/x)^{Cp/Cv} = M\ d^2x/d\theta^2 + \tau_1 + \tau_0 e^{-\beta(\theta)} \tag{3.57}$$

Equation 3.57 can be solved (probably numerically) and results used to examine data and make predictions (Thomas, C.T. 1992). Note that the assumption in Equation 3.46 may be replaced by assuming that the force from the rotating crankshaft acting on the piston rod is a function of time according to the specific design, position of the piston rod, and so on, so that Equation 3.57 may be tested with variations of functions as input since the piston motion will be sinusoidal in nature (EPI 2014). In this case, Equation 3.57 becomes

$$P_4\left(\frac{x_f}{x}\right)^{Cp/Cv} = M\ d^2x/d\theta^2 + \tau_1 + \tau_0 e^{-\beta(\theta)} + \text{forcing function to result in sinusoidal motion} \tag{3.58}$$

By inspection, the forcing function should be of the form $dx/d\theta$ and one to try first is $k\ dx/d\theta$. Since there is a power term, C_p/C_v, in the equation, an analytical solution is unlikely. In any case, Equation 3.58 may be written

$$P_4\left(\frac{x_f}{x}\right)^{Cp/Cv} = M\ d^2x/d\theta^2 + \tau_1 + \tau_0 e^{-\beta(\theta)} + k\ dx/d\theta \tag{3.59}$$

The forcing function shown in Equation 3.59 is related to the crankshaft angular velocity (ω) as described in the literature (Wikipedia 2014; EPI 2014). If we were to assume that c_p/c_v is unity merely to check the mathematics, we would find that the solution to the differential equation shown as Equation 3.59 will have cosine and sine terms along with exponential terms so that its form will fit the sinusoidal variation we expect to see (Churchill, R.V. 1941; Rainville, E.D. 1958). Therefore, this form is appropriate to test on actual data to see what may be said about relating terms such a pressure, piston measurements, and crankshaft rotational parameters on optimizing friction reduction. This is a fruitful area for research.

Experiments can be carried out to assess the values for shear or friction. Note when θ is the final time (dictated by crankshaft rotational speed), then the final distance is known (the extent of piston travel), the mass of the piston is known and so are the initial conditions that follow combustion. Therefore, with several experiments, then unknowns can be determined using statistical analyses. This can provide insights into frictional terms and effects on them. The pressure rise from combustion powers piston travel. The piston travel is controlled to stop when expansion is over based on the rotation of the crankcase and connection to the piston rod.

The analyses in this chapter and equations represent variables that can be examined in relation to fuel use and materials in the cylinder-piston-ring seals as described in order to see about reducing energy usage as a consequence of using ethanol in fuel mixtures. Examining facets of variables and implications allows one to postulate (and later experimentally check for) small changes that in total may result in significant positive contributions for ethanol use. Looking at the sum total of factors in all aspects is a key objective of this book since the author believes the sum total of small changes will result in large impact on pointing to factors that result in the maximum value of ethanol usage. It is better not to assume that there will be one "magic" step to dramatically optimize the situation. It is always possible some discovery will be made that will change matters suddenly. Perhaps, it is possible that by such step-wise optimization that such discoveries will made because of the focused attention of researchers on multi-step improvements.

APPENDIX 3A: DERIVATION OF THE EQUATION FOR ISENTROPIC EXPANSION OR COMPRESSION FOR AN IDEAL GAS

Starting with

$$\left(\frac{P}{P_1}\right)=\left(\frac{T}{T_1}\right)^{Cp/R} \tag{3.12}$$

For an ideal gas,

$$PV = nRT \tag{3.2}$$

Starting with Equation 3.12, we substitute for T from the perfect gas Equation 3.2, and similarly for T_1, and obtain

$$\frac{P}{P_1}=\frac{(PV/nR)^{Cp/R}}{(P_1V_1/nR)^{Cp/R}} \tag{3.13}$$

The nR cancels out from Equation 3.13 resulting in

$$\frac{P}{P_1}=\frac{(PV)^{Cp/R}}{(P_1V_1)^{Cp/R}} \tag{3.13a}$$

Divide the pressure terms on the left-hand side of the equation by the pressure terms raised to the Cp/R power shown on the right-hand side of Equation 3.13a

$$\frac{[P/P^{Cp/R}]}{[P_1/P_1^{Cp/R}]} = \frac{V^{Cp/R}}{V_1^{Cp/R}} \tag{3.13b}$$

Reexpressing exponents on the left-hand side, Equation 3.13b can be written

$$\frac{P^{1-Cp/R}}{P_1^{1-Cp/R}} = \frac{V^{Cp/R}}{V_1^{Cp/R}} \tag{3.13c}$$

Rewriting the exponents on the left-hand side of Equation 3.13c

$$\frac{P^{R/R-Cp/R}}{P_1^{R/R-Cp/R}} = \frac{V^{Cp/R}}{V_1^{Cp/R}} \tag{3.13d}$$

We can rewrite the exponents on the left-hand side of Equation 3.13d again and show the result as

$$\frac{P^{(R-Cp)/R}}{P_1^{(R-Cp)/R}} = \frac{V^{Cp/R}}{V_1^{Cp/R}} \tag{3.13e}$$

$$\text{Since } C_p - C_v = R \text{ for an ideal gas, then } R - C_p = -C_v \tag{3.13f}$$

Substituting this result in the numerator of the exponent on the left-hand side of Equation 3.13e results in

$$\frac{P^{-Cv/R}}{P_1^{-Cv/R}} = \frac{V^{Cp/R}}{V_1^{Cp/R}} \tag{3.13g}$$

Raising both sides of Equation 3.13g to the reciprocal of Cv/R (1/Cv/R) power results in

$$\frac{P^{-1}}{P_1^{-1}} = \frac{V^{(Cp/R)/(Cv/R)}}{V_1^{(Cp/R)/(Cv/R)}} \tag{3.13h}$$

Cancelling the R terms in the exponents of the right-hand side of Equation 3.13h results in Equation 3.13i

$$\frac{P^{-1}}{P_1^{-1}} = \frac{V^{Cp/Cv}}{V_1^{Cp/Cv}} \tag{3.13i}$$

Rewriting Equation 3.13i to simplify it results in

$$\frac{P^{-1}}{P_1^{-1}} = \left(\frac{V}{V_1}\right)^{Cp/Cv} \tag{3.13j}$$

Inverting both sides of Equation 3.13j results in Equations 3.14 and 3.15 as shown in the main text.

$$\left(\frac{P}{P_1}\right)=\left(\frac{V_1}{V}\right)^{Cp/Cv} \tag{3.14}$$

APPENDIX 3B: DERIVATION OF THE HEAT OF COMBUSTION (HEATING VALUE) OF THE ETHANOL–GASOLINE MIXTURE

Ethanol and gasoline can be blended in different proportions (Curtis, S. et al. 2008). This blending is done by volume (DeSimio, L. et al. 2012). A typical ratio is designated E15–15% ethanol and 85% gasoline (Office of Energy Efficiency and Renewable Energy 2003). The heat of combustion of pure anhydrous ethanol is 12,200 BTU/lb or 6764 kcal/kg (Hodgman, C.D. et al. 1956, p. 1784). The heat of combustion of gasoline (assuming a mean value near that of heptane) is 19,850 BTU/lb or 11,005 kcal/kg (Curtis, S. et al. 2008). The density of pure anhydrous ethanol is 6.836 lbs/gal (0.819 kg/L). The density of gasoline, considering heptane as the average, is 6.152 lbs/gal (0.736 kg/L). The heat of combustion of ethanol on a weight basis is considerably less than that of gasoline as indicated (a disadvantage on the basis that one wants to put as much energy per unit weight as possible into the fuel). In terms of optimization, maybe that is not such a disadvantage once all factors are considered as discussed in this book.

The energy density per unit weight difference narrows when one considers that ethanol has a density that exceeds that of the heptane based average gasoline. Multiplying the heat of combustion in BTU/lb or kcal/kg by the density in lbs/gal or kg/L gives the heat of combustion in BTU/gal or kcal/L. For ethanol that is 83,399 BTU/gal. For gasoline, the heat of combustion is 122,117 BTU/gal. The weighted mean heat of combustion (heating value) for E15 on this volumetric basis is 116,309 BTU/gal. The mean density of the mixture is 6.2455 lbs./gal. On a weight basis again, the heating value is 18,600 BTU/lb or 10,312 kcal/kg. This is about 5% below that of pure gasoline.

APPENDIX 3C: DERIVATION OF THE EQUATION SHOWING THE DEPLETION OF FUEL AS A FUNCTION OF TIME

Starting with

$$\frac{dC_{fuel}}{d\theta}=-\alpha\, k\, C_{fuel}^{\;2} \tag{3.24}$$

Rewriting Equation 3.24,

$$\frac{dC_{fuel}}{C_{fuel}^{\;2}}=-\alpha\, k\, d\theta \tag{3.24a}$$

Integrating this expression (see Zegarelli, M. 2008, p. 60 of this example representative of excellent books for persons wishing to learn a new area).

$$-\frac{1}{C_{fuel}} = -\alpha \text{ k } \theta + \text{the constant of integration} \quad (3.24b)$$

At time zero ($\theta = 0$), $C_{fuel} = C_{fuel0}$, the initial fuel concentration in the clearance volume. Therefore,

Substituting into Equation 3.24b

$-1/C_{fuel0} = 0 +$ the constant of integration, so the constant of integration is given by

$$\text{Constant of integration} = -\frac{1}{C_{fuel0}} \quad (3.24c)$$

Substituting Equation 3.24c into Equation 3.24b, we obtain

$$-\frac{1}{C_{fuel}} = -\alpha \text{ k } \theta - \frac{1}{C_{fuel0}} \quad (3.24d)$$

Rearranging Equation 3.24d to solve for C_{fuel}, we obtain

$$C_{fuel} = \frac{C_{fuel0}}{(1+\alpha \text{ k } C_{fuel0}\theta)} \quad (3.25)$$

APPENDIX 3D: INTEGRATION OF THE CHANGE IN TEMPERATURE WITH TIME TO DETERMINE THE TEMPERATURE DUE TO COMBUSTION

Starting with

$$\frac{dT}{d\theta} = \frac{[\Delta H_{comb} \alpha \text{ k } C_{fuel0}]}{\{[(1+\alpha \text{ k } C_{fuel0}\theta)^2][(1+\alpha)C_v]\}} \quad (3.30)$$

For clarity, let Z represent the grouping $\{[\Delta H_{comb} \alpha \text{ k } C_{fuel0}]/[(1 + \alpha)C_v]\}$
Therefore,

$$\frac{dT}{d\theta} = \frac{Z}{[(1+\alpha \text{ k } C_{fuel0}\theta)^2]} \quad (3.30a)$$

Further let

$$U = 1+\alpha \text{ k } C_{fuel0}\theta \quad (3.30b)$$

to use the variable substitution technique (Zegarelli, M. 2008, p. 118)

Therefore, differentiating this new variable results in Equation 3.30c

$$dU = \alpha\ k\ C_{fuel0} d\theta \tag{3.30c}$$

Expressing $d\theta$ in terms of dU,

$$d\theta = \frac{dU}{(\alpha\ k\ C_{fuel0})} \tag{3.30d}$$

Substituting Equations 3.30b and 3.30d into Equation 3.30a results in

$$(\alpha\ k\ C_{fuel0}) dT/dU = Z/U^2 \tag{3.30e}$$

Rewriting Equation 3.30e,

$$(\alpha\ k\ C_{fuel0}) dT = Z\left(\frac{dU}{U^2}\right) \tag{3.30f}$$

This equation is readily integrated to yield

$$(\alpha\ k\ C_{fuel0}) T = Z\left(\frac{-1}{U}\right) + \text{constant} \tag{3.30g}$$

Replacing U with Equation 3.30b in Equation 3.30g

$$(\alpha\ k\ C_{fuel0}) T = Z\left[-\frac{1}{(1+\alpha\ k\ C_{fuel0}\theta)}\right] + \text{constant} \tag{3.30h}$$

When $\theta = 0$, $T = T_2$, the initial temperature prior to combustion. Substituting into Equation 3.30h at $\theta = 0$, $(\alpha\ k\ C_{fuel0})T_2 = -Z + \text{constant}$, so the constant is expressed as

$$\text{Constant} = (\alpha\ k\ C_{fuel0}) T_2 + Z \tag{3.30i}$$

Substituting Equation 3.30i into Equation 3.30h

$$(\alpha\ k\ C_{fuel0}) T = Z\left[-\frac{1}{(1+\alpha\ k\ C_{fuel0}\theta)}\right] + (\alpha\ k\ C_{fuel0}) T_2 + Z \tag{3.30j}$$

Combining terms involving Z results in

$$(\alpha\ k\ C_{fuel0}) T = Z\left[1-\frac{1}{(1+\alpha\ k\ C_{fuel0}\theta)}\right] + (\alpha\ k\ C_{fuel0}) T_2 \tag{3.30k}$$

Dividing through by $\alpha\ k\ C_{fuel0}$ results in

$$T = Z\,[1 - 1/(1+\alpha\ k\ C_{fuel0}\theta)]/(\alpha\ k\ C_{fuel0}) + T_2 \tag{3.30l}$$

Replacing Z by its definition $\{[\Delta H_{comb}\ \alpha\ k\ C_{fuel0}]/[(1 + \alpha)C_v]\}$ in Equation 3.30l results in

$$T = \left\{\frac{[\Delta H_{comb}\ \alpha\ k\ C_{fuel0}]}{[(1+\alpha)\ C_v]}\right\} \frac{[1 - 1/(1+\alpha\ k\ C_{fuel0}\theta)]}{(\alpha\ k\ C_{fuel0})} + T_2 \quad (3.30m)$$

Cancelling the $\alpha\ k\ C_{fuel0}$ term from the numerator and the denominator of the first term m the righ-hand-side of Equation 3.30m results in

$$T = \{[\Delta H_{comb}]/[(1 + \alpha)\ C_v]\}\ [1 - 1/(1 + \alpha\ k\ C_{fuel0}\theta)] + T_2$$

and rewriting this slightly results in Equation 3.31

$$T = T_2 + \left(\frac{\Delta H_{comb}}{[(1+\alpha)\ C_v]}\right)\left\{1 - \left[\frac{1}{(1+\alpha\ k\ C_{fuel0}\theta)}\right]\right\} \quad (3.31)$$

APPENDIX 3E: COMPARISON OF THE ASSUMPTION OF A PERFECT GAS VERSUS A GAS FOLLOWING VAN DER WAALS EQUATION TO PREDICT THE PRESSURE AFTER COMBUSTION

As indicated in the main body of the text (Section 3.52), the pressure following combustion is predicted to be approximately 60 atm or 60 bar using the Perfect Gas Law. The equation of state may not be a Perfect gas exactly (Bodner Research Web 2014). It may (or may not) be better represented by using van der Waals equation as indicated in this reference. The expression for van der Waals equation is given by

$$\left(P + \left(\frac{n}{V}\right)^2 a\right)(V - nb) = nRT \quad (3E.1)$$

In this expression n is the moles of gas, V is the volume, a is the van der Waals constant to adjust pressure attained to a lower value, b is the van der Waals constant to allow correcting the volume for the molecular size of the gas, R is the gas constant and T is temperature.

In the initial conditions before compression, the pressure is 1 atm or 1 bar and the temperature is 297 K for one mole of gas. The volume per mole is computed to be 24.35 liters per mole using the Perfect Gas equation. After adiabatic compression, where the pressure attained is 21 atm, the volume is reduced to 2.8 liters per mole.

The combustion reaction then ensues raising the temperature to 2054 K. Assuming nitrogen as the representative gas, the value for a is 1.37 atm $(\text{liter/mole})^2$ and that of parameter b is .0387 liter/mole. In the equation, substituting, we obtain

$$(P + 0.1789)\ (2.73) = (1)\ 0.082)\ (2054)$$

or $P = 61.5$ atm, indicating that the van der Waals equation gives about the same results as the Perfect Gas equation. This assumes the volume per mole is unchanged.

If the volume per mole changes much less than expected, not predicted by the Perfect Gas law or by van der Waals equation, then the pressure elevation will be less. Suppose the volume only is reduced by about 50% during compression in reality. After all, the van der Waals equation is also a model assumption. Then the pressure elevation only reaches about 14 atm, not 21 atm. This points out the need to get actual measurements of pressure, temperature, and specific volume to accurately interpret events in the internal combustion cylinder.

REFERENCES

Ahmed, N.S. and Nassar, A.M. 2011. Tribology—Lubricants and lubrication. In Dr. Chang-Hung Kuo (Ed.), *Lubricating Oil Additives*. Egyptian Petroleum Research Institute, Chapter 10. ISBN: 978-953-307-371-2. http://cdn.intechopen.com/pdfs-wm/21936.pdf.

Andersson, Peter, Tamminen, Jaana, and Sandstroem, Carl-Erik. 2002. *Piston Ring Tribology. A Literature Survey*. VTT Research Notes 2178. VTT Industrial Systems. Espoo, Finland.

Balich, G.W. and Ashenbach, C.R. 2004. The Gasoline 4-Stroke Engine for Automobiles. Department of Aerospace and Mechanical Engineering. University of Notre Dame. Notre Dame, IN. http://www3.nd.edu/~msen/Teaching/DirStudies/Gas4Stroke.pdf.

Bodner Research Web. 2014. *Deviations from the Ideal Gas Law*. http://chemed.chem.purdue.edu/genchem/topicreview/bp/ch4/deviation.php

Brooke, L. 2015. Slick solutions for friction reduction. *Automotive Engineering*. Vol. 2, No. 3. P. 24. SAE International, Mechanicsburg, PA.

By, A., Kempinski, B., and Rife, 1.M. 1981, *Knock in Spark Ignition Engines, SAE Papcr No. 810147*. Society of Automotive Engincers. Warrendale, PA.

Callen, H.B. 1960. *Thermodynamics*. John Wiley & Sons, Inc. New York, NY.

Carden, P. 2010. Calculation of friction in high performance engines. Ricardo Software European User Conference. Powerpoint presentation. Ricardo, Millbrook, UK. RD.10/174701. http://www.ricardo.com/

Cesur, I., Ayhan, V., Parlak, A., Savas, O., and Aydin, Z. 2014. *The Effects of Different Fuels on Wear between Piston Ring and Cylinder*. Advances in Mechanical Engineering. Volume 2014, Article ID 503212, http://dx.doi.org/10.1155/2014/503212.

Churchill, R.V. 1941. *Fourier Series and Boundary Value Problems*. McGraw-Hill Book Company, New York.

Curtis, S., Owen, M., Hess, T., and Egan, S. 2008. Effect of Ethanol Blends on a *Spark Ignition, 4-Stroke, Internal Combustion Engine*, Brigham Young University, Provo Utah.

DeSimio, L., Gambino, M., and Iannaccones, S. 2012. *Effect of Ethanol Content on Thermal Efficiency of a Spark-Ignition Light Duty Engine*. Instituto Motori. Italian National Research Council. ISRN Renewable Energy. Article ID 21919703. http://dx.doi.org/10.5402/2012/219703.

EPI. August 2014. Piston Motion Basics. http://www.epi-eng.com/piston_engine_technology/piston_motion_basics.htm.

Flagan, R.C. and Seinfeld, J. 1988. Internal combustion engines. In *Fundamentals of Air Pollution Engineering*, Prentice-Hall, Inc., Englewood Cliffs, NJ, Chapter 4.

Ferrante, A. and Elghobashi, S. November 2005. Reynolds number effect on drag reduction in a microbubble-laden spatially developing turbulent boundary layer. *J. Fluid Mechanics*. Vol. 543. P 93–106. http://dx.doi.org/10.1017/S0022112005006440.

Gangopadhyay, A., McWatt, D.G., and Zdrodowski, R.J. October 19, 2012. Engine Friction Reduction through Surface Finish and Coatings. *DEER Conference*. Ford Research and Advanced Engineering. Dearborn, MI.

Heghes, C.I., Sept. 2006. C_1–C_4 Hydrocarbon Oxidation Mechanism, Dissertation. Doctor of Natural Sciences. Rupertus Carola University, University of Heidelberg. Germany.

Hodgman, C.D., Weast, R.C., and Selby, S. 1956. *Handbook of Chemistry and Physics*, 38th Edition. Chemical Rubber Publishing Company. Cleveland, OH.

Honda Motor Sales. 2007. Honda Accord Sedan, Model XE, Engineering Specifications.

Johnston, R.M., Brockett, W.A., Bock, A.E., and Keating, E.L. 1992. *Elements of Applied Thermodynamics*. 5th Edition. Naval Institute Press. Annapolis, MD.

Karloef, L., Torgersen Axell, L., and Slotfeldt-Ellingsen, D. September 30, 2005. *Why is (sic) Ice and Snow Slippery? The Tribo-Physics of Sking*. Swix Report AS. SWIX Research and Development. Lillehammer, Norway.

Kennedy, M., Hoppe, S., and Esser, J. May, 2012. *Development Friction*. MTZ Worldwide 0512012. Vol. 73. P. 40. Federal-Mogul.MTZ_CarboGlide_Eng_Apr12.pdf.

Knothe, G. August, 2012. *Biodiesel Composition and Fuel Properties*. power point presentation. USDA/ARS/NCAUR. Peoria, Illinois. biodieselcomposition-gerhard-knothe-2012-08.pdf.

Kuo, P.S. 1996. Cylinder pressure in a spark-ignition engine: A computational model. *J. Undergrad. Sci.* Vol. 3. P.141–145.

Levenspiel, O. 1999. *Chemical Reaction Engineering*. 3rd Edition. John Wiley & Sons, New York, NY.

Martinez, I. 1995. Combustion Kinetics. http://webserver.dmt.upm.cs/~isidoro/bk3/cis/Combustion%20Kinetics.pdf.

Nbu-Nada, E., Al-Hinti, F., Al-Sarkhi, A., and Akash, B. March 2008. Effect of piston friction on the performance of the SI engines: A new thermodynamic approach. *Journal of Engineering for Gas Turbines and Power*. Vol. 130, P. 022802–1.

Nolan, P.J. June 1993. *Fundamentals of College Physics*. 5th Edition. William C. Brown Publishers, Dubuque, IA.

Noller, C.R. 1958. *Textbook of Organic Chemistry*. 2nd Edition. W.B. Saunders Company, Philadelphia, PA.

Office of Energy Efficiency and Renewable Energy. 2003. *Freedom CAR and Vehicles Technologies Program*. U.S. Department of Energy. August 2003. http://energy.gov/eere/vehicles/vehicle-technologies-office.

Postma, R.W. 2004. *Force and Torque Margins for Complex Mechanical Systems*. http://www.esmats.eu/amspapers/pastpapers/pdfs/2004/postma.pdf.

Rainville, E.D. 1958. *A Short Course in Differential Equations*. 2nd Edition. The Macmillan Company, New York.

Ratiu, S., Nicolae Popa, G., and Alexa, V. Jan. 2009. Monitoring of the pressure inside the cylinder for an internal combustion engine. *WSEAS Transactions on Circuits and Systems*. Vol. 8. Issue 1.

Reid, R.C., Prausnitz, J.M., and Poling, B.E. 1987. *The Properties of Gases and Liquids*. 4th Edition, McGraw-Hill, New York.

Shenoy, P.S. and Fatemi, A. 2006. Dynamic analysis of loads and stresses in connecting rods. *Proc. I. Mech E* Vol. 220. Part C. p. 615: J Mech Eng Sci. DOI 10.1243/09544062JMES105.

Sinnamon, J.F. and Tielking, J.T. October, 1974. *Hydroplaning and Tread Pattern Hydrodynamics*. Report No. UM-HSRI-PF-74-10. Highway Safety Research Institute. University of Michigan. Ann Arbor, Michigan.

Sivaramakrishnan, R., Su, M.C., Michael, J.V., Klippenstein, S.J., Harding, L.B., Ruscic, B. September 2010. Rate constants for the thermal decomposition of ethanol and its biomolecular reaction with OH and D: Reflected shock tube and theoretical studies. *J. Phys. Chem. A*. Vol 114(35), P. 9425–39.

Smedley, G. May, 2004. *Piston Ring Design for reduced Friction in modern internal combustion Engines*. MS dissertation. Department of Mechanical Engineering, Massachusetts Institute of Technology, Cambridge, MA.

Thomas, C.T. 1992. *Heat Transfer*. Prentice-Hall, Englewood Cliffs, NJ.

Toyota Motor Sales U.S.A., Inc. 2014. *Emissions#1-Combustion Chemistry*. http://www.princeton.edu/ssp/64-tiger-cub-1/64-data/combustion-chemistry.pdf.

Wei, Nasha, Fengshou, Gu, Guoxing, Li, Wang, Ti, and Ball, Andrew. July 13–17, 2014. Characterizing (sic) the Friction and Wear between the Piston Ring and Cylinder Liner based on Acoustic Emission Analysis. *The 21st International Congress on Sound and Vibration*. Beijing, China.

West, B.H., Lopez, A.J., Theiss, T.J., Graves, R.L., Storey, J.M., and Lewis, S.A. 2007. *Fuel Economy and Emissions of the Ethanol-Optimized Saab 9-5 Biopower*. Report 2007-01-3994. Oak Ridge National Laboratory. Oak Ridge, TN.

Westley, F. April, 1980. Table of Recommended Rate Constants for Chemical Reactions occurring in Combustion. National Bureau of Standards. U.S. Dept of Commerce. Washington, D.C.

Wikipedia. October. 2014. *Piston Motion Basics*. http://en.wikipedia.org/wiki/piston_motion_equations. Last modified on September 1, 2016.

Zegarelli, M. 2008. *Calculus II for Dummies*. Wiley Publishing Company, Hoboken, NJ.

4 Computer-Based Optimization of the Ethanol Process

4.1 INTRODUCTION

The ethanol process must achieve adequate economic status in order to be the basis of a successful commercial enterprise. This can best be achieved by a process that is controlled and optimized using computers. The process should be operated under control so that each aspect can be tuned to operate as best as it can.

Similarly, economic aspects dictate that specification and supply of raw materials, particularly feedstock, should be managed through the assistance of computers. This should preferably include management, planning, and purchasing of feedstock using advance arrangements.

Process control includes optimization of the method and interaction of components used to make ethanol. Personnel are intimately involved with the control process. Based on a paper, the author wrote a report on how people are part of process batch delays and the speed at which process batch sequences are carried out (Forrester, J.W. 1968). In the past, the author's experience included developing machine language software to operate a discrete analyzer to process samples for blood component measurements (Goldstein, W.E. et al. 1975). The device was constructed and software designed so that samples would be processed and analyzed at a rate totally relaxed in response to the number of the time sequences required for each instrument function, the sample type such as testing for blood glucose and urea, and the frequency at which samples are presented to this system. This is a form of control.

A second example, perhaps more germane to the ethanol process, is the development of the pilot plant computer process control system to optimize fermentation processes so that they were technically and economically optimal for introduction into manufacturing. This work was done at the pilot plant scale through 1000 L size. Recipes were written and carried out using computer-based monitoring of remote sensors and then actuation of equipment such as automatic valves, steam traps, stirrers, plus heating and cooling elements using visual icons on a computer screen (Schlager, S.T. et al. 1980). Each icon represented a host of functions that could be selected as part of a recipe. The overall control and initiation of the recipe and data management involved using a supervisory computer that acted at a higher level than the process control computer. A schematic of this system is presented as Figure 4.1. As shown in Figure 4.1, fermentors were under the control of process control system (Beckman MV-8000), which was under the supervisory control using software provided by an experienced party (Rinteko Oy of Finland). This software was installed

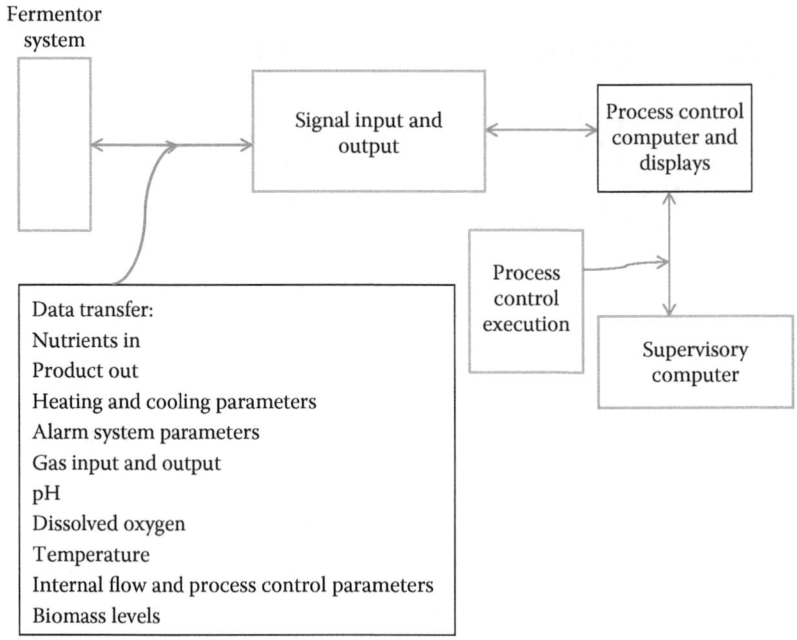

FIGURE 4.1 Pilot plant computer process control system.

in a PDP 11-34 digital computer (Digital Equipment Corporation) with data storage provided by the company mainframe (IBM). The system is described further to a limited extent in the literature (Goldstein, W.E. 1987).

The equipment was used to find best ways to make products and develop technology that could be scaled up to manufacturing. Ways were found to save energy and associative costs plus improve yields and reduce time in processing. Getting this system debugged and functional was not easy as many problems arose to be solved that required new solutions. It took time to get this operational such that the patience of our management was severely tested.

This example could well apply to the ethanol fermentation as a research and process development program that would be interactive with manufacturing and incorporate modern equipment. It would be based on the same principles developed about 30 years ago in the facility staffed by personnel that the author had the good fortune to lead, and with whom he closely worked.

A next step for the ethanol process would be to apply and formulate equations for control of the ethanol process that could be tested at pilot plant scale and then introduced into manufacturing replicating justifiable control functions. The equations would be coded into algorithms that represent the process. The computer system involved would access data from instruments, and software and a supervisory system would be used to decide to control certain aspects to try and optimize the situation. The objective is to manufacture more products for the least cost.

This is, of course, a significant undertaking. The investment would be justified by the cost savings that would result and the business that would be developed for a job well done. For the personal situation described in the author's past, during the 1980s, the pilot plant equipment (including fermentors needed), the process control computer, the supervisory computer, and the software needed were designed and specified. As the project proceeded, some of the software had to redone and developed anew due to matters not anticipated, including undercapacity wiring. At the beginning, a system and design was specified, costs developed, and a request for funds submitted to corporate management. This request was for approximately $2 million, a substantial sum today, and significantly more substantial in that past time. This particular project was not in the company's experience. On Christmas day, circa 1980, the author was called into the office of the Chief Operating Officer (COO) of the corporation. The COO handed the author a cashier's check for the total and asked a question that was simple with implications. Would this work? The author told him it would. It took a year and a half to show that it did indeed work as promised.

4.2 SPECIFICATIONS FOR CONTROL OF THE ETHANOL PROCESS

The same experience analogously applies to the ethanol process. It is necessary to set up excellent microbiology and analytical laboratories. The process is planned so that analytical measurements can be done online. The process is simulated at bench scale and at pilot scale so that theoretical (chemical engineering modeling) and experimental practices can be implemented and tested at small scale before technologies are implemented at the manufacturing scale. This saves significant amounts of money since it minimizes (but does not prevent) mistakes made at the large scale.

The ethanol process is based on propagating a specific yeast and providing conditions conducive for it to produce ethanol. Figure 4.2 shows a schematic for the ethanol process. Carbohydrate feedstock such as extracts of lignocellulose or ground corn are processed through a converter where enzyme systems break the complex carbohydrates down to simple carbohydrates such as glucose (including xylose). This stream is mixed with a yeast biomass recycle stream feeding an ethanol fermentation where the ethanol is formed by the yeast from the carbohydrate and other nutrients. The ethanol then passes through a separator (centrifuge, hydroclones, or another type such as settler). The ethanol is pumped off as a product, the yeast biomass is separated into a yeast biomass stream that is a by-product of the process to be sold, and a portion of the biomass is recycled as indicated.

Depending on the yeast, the biomass may be further propagated aerobically to compensate for the biomass withdrawn as a by-product in a separate stage as indicated. In turn, it may be propagated in the fermentor in an aerobic section that precedes the anaerobic section where ethanol is produced. It may be that yeast can readily switch from anaerobic to aerobic to anaerobic conditions quite readily with little delay or stagnation (Curry, J. 2009). This step is included to show that the matter must be considered for the individual specific process. Certainly, this changeover is an important aspect of control governed by the yeast characteristics, and conditions such as dissolved oxygen, pH, and other nutrient content (Trafton, A. 2014).

Besides process flows and variables as indicated in Figure 4.2 (an overall schematic of the ethanol process), quantities need to be measured since they are critical to the process. Some of these can be measured as part of the computer control network

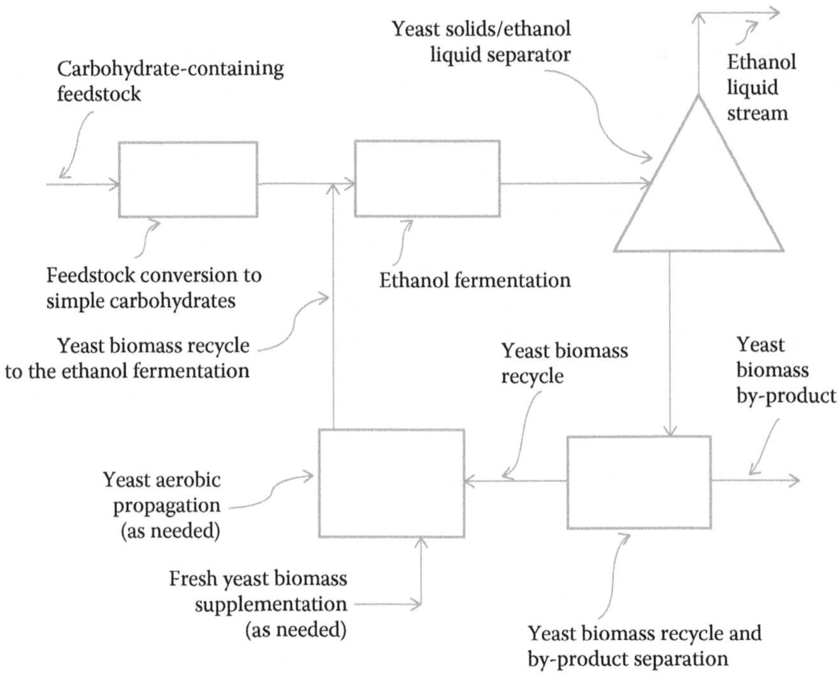

FIGURE 4.2 Ethanol process flow chart.

and others will be measured off-line (not in the computer network but entered in later as manual data input). Table 4.1 shows key measurements to be made, ways to measure the entity, and presents actions based on events, plus describes the utility of the data (Acumedia. 2011; Gresham, D. et al. 2008; Hanson, P. 2008; Heist, P. 2008; Martínez-Moreno, R. et al. 2012; Ortiz-Julien, A. et al. 2002; Saldanha, A.J. et al. 2004; Vieira, E.D. et al. 2013).

As indicated in Table 4.1, the rate of the feedstock stream and its composition is important since that is the starting point for production. One needs to know the carbohydrate composition since if the purity is off, productivity will obviously suffer. Measurements of this stream composition may be typically made by high-performance or high-pressure liquid chromatography (HPLC) (Cooley, R.; Cecil, S. 2011; Hall, G. and Reuter, W.M. 2007). In HPLC, the liquid is pushed through a tightly packed bed in a down flow direction under high pressure. These systems can be provided online so that if it is in a facility, it is better since one will have results with much less delay than sending the sample to the laboratory. However, samples should still be pulled and sent to the laboratory to check them against the production streams. They should match within an error known for the facility. If they do not match, one investigates to find out why this happened. In production, many things happen that appear unusual (maybe crazy) until investigations are carried out. Selecting samples and checking the process are important and must occur while one is developing their online system enough to trust it.

TABLE 4.1
Yeast Nutrients and Process Factors

Process Factor	Measurement Method		References
Feedstock flow rate	Flow meter		
Feedstock density	Densitometer		
Oxygen concentration in biomass propagation	Sterilizable electrode		
Oxygen concentration in ethanol fermentor	Sterilizable electrode		
Oxygen concentration in biomass propagation	Sterilizable electrode		
Oxygen concentration in ethanol fermentor	Sterilizable electrode		
Carbon dioxide concentration in off-gas	Mass spectrometer	Manual method lab and pilot plant only	
Oxygen concentration in off-gas	Mass spectrometer	Manual method lab and pilot plant only	
Carbon dioxide concentration in off-gas from biomass propagator	Mass spectrometer	Manual method lab and pilot plant only	
Oxygen concentration in off-gas from biomass propagator	Mass spectrometer	Manual method lab and pilot plant only	
Carbon dioxide concentration in off-gas from ethanol fermentor	Mass spectrometer	Manual method lab and pilot plant only	
Oxygen concentration in off-gas from ethanol fermentor	Mass spectrometer	Manual method lab and pilot plant only	
Nitrogen concentration in off-gas from biomass propagator	Mass spectrometer	Manual method lab and pilot plant only	

(*Continued*)

TABLE 4.1 (*Continued*)
Yeast Nutrients and Process Factors

Process Factor		Measurement Method				References
Nitrogen concentration in off-gas from ethanol fermentor		Mass spectrometer	Manual method lab and pilot plant only			
Other gasses in off-gas from propagator		Mass spectrometer	Manual method lab and pilot plant only			
Other gasses in off-gas from ethanol fermentor		Mass spectrometer	Manual method lab and pilot plant only			
Yeast settling, floc size, diameter, visual images		Grad samples by laboratory	Lab and pilot plant special meter			
Ethanol flow rate in streams and effluent		Meter	Meter in pilot plant			
Ethanol concentration and other alcohols		Online HPLC	Manual lab HPLC			
Ingredient	**Source**	**Source Concentration (mg/L)**	**Source Concentration (g/L)**	**Measurement Method**		
Glucose			300	Online HPLC	Manual lab HPLC	Trafton, A. (2014)
Glucose			10	Online HPLC	Manual lab HPLC	Gresham, D. et al. (2008)
Maltose				Online HPLC	Manual lab HPLC	
Maltotriose				Online HPLC	Manual lab HPLC	
Maltotetraose				Online HPLC	Manual lab HPLC	
Sucrose			250	Online HPLC	Manual lab HPLC	Ortiz-Julien, A. et al. (2002)
Sucrose			150	Online HPLC	Manual lab HPLC	

(*Continued*)

TABLE 4.1 (*Continued*)
Yeast Nutrients and Process Factors

Ingredient	Source	Source Concentration (mg/L)	Source Concentration (g/L)	Measurement Method			
Sucrose hydrol			280	Online HPLC	Manual lab HPLC		Martínez-Moreno, R. et al. (2012)
Nitrogen	Urea	300–500		Online nitrogen indicator	Nitrogen indicator	Nitrogen indicator	Heist, P. (2008)
Nitrogen	Variety aa	5–200 N		Online nitrogen indicator	Nitrogen indicator	Amino acid analysis	Martínez-Moreno, R. et al. (2012)
Nitrogen	DAP	10–60 N		Online nitrogen indicator	Nitrogen indicator	By outside lab	Ortiz-Julien, A. et al. (2002)
Nitrogen	NH_4Cl	500		Online nitrogen indicator	Nitrogen indicator	Total nitrogen	Vieira, E.D. (2014)
Phosphate	KH_2PO_4	13.3		Online phosphate indicator	Manual phosphate	By outside lab	Saldanha, A.J. et al. (2004)
Phosphate	KH_2PO_4	10			Anion exchange chrom		Gresham, D. et al. (2008)
Phosphate	KH_2PO_4	5			Manual phosphate		Vieira, E.D. (2014)
Sulfate	$(NH_4)_2SO_4$	3		Online sulfate indicator	Anion exchange chrom		Saldanha, A.J., et al. (2004)
Sulfate	$(NH_4)_2SO_4$		5				Gresham, D. et al. (2008)
	$MgSO_4$		1				Vieira, E.D. (2014)
	$MgSO_4$		0.5				Gresham, D. et al. (2008)

(*Continued*)

TABLE 4.1 (*Continued*)
Yeast Nutrients and Process Factors

Ingredient	Source	Source Concentration (mg/L)	Source Concentration (g/L)	Measurement Method		
Uracil						Hanson, P. (2008)
Adenine						Hanson, P. (2008)
Potassium	KOH		Elevated	Online potassium indicator	Manual potassium	Trafton, A. (2014)
Potassium	KCl	100			Cation exchange chrom	Vieira, E.D. (2014)
Vitamins and Supplements						
Yeast extract	Vitamin, nitrogen, factors		6	None	Certificate of analysis	Vieira, E.D. (2014)
Yeast extract	Vitamin, nitrogen, factors		3–5		Outside lab and supplier	Ortiz-Julien, A. et al. (2002) Acumedia (2011)
Calcium chloride		100		None	Manual lab cation exchange	Gresham, D. et al. (2008)
Sodium chloride		100		None	Manual lab cation exchange	Gresham, D. et al. (2008)
Boric acid		0.5		None	Micronutrient certificate of analysis by supplier	Gresham, D. et al. (2008)
Copper sulfate		0.04				Gresham, D. et al. (2008)

(*Continued*)

TABLE 4.1 (*Continued*)
Yeast Nutrients and Process Factors

Vitamins and Supplements	Source	Source Concentration (mg/L)	Source Concentration (g/L)	Measurement Method	
Potassium iodide		0.1			Gresham, D. et al. (2008)
Ferric chloride		0.2			Gresham, D. et al. (2008)
Manganese sulfate		0.4			Gresham, D. et al. (2008)
Sodium molybdate		0.2			Gresham, D. et al. (2008)
Zinc sulfate		0.4			Gresham, D. et al. (2008)
Biotin		0.001			Gresham, D. et al. (2008)
Calcium pantothenate		0.2			Gresham, D. et al. (2008)
Folic acid		0.001			Gresham, D. et al. (2008)
Inositol		1			Gresham, D. et al. (2008)
Niacin		0.2			Gresham, D. et al. (2008)
p-Aminobenzoic acid		0.1			Gresham, D. et al. (2008)
Pyridoxine		0.2			Gresham, D. et al. (2008)
Riboflavin		0.1			Gresham, D. et al. (2008)
Thiamine		0.2			Gresham, D. et al. (2008)

Note: This table is a duplication of Table 2.1. This duplication is intentional so that readers could view the table in both chapters.

The next entry in Table 4.1 is the gasses (Ametek. 2014). As indicated, an important measure is the dissolved oxygen in fermentations (for both propagating yeast and for ethanol formation). The appropriate levels and patterns need to be established. They may be particular to a facility and perhaps have to vary with time. Also, if aeration is required as necessary to propagate yeast, this requires energy and agitation. The computer control system should be able to use models to help determine how to vary this to save electricity. In the ethanol fermentation, the theoretical environment is an absolutely anaerobic environment (void of oxygen). A certain minimal level of oxygen (through air introduction) must be maintained (for the health of the yeast) and it has to be controlled. The measurement of dissolved oxygen is done with a sterilizable oxygen electrode system in an entry port in the fermentors (Van London Company. 2015). Bench-scale and pilot plant units would be similarly equipped in order to be able to study the process, gain information, and then be able to advise on what is observed and measured.

Aerobic and anaerobic conditions are maintained by manipulation of the composition of gasses in the feed to the fermentors. In one case, air is used (which may be enriched with oxygen beyond normal atmospheric content of about 21% by volume if this is justified by results) or pure nitrogen gas can be fed using a nitrogen supply system to maintain an anaerobic environment. If the gas supply is significant enough, a facility could have an on-site air separation plant where a distillation process is used to separate nitrogen from oxygen for the purposes of supply to the ethanol plant.

As indicated in Table 4.1, the off-gas (exhaust gas) from the fermentors is monitored using a mass spectrometer system designed for this purpose (Ametek. 2014). This device provides the capability to obtain nitrogen, oxygen, carbon dioxide, and argon composition in the off-gasses. This then allows mass balances to be made on the utilization of oxygen in the fermentor. Furthermore, mass transfer rates can be monitored using the dissolved oxygen measurement. Also, the level of carbon dioxide is a way to help monitor the fermentations involving growth and maintenance of yeast and the formation of ethanol by the yeast.

The yeast used needs to be characterized through testing of settling characteristics, and obtaining a profile of individual yeast cell size, plus the apparent density and size distribution of the flocs. Photographic images should be obtained and stored for quality control purposes. This requires investment in a good-quality light microscope fitted with a camera. The microscope should allow the yeast cells to be observed under different kinds of light and variations. These include visible light, phase contrast, and different wavelengths. The ability to observe cells and cell flocs using fluorescence imposed on the sample is also relevant for classifying the yeast (Leica Microsystems. 2016; Zeiss. 2016). The microscope would be interfaced to the computer system so that images could be studied, compared, and archived.

The laboratory equipment should include a bioreactor assembly that can be used to monitor the fermentation at bench scale. This can be a system that can operate in batch, fed batch, and continuous mode (Applikon Biotechnology. 2016). The information from this can be used as a bench mark for comparison with the pilot system yeast cells and cells in production. Therefore, a research intensive facility is recommended since the rewards from this will be substantial.

The next category in Table 4.1 is nitrogen. It may be supplied in all kinds of forms. The organism likely finds ways using secreted enzymes to break the nitrogen supply down to simplest forms such as NH^{4+} or NH_3, or NO_2. Therefore, the special measurement device being proposed will measure these simpler molecules and refer that back to the nitrogen supply that may be in yeast extract, soy isolate, or amino acid. Success in this might allow the use of HPLC as an online measurement in production. Another possibility may be a probe (Taranovich, S. 2012; WTW Laboratory and Field Instrumentation. 2016). The measurement should be duplicated in the analytical lab and also in the pilot plant being recommended in this book.

Table 4.1 indicates that particular anions such as phosphate and sulfate and cations such as ammonium and potassium be measured using ion chromatography. Micronutrient supplements should be obtained from a respected vendor and have a suitable certificate of analysis. Critical items such as vitamins might be checked at random using a special services laboratory for comparison to the values in the certificate of analysis. A special service laboratory should be retained for work so that they can be called upon as necessary. A laboratory with the capabilities to fulfill all of the supplemental and selective process needs may be hard to identify and so the search for this laboratory should be done early.

Note that it is recommended that bench-scale and pilot-scale systems be established and be running to check results in production and look for drifts in data, malfunctions, and so on. The small systems are approximate scale down models of production where their design is quite particular to the particular facility and its practices. As the process proceeds and begins to consistently hum, the need for laboratory and pilot plant work becomes less and less. It is important to keep such bench-scale and pilot plant work going in order to look and anticipate problems before they occur. Well-managed facilities do this properly and exhibit results in their revenue, cost control, and profits. The computer control and monitoring, once established, will keep the facility on track and performing well.

In manufacturing, personnel are totally preoccupied and focused on keeping the operation running. They have little time to think about nuances, process development, and improvements. It is often up to the research and development people manning the analytical laboratory bench-scale unit and pilot unit to think about nuances, process development, and improvements. This is so if the parties have developed interactive and cooperative relationships where the parties know they can call each other for help. Researchers and process development personnel need to find ways to introduce changes in usually an evolutionary manner and thereby convince production to see the benefit of incorporating findings. It is recognized that findings in production often are unique and may not be seen in laboratory or pilot plant mimicking units. When this happens, what production sees must be learned by research and process development so that they can study and replicate the work in the laboratory and pilot plant. Often this is tough to accomplish.

Success in such activities requires establishing a great working relationship. It is not humanly possible to think of all things at once. Everyone has to contribute.

Management has to make this conducive by the environment they establish and the mutual respect they engender besides the reward system. This does not have to be a "nirvana" environment. It can be followed up day to day as a practical endeavor. If the culture effectively kills destructive internal politics (that does little more than harm productivity), then the company has a fighting chance to realize objectives.

This is important in the science of ethanol opportunity since there are barriers to overcome. People make this happen, not a computer system or the process equipment. It depends on what people do in their activities and with information gained.

4.3 APPLICATION OF COMPUTER-BASED PROCESS CONTROL TO ETHANOL

The modern adaptation of the experience of the author and colleagues in this area (circa 1980) may be the HMI/SCADA or Human Machine Interface/Supervisory Control and Data Acquisitions being installed in present days (Filice, G. 2016; Wikipedia. 2014). It seems this technology may well fit implementation to help ensure success for ethanol production and supply ventures).

The ethanol manufacturing facility and business has hundreds of variables that need to be controlled to ensure the supply of a quality product that results in a profitable business. These start at the macrolevels with ethanol production rate, feedstock supply rate, production of quality yeast biomass, and so on, and then extend to individual process areas, equipment, and key valves that are certainly not minutiae. The author remembers an incident where a chemical used to clean out boiler tubes first showed up as strange colors in a product, as a consequence of a pipe rotting out due to a replacement of a section of stainless-steel pipe with carbon steel, with the carbon steel developing leaks. The product turned colors affecting millions of dollars of shipping. This problem had to be solved immediately as the product could not be shipped. As it happens, the author remembered a brochure on the chemical that said it turned colors as a function of pH. We solved the problem and cleaned up the product so people could go home and spend Easter Sunday with their family. The section of pipe was only a few feet long. Big problems start in small ways.

Optimizations also include suppliers providing what they say they will ship and what may happen when something is out of specification and no one catches this resulting in, for example, a bad facility problem. It also extends to work interferences and labor problems in a facility where people are unhappy and purposely slow their work down. It also extends to terrorism and cyber-attacks in this modern age. This is best managed by paying attention to key aspects and to sound management and common sense to achieve objectives. To help in this regard, one establishes modern control systems, technical and business, to help set a baseline for guidance and control based on the best computers available, the mental capacity of you and your colleagues.

The HMI/SCADA systems, properly specified and implemented, can provide what is needed as a basis if they are used as they should be used. These systems consist of the following:

- The ability to connect to plant devices and collect data systematically (SCADA)
- Ways for users (operators, planners, designers, process specialists, management) to visualize the process and what is happening (HMI)
- Ways for automatic, feedback, and feed forward techniques to control the process by use of specially designed software for supervisory control (SCADA)
- Ways to gather and securely store data with the software designed to ease the ability to study history and be able to readily use that information

- Ways to interface the supervisory system with business management to link control of the process to managing the business leading to profitability
- The ability to act independently and revert to manual operation under defined plans if the computer system fails and emergency measures are needed

The connections to the plant equipment from the supervisory and process control computers may be distributed by area to allow some autonomy by process sections. In any case, the data needed, whether it comes from the plant or a laboratory source, are acquired and stored systematically according to how it will be processed and used.

The process system should be visualized on displays where operators and others can query portions of the facility by clicking on icons to see what is happening. This can include status of alarms, operation of elements such as transmitters, valves, heat exchangers, pumps, etc. The supervisory computer has coded recipes to carry out functions. It also has software to impart process control according to algorithms and defined strategies.

Data acquired and actions taken are stored historically so the information can be assessed and studied for improvements as well as analysis of problems, both past and present. This would be in the supervisory computer area. This is an important area since this can tend be a source of misguided comfort since everything is being saved. The system is useless if the data are not accessed and studied as a defined assignment. One needs to do this to make sure one can get to such data and analyze it without some impediment. Such an impediment would mean a mis-design in the system.

The system needs to be accessible in terms of key functions important to maintaining the profitability of the business. Of course, it is important to plan for emergencies. It is important to be able to operate the facility manually and make product in cases of emergency that can be for any number or reasons.

4.4 OPTIMIZATION EXAMPLE

For any facility, a multitude of technical and business problems will occur that are connected. Some of these may be amenable to mathematical optimization analysis. Others may be solvable with good strategic thinking. Both are important for the problems that are conducive to mathematical analysis. Such problems may involve hundreds of variables. The situation is complex but should be studied in advance in an anticipatory manner, since valuable information may result that will put your organization ahead of the competition and maybe those plain opposed to ethanol for their own reasons, logical and illogical. The mathematics will end up involving multidimensional differential equations and matrix algebra (Cooke, D.).

Once the equations are formulated, computers and various packages can be used. These can be installed on a supervisory computer where results can be used to tweak the process once the matter is understood sufficiently. One would check this out in the pilot plant. It should be obvious by now that the author believes such work is immensely invaluable and has evidence in his career and in the career of close associates that this works favorably in many circumstances.

Consider a basic example of optimization of the ethanol process that involves the effect of a micronutrient on yeast. A steady amount of micronutrient is necessary to keep the yeast actively growing or maintained while too much of the micronutrient becomes inhibitory or toxic to the yeast. This is expressed in

$$Y_{gr} = \frac{\mu}{(1 + k_1/C_i + k_2C_i)} \tag{4.1}$$

where Y_{gr} is the yeast growth rate in s^{-1}, μ is the maximum specific growth rate of the yeast in s^{-1}, k_1 is the constant for activity enhancement of yeast due to the micronutrient in g/cm^3, C_i is the concentration of the micronutrient in g/cm^3, and k_2 is the constant for inhibition of yeast due to the micronutrient in g/cm^3.

We want to find out the limits of concentration and the best concentration of micronutrient to use. Note that k_1 and k_2 are fixed as characteristics of the particular yeast strain being used. If these characteristics are undesirable, then we would have to search for (or develop) a new yeast strain with characteristics desired making sure other characteristics have not changed.

In order to find limits or what is known as extrema, Equation 4.1 is differentiated with respect to C_i with other variables held constant, making this a partial differentiation (Miller, K.S. 1953), resulting in

$$\frac{\partial Y_{gr}}{\partial C_i} = \frac{-\mu(-k_1/C_i^2 + k_2)}{(1 + k_1/C_i + k_2C_i)^2} \tag{4.2}$$

This derivative is set to zero since that is where a maxima (or a minima) is to occur since the point where that occurs has a constant C_i value by definition (Perry, J.H. 1950). This results in

$$\frac{\partial Y_{gr}}{\partial C_{iextrema}} = \frac{-\mu(-k_1/C_i^2 + k_2)}{(1 + k_1/C_i + k_2C_i)^2} = 0 \tag{4.3}$$

Simplifying this equation, terms cancel out, resulting in

$$\frac{\partial Y_{gr}}{\partial C_{iextrema}} = \frac{-k_1}{C_i^2 + k_2} = 0 \tag{4.4}$$

Solving, it can be seen that at this extrema,

$$C_i = \sqrt{k_1/k_2} \tag{4.5}$$

The sign on the square root is positive since C_i is a positive quantity.

We can show that this extrema is a maximum by differentiating $\partial Y_{gr}/\partial C_i$ (Equation 4.2) to obtain $\partial^2 Y_{gr}/\partial C_i^2$ and showing it is always a negative quantity. We can also readily show this by inserting numbers into Equation 4.3 and showing that Equation 4.5 is indeed a point where the yeast growth rate is at its maximum assuming that Equation 4.1 is a valid representation of the situation.

Therefore, we would never want to increase C_i beyond the value of $\sqrt{k_1/k_2}$ since beyond this value the yeast growth rate will be reduced and this action also presumably involves more cost. Similarly, unless there is some problem, we would not want the value of C_i to be much below $\sqrt{k_1/k_2}$ as that would lower the yeast growth rate.

For practical reasons, this micronutrient may be provided as part of a standard mixture from a supplier. Therefore, it may be impractical to change it in any case. Experiments would have to be done to determine k_1 and k_2. One may do this as part of a general optimization study where studying in general how the growth rate of the yeast being used varies with micronutrient composition if that was shown to be important. In this case, assuming this model is acceptable, the general form being studied would be expressed using

$$Y_{gr} = \frac{\mu}{(1 + k_1/C_i + k_2C_i + k_3/C_i + k_4C_i + k_5/C_i + k_6C_i + \cdots)} \tag{4.6}$$

This becomes a more significant optimization problem theoretically and experimentally. However, there might be considerable value in carrying out work of this type since it can lead to better economics.

Often multiple relationships are involved that would be solved simultaneously that lead to more completed optimizations of several variables at once. An example lies in a case where one is optimizing the yeast's growth rate and also the utilization of feedstock in order that the variation in both in operation can be minimized since that can negatively affect productivity and perhaps cause the yeast to deteriorate more quickly. In a continuous process, where a portion of the yeast is recycled to be reused to operate at higher yeast biomass levels, changing the proportionate amount of yeast biomass that is recycled may help reduce the variation in the reaction analogous to what is practiced in waste treatment (Blanch, H.W. and Clark, D.S. 1997).

Investing in an HMI/SCADA system and undertaking research activity directed to improving plant economics through a modeling program will be immensely valuable. This combination will provide insights into the process leading to best strategies for process control. One has to undertake this activity for their facility and its particular nuances. The research would be conducted at bench and pilot scales and the results transferred into production once it is expected that it will be beneficial.

REFERENCES

Acumedia. 2011. Yeast Extract (7184). Accumedia Manufacturing, Inc. http://www.neogen.com/

Ametek. 2014. Direct Measurement of Respiratory Quotient (RQ) in Fermentation Processes. Ametek Process Instruments Application Note. http://www.ametekpi.com/.

Applikon Biotechnology. 2016. Cultivation Systems from Discovery to Production. http://www.applikon-bio.com.

Blanch, H.W. and Clark, D.S. 1997. *Biochemical Engineering*. Marcel Dekker, New York. p. 325.

Cecil, S. 2011. *Fuel Ethanol: Fermentation Analysis by HPLC*. Sigma Aldrich. Reporter US Volume 26.3. http://www.sigmaaldrich.com.

Cooke, D. *Multi-Variable Calculus and Optimization*. EC2040_T3.pdf. Trinity College, London.

Cooley, R. On-Line HPLC as a Process Analytical Technology (PAT) for Controlling Product Collection from Process-Scale Chromatography Columns. Dionex Corporation. http://www.dionex.com/.

Curry, J. 2009. Yeast and Fementation. Dukes of Ale. BJCP Preparation Course. http://www.mgriesmeyer.com/doatest/bjcp/yeastandferentation.pdf.

Filice. G. 2016. The Smart Operator: Super HMI?SCADA: Learning from Large Systems. General Electric. http://www.ge.com/digital/.

Forrester, J.W. 1968. Industrial dynamics—After the first decade. *Management Science*. 14(7). p. 398–415.

Goldstein, W.E. 1987. Enzymes as bulk products. In *Basic Biotechnology*. Eds. J. Bu'Lock and B. Kristiansen. Chapter 14. Academic Press, London. p. 395.

Goldstein, W.E., Rogers, R.W., and Genshaw, M.A. 1975. Interactive computer control as applied to clinical analyzers. *American Laboratory*. 7(9). p. 47–55.

Gresham, D., Desai, M.M., Tucker, C.M., Jenq, H.T., Pai, D.A., Ward, A., DeSevo, C.G., Botstein, D., and Dunham, M.J. 2008. The repertoire and dynamics of evolutionary adaptations to controlled nutrient-limited environments in yeast. *PLoS Genetics*. 4(12). p. e1000303. doi: 10.1371/journal.pgen.1000303.

Hall, G. and Reuter, W.M. 2007. HPLC Analysis for the Monitoring of Fermentation Broth During Ethanol Production as a Biofuel. Liquid Chromatography Application Brief. PerkinElmer. http://www.perkinelmer.com/.

Hanson, P. 2008. Introduction to *Saccharomyces cerevisiae*. The Vegetative Life Cycle of Yeast. http://faculty.bsc.edu/phanson/yen/hanson_introduction.pdf.

Heist, P. 2008. A guide to successful yeast propagation. *Ethanol Producer Magazine*.

Leica Microsystems. 2016. Obese and Slim Yeast Cells. Microscopic Insights into Cellular Lipid Metabolism. http://www.leica-microsystems.com/science-lab/obese-and-slim-yeast-cells/.

Martínez-Moreno, R., Morales, P., Gonzalez, R., Mas, A., and Beltran, G. Biomass production and alcoholic fermentation performance of *Saccharomyces cerevisiae* as a function of nitrogen source. 2012. Federation of European Microbiology Societies, *Yeast Research*. 12. p. 477–485.

Miller, K.S. 1953. *Partial Differential Equations in Engineering Problems*. Prentice-Hall, Inc., New York.

Ortiz-Julien, A., Dulau, L., Katchmer, J., Molinari, S., and Marides, A. 2002. Nutrients for the Alcohol Fermenation. Lallemand. http://www.lallemand.com/

Perry, J.H. Ed. 1950. *Chemical Engineers' Handbook*. 3rd Ed. McGraw-Hill Book Company, New York. p. 82.

Saldanha, A.J., Brauer, M.J. and Botstein, D. 2004. Nutritional homeostasis in batch and steady-state culture of yeast. *Molecular Biology of the Cell* 15. p. 4089–4104.

Schlager, S.T., Schwartz, A., Goldstein, W.E. et al. 1980. System for computer control and monitoring of a fermentation process system. Biotechnology Group, Miles Laboratories, Inc., Division of Bayer, A.G.

Taranovich, S. 2012. Optical *In Situ* Glucose Sensor. Stratophase, Ltd. http://stratophase.com.

Trafton, A. 2014. New Approach to Boosting Biofuel Production. MIT News. http://news.mit.edu/2014/yeast-ethanol-biofuel-production-1002.

Van London Company. 2015. Steam–Sterilizable Oxygen Electrode. http://www.vlpc.com/default/index.cfm/dissolvedoxygen.

Vieira, E.D., Andrietta, M., da, G.S., Andrietta, S.R. 2013. Yeast biomass production: A new approach in glucose-limited feeding strategy. *Brazilian Journal of Microbiology* 44(2). p. 551–558.

Wikipedia. 2016. *SCADA*. https://en.wikipedia.org/wiki/SCADA. Last modified on September 22, 2016.

WTW Laboratory and Field Instrumentation. 2016. IQ sensors for the determination of the nitrogen parameters ammonium, nitrate and nitrite. http://www.wtw.com/.

Zeiss. 2016. Cell Observer. Efficient Observation of Living Cells. http://www.zeiss.com/microscopy.

5 Raw Material Resources to Produce Ethanol

5.1 PLANT RAW MATERIAL RESOURCES

5.1.1 Introduction to Use of Plant Raw Materials to Manufacture Ethanol

There are several objectives concerning the selection and use of raw materials (plant resources) for the bioproduction of ethanol (producing ethanol in fermentation vessels) and processing ethanol for applications, the main one being to be a component of a fuel for automobiles. Ethanol is presently blended with gasoline for use in automobiles. This blending is required by federal statute (Renewable Fuels Association. 2014a,b).

One thing to note is that the ethanol is made from the raw material resource by a select microorganism, usually yeast. Yeasts are not all the same, and certainly, there is a specific yeast for optimal ethanol production from a specific raw material.

The raw materials for producing ethanol by a biological route come from many sources. If one particular raw material is used, the yeast will be developed to use that material, even though it may work fine for alternative raw materials.

Examples of feedstock raw materials include corn, sugarcane, sugar beet, potato, rice, wheat, cassava, sorghum, oats, and barley. Various sources of lignocellulose include foliage, wood, trash, garbage, sugarcane and beet sugar plant materials, and plant materials that are not the main food product of crops such as those associated with corn, wheat, potato, rice, cassava, oats, and barley. Several components of a plant resource can, therefore, be resources of materials used to make ethanol if they are not used in other ways as products. Examples of this are the leaves and fibers in the corn plant as well as the lignocellulose entities associated with oil in the corn plant. Since the oil and fibers are also products (or by-products, also termed coproducts) and since these are also sold, one has to be concerned about providing these coproducts in the best way and form possible without altering the food product. For example, the oil from the corn plant must have the purity and composition and stable characteristics to satisfy the industries to which it is sold.

The fiber and what is associated with the fiber, such as protein, again within the corn plant, for example, must satisfy the needs of the animal feed market with this market's own demands and needs. Basically, the raw material is a multiproduct source. Matters are further complicated since materials formed in the processing of the raw material (e.g., corn) to ethanol can be recycled back and reprocessed. This recycle and blending with the ethanol/coproduct production facility creates new materials that result in new and modified forms, and at times, results in obtaining additional ethanol out of the plant raw material resource (raising the yield). Besides producing valued products, the products derived from the facility must get their due attention as costs for all steps in the facility must be minimized so the facility is economically successful. It should be evident that this is a considerable challenge.

The list of complications goes on—one being that a by-product produced and sold may have a specialized form and specific properties related to that form. The form may end up spawning an industry such as the supply of pellets (Portz, T. 2016).

There are many objectives in choosing a plant resource. One is that the plant resource is economical to use and that its use does not cause a problem in resource allocation (e.g., using corn that is a source of food to make ethanol thereby lessening food corn supplies and raising prices). They are ways around this challenge. However, this is another matter to consider.

The plant raw material resource should be plentiful in the area where the facility to produce the ethanol is located, plus the cost to harvest and deliver that material at allowed quality should be acceptable.

All of the plant raw material resource to be used is not converted to ethanol, and may result in valuable by-products, or may end up as waste or as a source of energy. Some of this plant raw material is converted, and some may just pass through the process. There are losses that occur due to nonethanol by-products being made unless these losses result in valuable products.

Some of these by-products just do not get converted at all. Discarding these by-products is a poor option since there is a cost to process them that one wants to retrieve. Preferably, value for by-products can be created to increase profitability. Making sure that profit is generated from everything is important so the facility fulfills its objectives and is economically successful.

The portion of the plant resource that is not converted to anything needs to be in a state where it can be converted or used to make something of value and processed and transported at a minimal cost. If the nonconverted portion of the plant resource will remain that way, then it should be in a state most favorable for its end use. For example, if the end use of the accumulated biomass (yeast and other entities) needs to meet certain specifications such as protein content for animal feed, then the process should be designed to meet that aim.

It may be that clean separation of such by-products is important for their purity in end use. That being the case, then the process needs to be designed to accomplish that task. This may also mean that the plant resource raw material should be selected so that it already has properties desirable for such clean separation (Harmsen, P.F.H. et al. 2010). That can be a factor in growing and using the plant resource. Even though the primary objective is to make in-specification ethanol, all of the factors mentioned must be considered to maximize profitability and have best chances for success against a variety of competitive products and methods.

Sometimes the conversion product formed is undesirable in the environment and may pose a health hazard. That also has to be considered since an environmental constraint can render a plant resource and processing method inadequate to fulfill objectives.

There are many crop sources of raw materials for use in ethanol production. For obvious reasons of planning and security, these should be cultivated or planned-for crops, not simply sourced from the wild. Can a crop be selected to be targeted for ethanol production? This is actually a complex question. Let us start by examining what might occur (and has occurred) in the regions of the world.

5.1.2 Preliminary Assessment of the World's Regions to Provide Raw Materials to Manufacture Ethanol

Given there is interest and agreement within a country and region that it is useful to produce ethanol there for a variety of good reasons (such as creating a cash flow to the region and developing/accessing technology to sustain such production), then each country can potentially develop a strategy and plan to enter this area. This is a preferred way to look at this since certain crops are best produced in select areas of the world for a number of reasons. One reason is climate, another is prevalence of crop disease, and another is cultural history and experience.

Any country anywhere can develop the means to produce ethanol for its own commercial gain for internal use and for export if the natural resources are available (or it has the ability and means to develop technology for this end by cooperating with another nation that has such material resources). Furthermore, the effort involved in this will naturally create technology and experience important for that particular nation.

In this section, each region of the world is addressed in regard to its capability to produce ethanol. The countries, and regions in which the country is located, plus the amount and type of crop produced in that country with potential to be used to make ethanol through biological means (processing through yeast fermentation) are presented in Table 5A.1 in Appendix 5A. The countries in each specific region are covered explicitly according to records on their capability and desire to produce crops of a certain type. The regions (and subregions) of the world so addressed (in alphabetical order) are Africa, Asia, the Atlantic Ocean region, the Caribbean, Central America, Europe, the Indian Ocean region, North America, the Pacific Ocean region, and South America.

The crops addressed as a set that have the potential to be used as a raw material to produce ethanol (or contribute to the product commercially) are sugarcane, corn, rice, wheat, potato, cassava, sugar beet, sorghum, oats, barley, and soybean. Only some of these crops can be produced in a practical manner in a particular region or country.

What should be done of course is to focus on the crop or crops best for that country from agricultural, economic, cultural, and technology aspects. Since these crops provide important foods, then if it is desired to use them to produce ethanol as well, sufficient crops must be grown to cover both the food and ethanol aspects. This growth is based on land area and/or yield increase to use the crop for food and for feedstock to make ethanol.

If the production of the ethanol may be undesirable because of the ambiance and pure natural beauty of the locale, then ways must be found to produce the ethanol commercially without disturbing such ambiance and beauty. This has to be part of the strategy and plan for a particular country to obviously avoid disturbing the natural attractiveness and heritage of a locale. Disturbing such beauty is a bad tactic since it will create opposition plus be intrinsically undesirable from a cultural standpoint.

5.1.2.1 Africa

Africa consists of five major regions (Western Africa, Central Africa, Eastern Africa, Northern Africa, and Southern Africa). Each region will be considered in terms of its potential or actual realization of its ability to provide raw materials needed to produce ethanol (Index Mundi. 2014a,b,c,d; Factfish. 2013a,b; UN Development Report. Maps of the World. 2015).

5.1.2.1.1 Western Africa

Starting in Western Africa, Figures 5.1 and 5.2 show the countries in this region and their capability to produce crops considered to have the potential to produce ethanol. In Figure 5.1, these include corn, cassava, and sorghum in Nigeria and in Ghana. Rice is produced at 4.8 billion kg in Nigeria. These are of course important food crops. However, they are important starch-containing sources with capability to produce ethanol.

The largest producing corn source in this region is Nigeria at 7.5 billion kg. The next largest producing source of corn is Ghana at 1.82 billion kg. The next largest corn producing source is Mali at 1.6 billion kg. For cassava, Nigeria can produce 53 billion kg, Ghana is able to produce 16 billion kg and Benin, 3.7 billion kg. For sorghum, Nigeria can produce 6.3 billion kg and Burkina Faso, 1.84 billion kg. The production can also be from the countries with the currently smallest production source (Gambia for corn, 8.3 million kg, Burkina Faso for cassava, 4.35 million kg, and Guinea-Bissau for sorghum, 14 million kg) (Wikipedia. 2013b; Factfish. 2013a; Index Mundi. 2014c; US Grains Council, Corn Refiners Association. 2015).

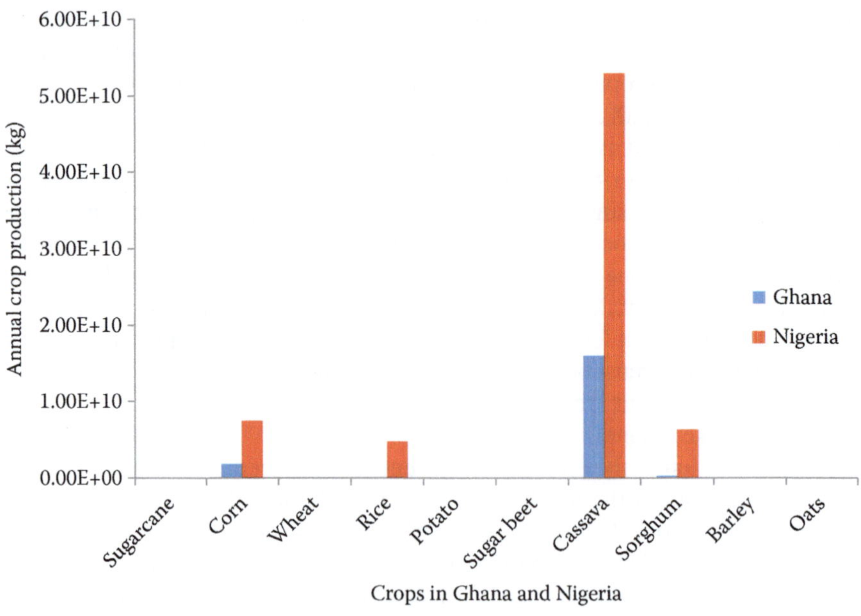

FIGURE 5.1 Production of starch-containing crops in Ghana and Nigeria.

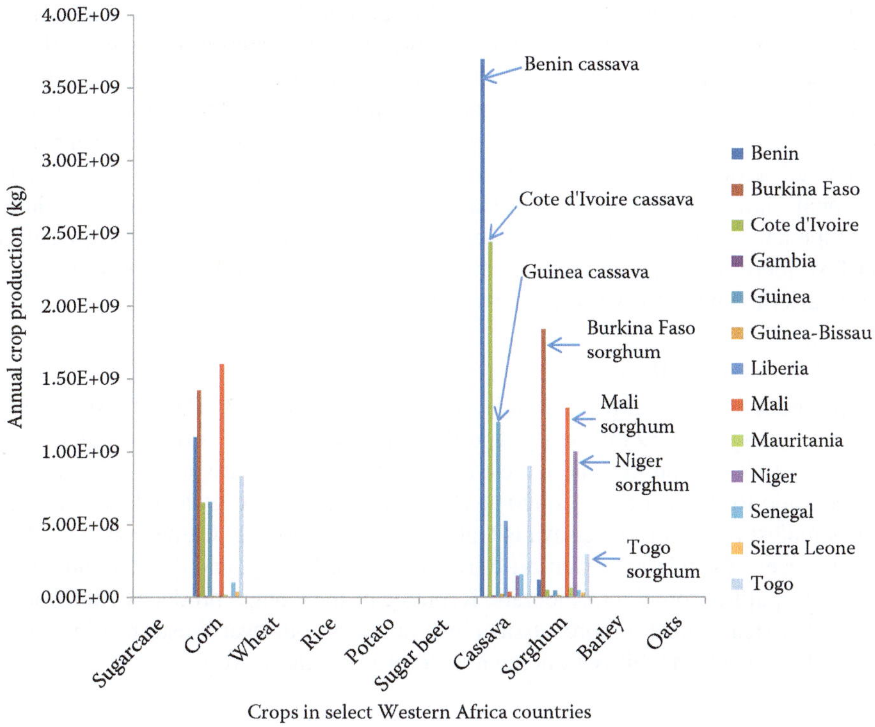

FIGURE 5.2 Starch-containing crops in select Western Africa countries.

If the country is experienced in several crop areas (no matter what production level), that is positive since that indicates the areas may benefit from cross-knowledge plus there may be an ability developed to switch from one to the other as a raw material source for ethanol for economic or other reasons. Development of the means and technology to switch from one source to the other is beneficial and certainly possible. The writer led efforts to accomplish this through adaptable fermentation technology to produce citric acid globally from a variety of starch sources in the past (Goldstein, W.E. 1978–1988).

Again, if diverting this capability to ethanol will negatively affect sources for food, this can be countered by developing ways to grow more or work with neighboring countries to be able to do what is necessary in a cooperative and optimally financial manner. Growing more crops, raising yields, and curbing crop disease plus working with neighboring countries are all helpful. For those needing more technology beyond their own, they can work with countries globally to get help and make this worthwhile for all parties concerned. If the nation has a history of producing such crops in their region, the knowledge is likely there. However, it has to be harnessed. The best way to counter political negatives is to develop ways to legitimately gain revenue in a way so all parties benefit.

Corn can be produced as animal feed as one type of variety and also for eating in the category sweet corn. The type of corn grown in the country, and distribution between sweet corn and that more typical of animal feed, or for processing, needs to be determined for each country. Each corn type will present its own challenges for breeding and for agronomy. Certain types may be grown in certain climates and there are likely best genetic varieties for production in the different countries indicated in Figure 5.2. There are already cooperatives in Africa fostering more combined action—Kenya Maize Consortium and Maize Breeders Network, the National Maize Association of Nigeria, and the Zimbabwe Seed Maize Association (Wikipedia. 2013b). According to International Institute of Tropical Agriculture (ITTA), there has been an increase and improvement in cultivation and growth of corn varieties suited to Western Africa (ITTA. 1997).

5.1.2.1.2 Central Africa

Crop production in Central Africa is indicated in Figure 5.3. The largest producer of corn is Angola (1.69 billion kg) and the smallest producer is the Republic of the Congo at 13.7 million kg. The largest producers of cassava are nearly the same, the Democratic Republic of the Congo at 16.5 billion kg and Angola at 17.4 billion kg. The smallest producer of cassava is Gabon at 315 million kg. Cameroon is the largest producer of sorghum in Central Africa at 1.15 billion kg, and the Republic of the Congo (separate from the Democratic Republic of the Congo) produces 7 million kg. Central Africa produces more cassava by a factor of four over Western Africa. Both areas show much capability in being able to produce these crops.

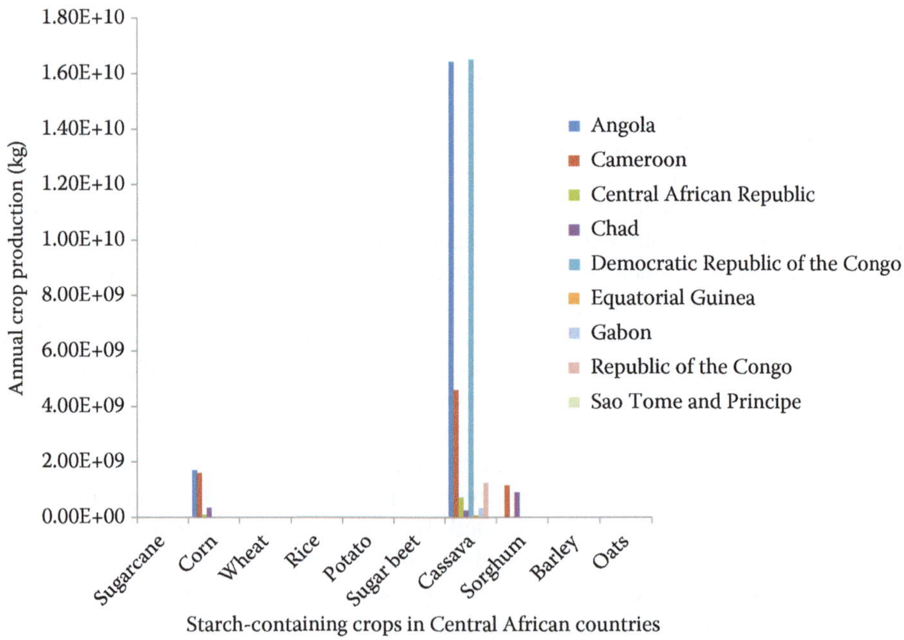

FIGURE 5.3 Starch-containing crop production in Central Africa.

5.1.2.1.3 Eastern Africa

In Eastern Africa (Figure 5.4), Ethiopia is the largest corn producer at 6.5 billion kg and Eritrea is the smallest at 20 million kg. Mozambique produces the most cassava (10 billion kg) and Somalia the least (90 million kg). Ethiopia produces the most sorghum in this region (4 billion kg) and Burundi the least at 32 million kg. In this region, additional crops that are being produced include potato in Malawi (3.26 billion kg) while Kenya is at 2.92 billion kg. Barley is also produced in this region in Ethiopia at 2.1 billion kg with the smallest quantity being produced in Zimbabwe at 60 million kg.

5.1.2.1.4 Northern Africa

In Northern Africa (as shown in Figure 5.5), there is a great variety of crops produced that contain starch for potential use in producing ethanol. Egypt produces the largest amount of corn annually (5.96 billion kg). Algeria produces 1 million kg per year of corn as the lower producer. Morocco is in-between at 200 million kg per year. Egypt produces 9.5 billion kg per year of wheat while the smallest producer of wheat shown on this chart is at 3.2 billion kg per year (Algeria). Egypt also produces 5.9 billion kg of rice as well as 4.5 billion kg per year of potato. Egypt produces 7.4 billion kg per year of sugar beet while Morocco produces 2.75 billion kg of sugar beet (FAO. 1999a,b; FAO Stats. 2011). Wheat and sugar beet appear in this northern

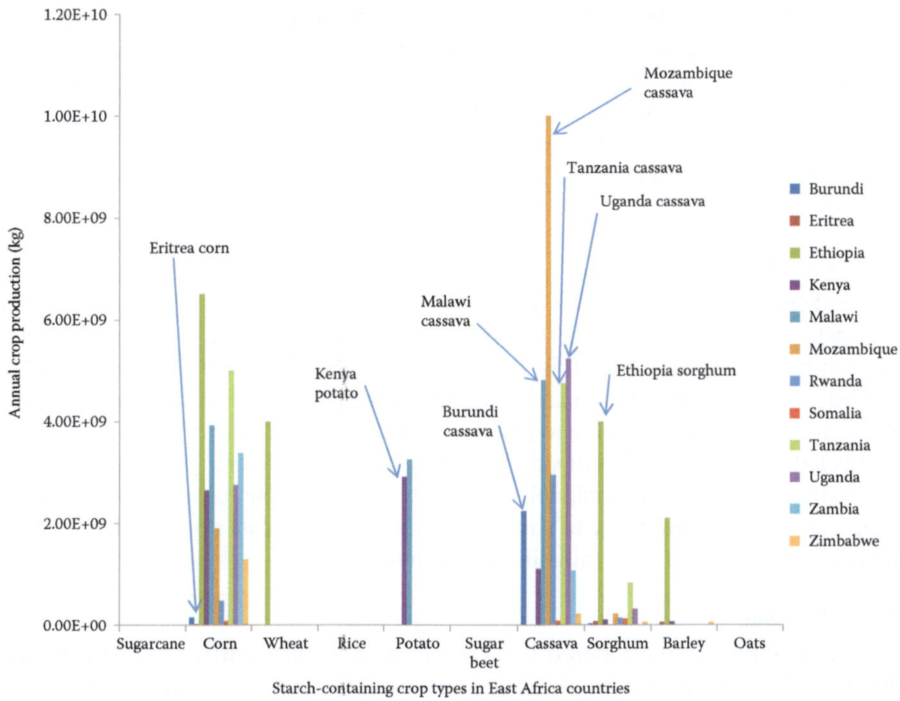

FIGURE 5.4 Annual production of starch-containing crops in East Africa.

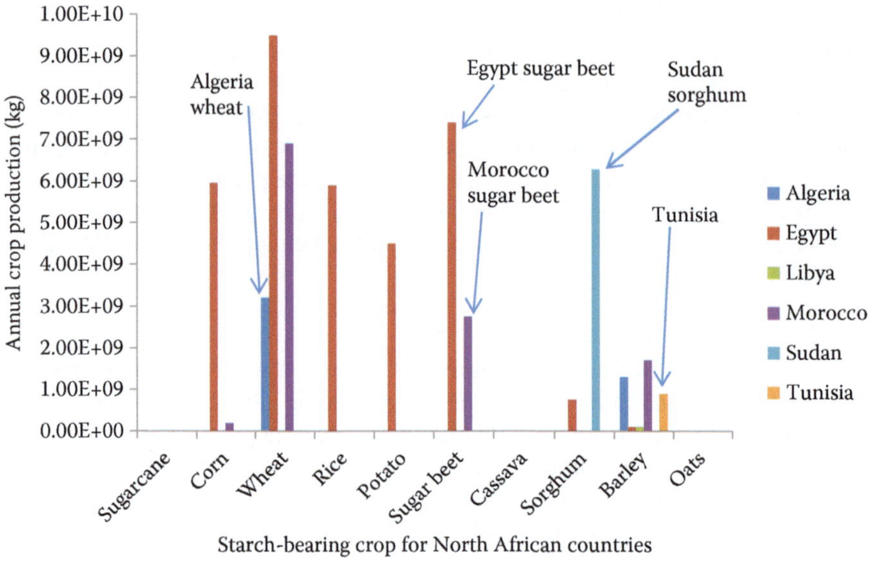

FIGURE 5.5 Production of crops containing starch in North African countries.

area of Africa (while not being shown in the Western, Central, and Eastern areas of Africa) pointing out the importance of climate and related factors in being able to produce crops in particular locales. The only entry for cassava is in the Sudan at 17.5 million kg. Sudan is able to produce sorghum at a level of 6.28 billion with Egypt trailing at 0.757 billion. Morocco produces barley at 1.7 billion kg per year while Egypt (0.108 billion kg) and Libya (0.1 billion kg) produce barley at nearly the same level. The diversification in Northern Africa to produce starch-containing crops is quite evident.

5.1.2.1.5 Southern Africa

In Southern Africa, as indicated in Figures 5.6 and 5.7, the Republic of South Africa is capable of producing corn at 11.3 billion kg per year and the lowest producer of corn in this region (Botswana) is at 35 million kg. South Africa can produce 0.158 billion kg per year of sorghum and 0.307 billion kg per year of barley (Index Mundi. 2014a,b,c,d). Swaziland is able to produce 1 million kg per year of sorghum.

Even though the amounts of starch-bearing crops of different types vary throughout Africa, the continent is capable of producing starch-containing crops of several types, extending from larger to smaller producers. Significant opportunity exists in this region because of the availability of such resources.

5.1.2.2 Asia

The regions of Asia considered to examine production in different countries are Western Asia, Southwestern Asia, Central Asia, Northeastern Asia, Southeastern Asia, and Southern Asia.

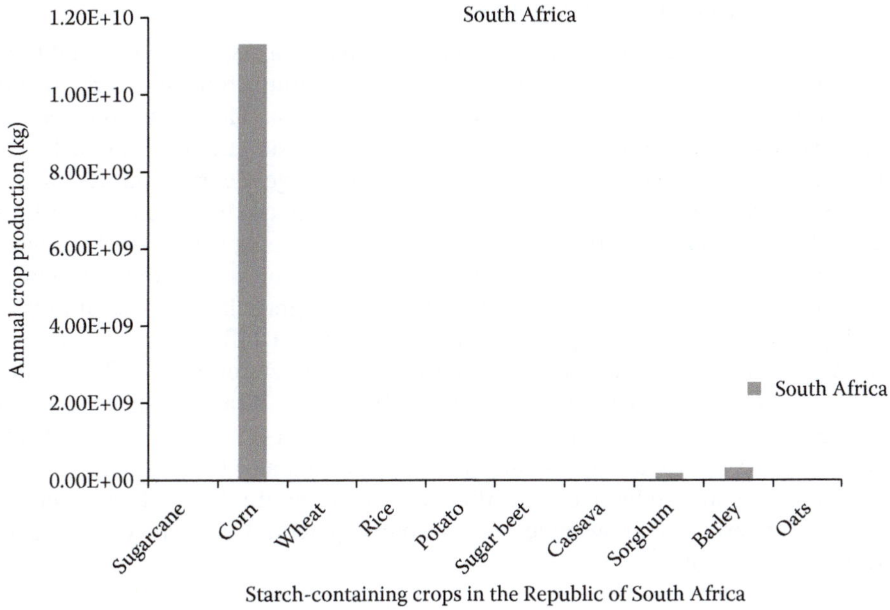

FIGURE 5.6 Production of starch-containing crops in the Republic of South Africa.

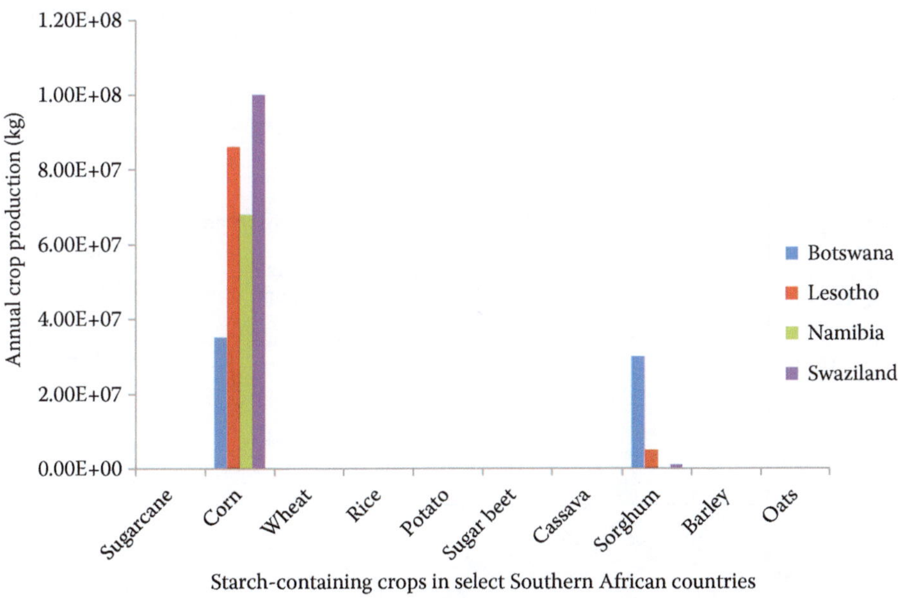

FIGURE 5.7 Production of starch-containing crops in select Southern Africa countries.

5.1.2.2.1 Western Asia (Middle East)

Figure 5.8 shows the countries in the region of Western Asia (also referred to as the Middle East). As indicated, the key starch-containing crops in this region are very similar to what was observed in Northern Africa. This is not unexpected since the regions are adjacent to each other at proximate latitudes. Interestingly, corn production is evident in four countries. The largest corn producer in this region is Turkey at 5 billion kg and the smallest producer is Jordan at 5 million kg. Turkey is the largest producer of wheat at 22.1 billion kg while Israel produces 0.45 billion kg of wheat as the smallest producer. Iran is the largest producer of potato at 5.4 billion kg while Turkey produces slightly less at 4.82 billion kg (Wikipedia. 2013c). Turkey produces sugar beet at 16.1 billion kg and Iran produces 3.8 billion kg as the lowest producer. Iran produces cassava at 23.9 billion kg. Saudi Arabia produces sorghum at 0.265 billion kg. Israel produces sorghum at 22.5 million kg. There are several barley producers, with the largest being Turkey at 3.95 billion and the smallest being Armenia at 0.16 billion kg. Several countries indicate productive capability in many crop areas, which is beneficial for them in terms of knowledge and of course potential capability to produce crop-based ethanol.

5.1.2.2.2 Southwestern Asia

The three countries in Southwestern Asia are Afghanistan, Azerbaijan, and Georgia (Figure 5.9). Afghanistan and Georgia both produce corn at 300 million kg per year, while Azerbaijan produces it at 160 million kg per year. The only other crop of significant production that may bear on ethanol is barley, produced at 600 million kg per year in Azerbaijan and at 400 million kg per year in Afghanistan. Ethanol therefore

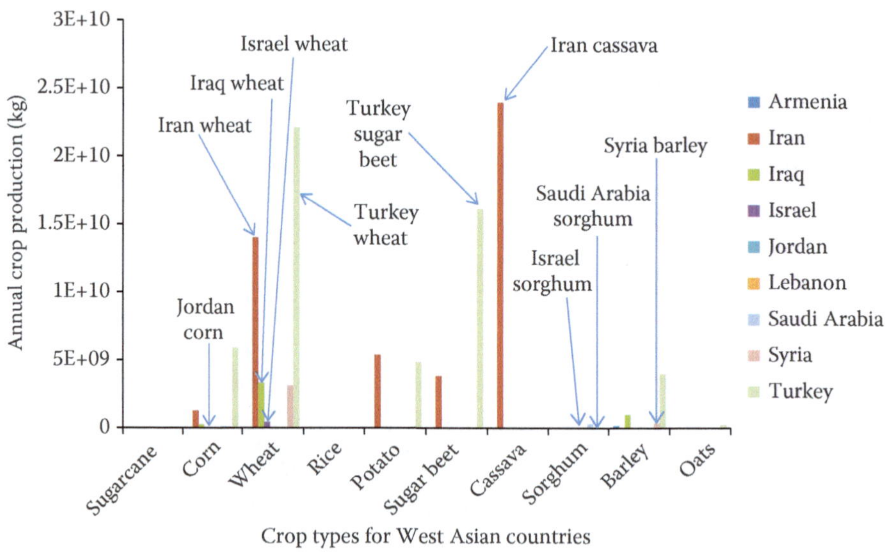

FIGURE 5.8 Starch-containing crop production for West Asian countries (Middle East).

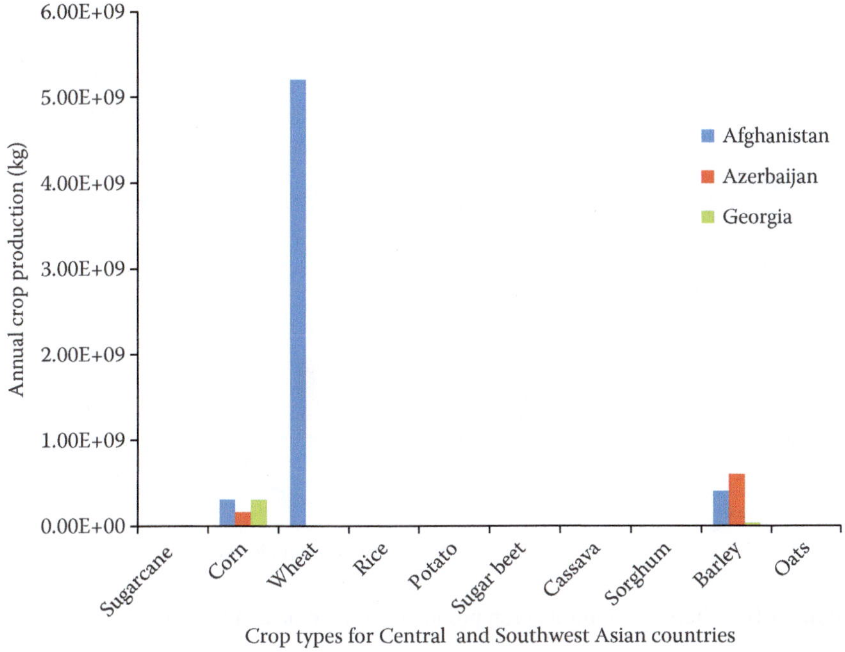

FIGURE 5.9 Annual production of starch-containing crops in Southwest Asia.

represents another opportunity for commerce in this region (given the political challenges can be addressed there in a manner so all parties gain from cooperation).

5.1.2.2.3 Central and North Central Asia

Figure 5.10 shows two countries (in Central and North Central Asia) with very large sugar/starch-containing production capability, China and Russia. China produces 126 billion kg per year of sugarcane so it can provide a large amount of sugar with potential for ethanol (Wikipedia. 2015). China produces 216 billion kg per year of corn and Russia, 11.3 billion kg per year of corn. China produces 122 billion kg of wheat per year while Russia produces 52.1 billion kg per year of wheat. China produces 204 billion kg per year of rice (Wikipedia. 2012). China produces 85.9 billion kg per year of potato while Russia produces 29.8 billion kg per year of potato. Russia produces 47.6 billion kg per year of sugar beet, ahead of China in this category which produces 10.7 billion kg per year. China produces 4.6 billion kg per year of Cassava and 2.7 billion kg per year of sorghum. Russia produces 20 billion kg per year of barley while China produces 1.55 billion kg per year. Russia produces 4.03 billion kg per year of oats and China 0.54 billion kg per year (Strychar, R. 2010). Both countries produce multiple crop categories and have diversified potential for ethanol production.

5.1.2.2.4 Northeast Asia

Figure 5.11 shows countries in Northeast Asia. North Korea has corn production at 1.93 billion kg per year. Japan's rice production is at 10.7 billion kg per year

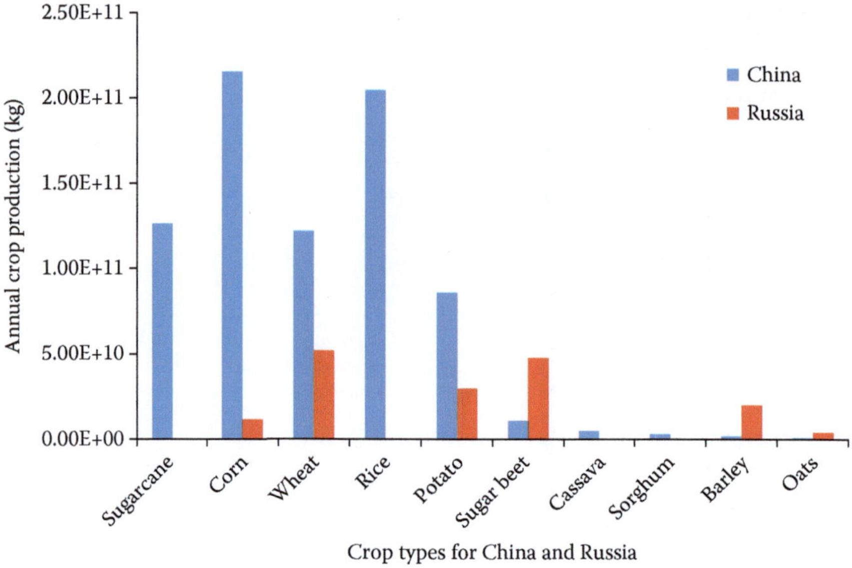

FIGURE 5.10 Starch-containing crop production for China and Russia.

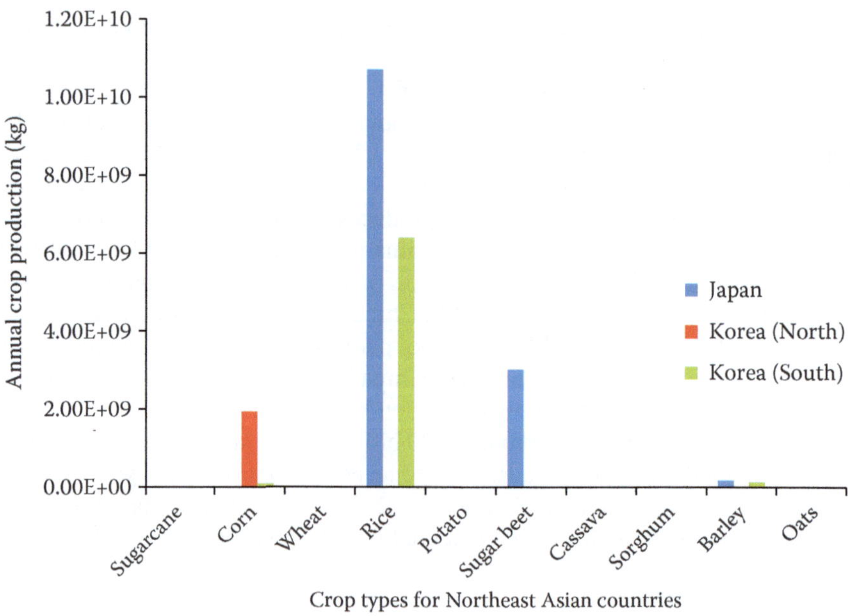

FIGURE 5.11 Starch-containing crop production for Northeast Asian countries.

while South Korea is at 6.4 billion kg per year. Japan's sugar beet production is at 3 billion kg per year. Japan's barley production is 1.7 billion kg per year while South Korea's barley production is at 1.31 billion kg per year. The capability in barley is much directed toward beverage alcohol. With this capability for this region and others, resources applied to producing industrial/fuel-grade alcohol should be easier to develop. North Korea has an opportunity to develop its corn crop and maybe other sources toward fuel/industrial alcohol, which can serve their country and people well if properly developed, preferably in cooperation with other countries.

5.1.2.2.5 Southern Asia

Figure 5.12 shows crop production that could be directed to ethanol for fuel/industrial use for countries in the southern region of Asia. India has large productive capability in several crops that can be directed to ethanol in part. Sugarcane (at 341 billion kg per year) is one of them. Pakistan can produce sugarcane at 6.38 billion kg per year. Cooperation between these countries in this area would be beneficial. India can produce corn at 22.5 billion kg per year. Bhutan can produce corn at 79 million kg per year as the lowest producer of this crop. India can produce wheat at 93.5 billion kg per year. Pakistan can produce 24.2 billion kg of wheat per year. India can produce 153 billion kg per year of rice. The lowest producer of rice on this chart is Nepal at 5.1 billion kg per year. Nepal could gain revenue for its country by perhaps expanding its rice production and using it to produce ethanol as a potential export product. India shows capability to produce 45 billion kg per year of potato. Nepal can produce 2.58 billion kg of potato (this can be another basis

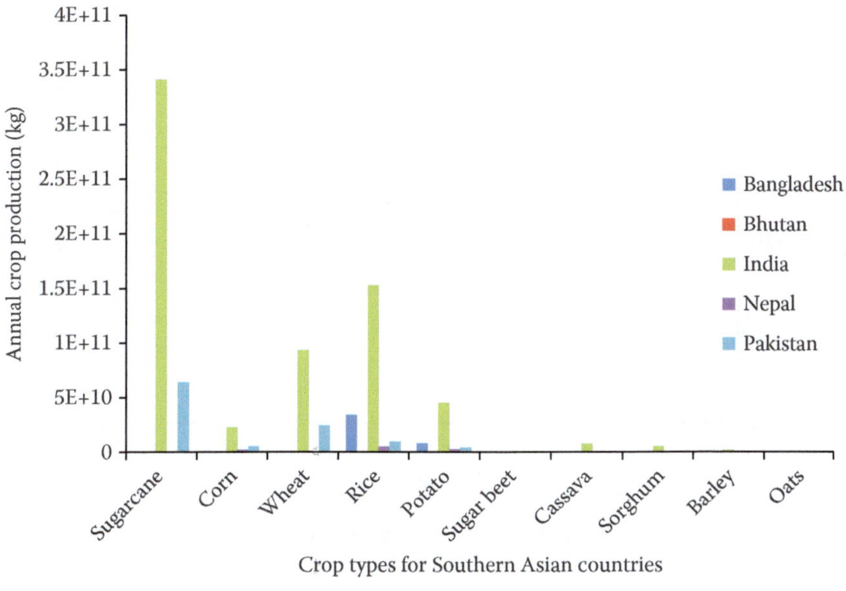

FIGURE 5.12 Starch-containing crop production for Southern Asia countries.

for their involvement in ethanol manufacture). India is able to produce cassava at 7.24 billion kg per year and can produce sorghum at 5 billion kg per year. Pakistan can produce sorghum at 145 million kg per year. India leads production in barley at 1.83 billion kg per year. The lowest barley producer is Bhutan at 5 million kg per year. Pakistan (265 million kg per year), Nepal (at 30 million kg per year), and Bangladesh (at 7 million kg per year) are in between the upper and lower figures for barley.

5.1.2.2.6 Southeastern Asia

Figure 5.13 shows crop production by country for Southeastern Asia. The largest sugarcane producer in this region is Thailand at 100 billion kg per year. The lowest sugarcane producer in this region based on this chart is the Philippines at 31.9 billion kg per year. The largest corn producer in this region is Indonesia at 9.4 billion kg per year. The lowest corn producer in this region according to this analysis and chart is Malaysia at 56 million kg per year. The largest producer of rice in this region is Indonesia at 69 billion kg per year. The lowest producer of rice is Cambodia at 9.3 billion kg per year. The largest producer of cassava is Thailand at 30.2 billion kg per year. The lowest producer of cassava is Malaysia at 81.7 million kg per year. This region has, of course, substantial production potential for ethanol.

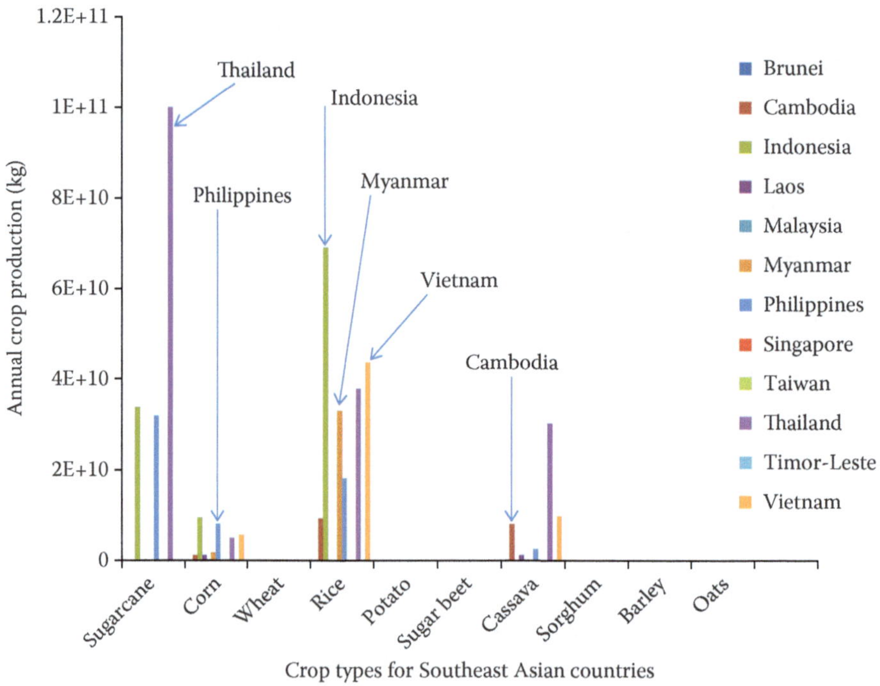

FIGURE 5.13 Starch-containing crop production for Southeast Asian countries.

5.1.2.3 Atlantic Ocean

This section covers those independent countries in the Atlantic Ocean where capability in certain sugar- or starch-bearing crops is noted in referenced data. Figure 5.14 shows data for Cape Verde, Grenada, and Sao Tome Principe. Cape Verde shows productive capability for corn at 1 million kg per year and cassava at 4.12 million kg per year. Sao Tome and Principe shows productive capability for cassava at 1.433 million kg per year, while Grenada can produce cassava at 210,000 kg per year. These are relatively smaller compared to other locales and these places are not necessarily close to each other to physically cooperate. However, they could still cooperate commercially by dealing with other parties if their crop production could be expanded to obtain the capability to produce ethanol perhaps elsewhere if not on their islands specifically.

5.1.2.4 Caribbean

Jamaica has the highest production of sugarcane in the Caribbean at 160 million kg per year (Figure 5.15). The French Antilles produce the least sugarcane in this region as per this figure at 20 million kg per year. Cuba produces the most corn in this region at 426 million kg per year. The Dominican Republic produces the least corn in this region at 45 million kg per year. Cuba has the highest cassava production in the region at 514 million kg/year. Haiti is next highest in cassava production at 418 million kg per year. A source claims that Cuba is trying to rejuvenate their sugarcane production (Sanchez, A.P. 2016). The Cayman Islands have the least cassava production in the region at 15,000 kg per year. Haiti produces the most sorghum in region at 90 million kg per year. The Dominican Republic produces 4 million kg per

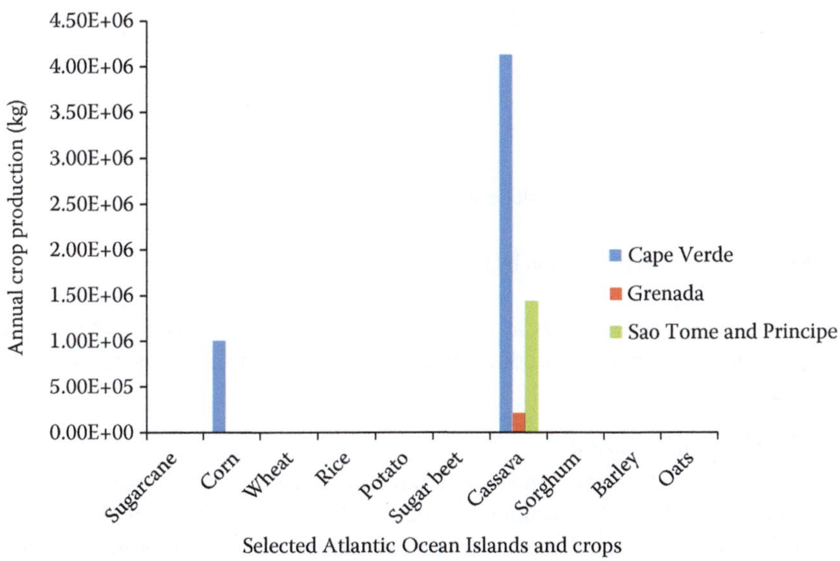

FIGURE 5.14 Starch-containing crops produced on selected Atlantic Ocean Islands.

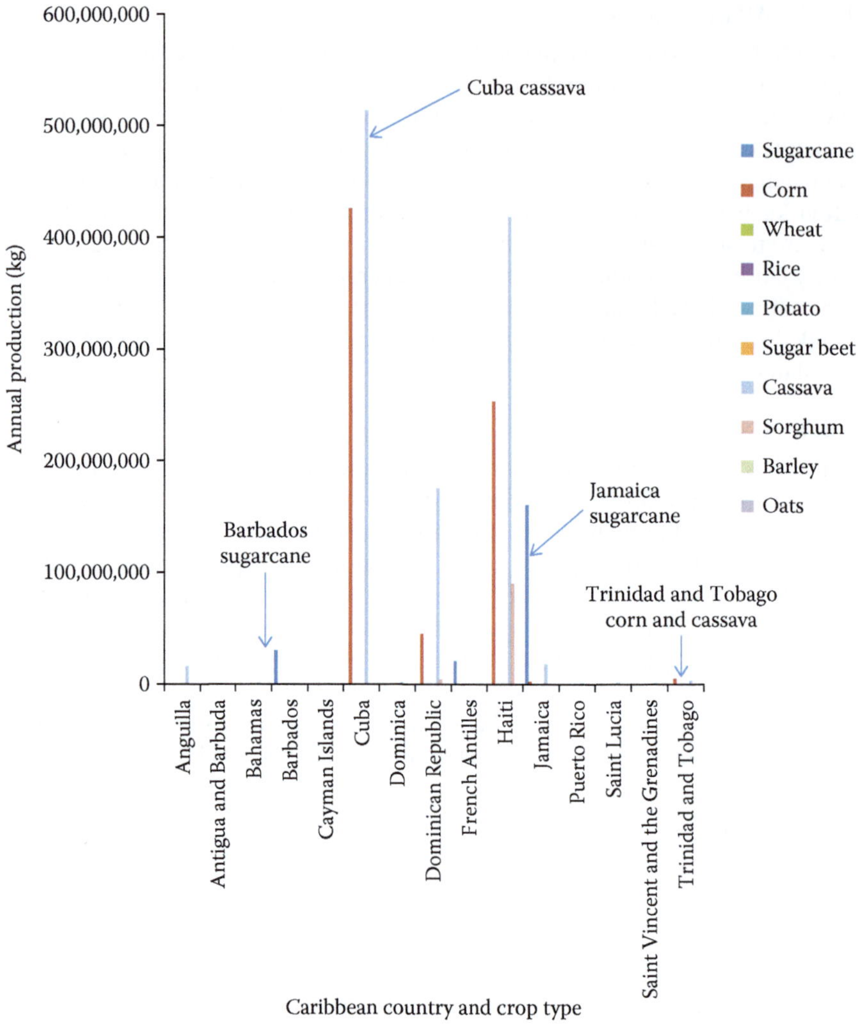

FIGURE 5.15 Annual production by Caribbean country and crop type.

year of sorghum. These islands are all relatively close together and could cooperate in crop production and even produce ethanol locally in one convenient locale in the Caribbean.

5.1.2.5 Central America

Figure 5.16 shows Central American countries and crops they grow suitable for use in supplying starch for ethanol production. Guatemala produces the most corn (1.64 billion kg per year). The next largest producer is El Salvador at 0.769 billion kilograms per year. The smallest producer is Costa Rica at 18 million kg per year. Nicaragua produces the most cassava (0.277 billion kg per year). The smallest producer of cassava is Panama at 21.5 million kg per year. El Salvador produces the

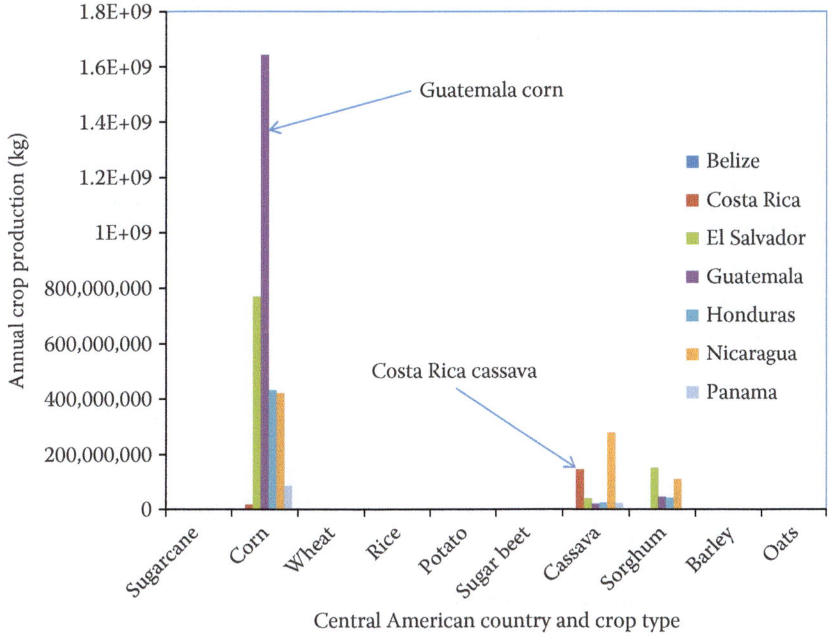

FIGURE 5.16 Central American crop production and type.

most sorghum (150 million kg per year). Honduras produces the least amount of sorghum shown (40 million kg per year). Once food needs are satisfied, all of these countries could have an opportunity to produce ethanol and perhaps cooperatively if they so choose to do so.

5.1.2.6 Europe

5.1.2.6.1 Southern Europe

Europe is divided into southern, northern, western, and eastern regions. In Southern Europe (Figure 5.17), the leading producer of corn is Serbia at 7.7 billion kg per year. Italy is next at 6.5 billion kg per year of corn. The smallest producer is Macedonia at 119 million kg per year.

Italy is the largest producer of wheat at 3.24 billion kg per year. The smallest producer of wheat is Portugal at 85 million kg per year. Italy is the significant producer of sugar beet at 3.25 billion kg per year. Italy also leads in the production of sorghum at 320 million kg per year. The smallest producer of sorghum is Macedonia at 60,100 kg of sorghum per year.

Italy is the largest producer of barley at 772 million kg per year. Portugal produces the least amount of barley shown at 30 million kg per year. Italy produces 320 million kg of oats per year. Slovenia produces 5 million kg per year of oats. The smallest producer could, for example, expand the production of the particular crop where they have the experience and resources and apply this to ethanol manufacture if they envision this as an opportunity.

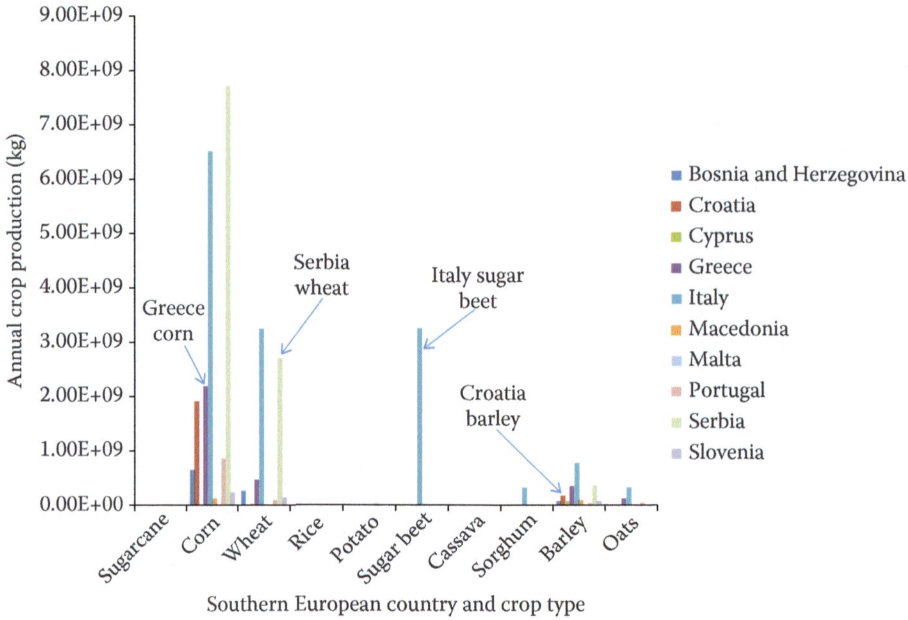

FIGURE 5.17 Production of starch-containing crops in Southern European countries.

5.1.2.6.2 Western Europe

The largest producer of corn in Western Europe is France at 15.1 billion kg per year (Figure 5.18). The next largest producer of corn in this region is Germany at 4.39 billion kg per year. The smallest producer of corn in this region from this chart is Switzerland at 146 million kg per year.

The largest producer of wheat in Western Europe is France at 38.6 billion kg per year. The next largest producer of wheat in this region is Germany at 25 billion kg per year. The smallest producer of wheat in this region is Luxembourg at 91 million kg per year.

The Netherlands is the sole significant producer of rice in this region at 1.34 billion kg per year. Several countries produce potato at almost the same level (France at 7 billion kg per year, the Netherlands at 6.77 billion kg per year, and Germany at 6.34 billion kg per year). Belgium produces 2.93 billion kg per year of potato. Several countries produce sugar beet, a viable potential source of sucrose for ethanol use as well as a food sweetener. France produces 38.1 billion kg per year of this crop, and Germany produces 25 billion kg per year. There is a drop in quantity produced by the next three countries (Netherlands at 6 billion kg per year, Belgium at 5 billion kg per year, and Austria at 3 billion kg per year). Sugar beet is an evident candidate since the sucrose once isolated can be directly used in the ethanol fermentation, thereby avoiding the intermediate conversion to form short oligosaccharides and glucose that is required for starch sources.

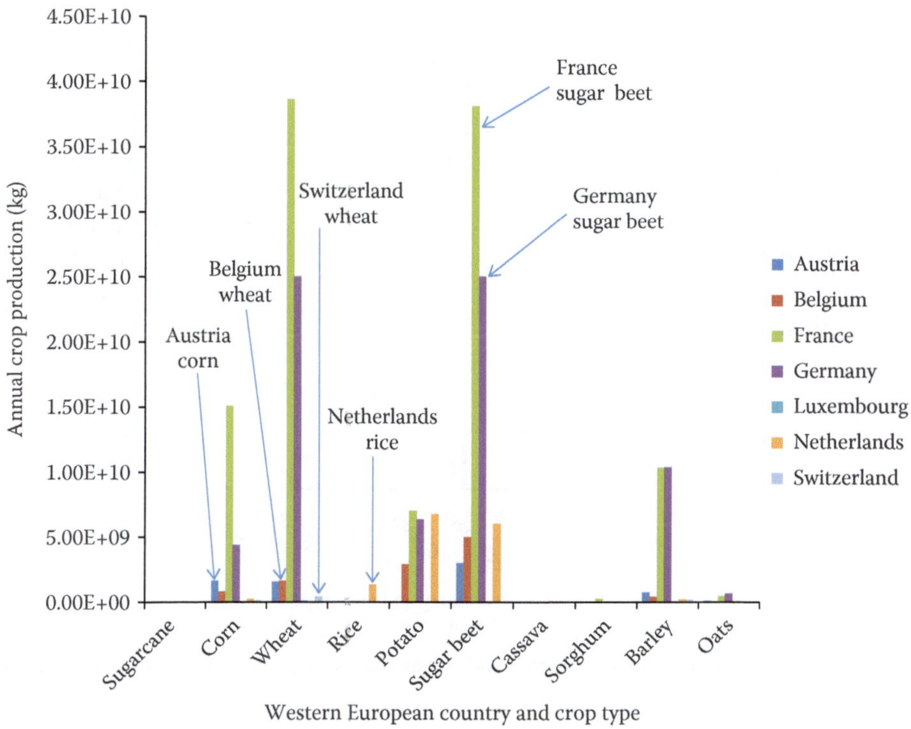

FIGURE 5.18 Production of starch-containing crops in Western European countries.

France (279 million kg per year) and Austria (10.2 million kg per year) each produce sorghum in this region as indicated by small values on the bar chart.

Germany and France produce almost the same amount of barley (1.03 billion kg per year). The smallest producer of barley is Belgium at 0.391 billion kg per year.

German produces 668 million kg per year of oats. France produces 450 million kg per year of oats, while Belgium and the Netherlands produce 34 and 6 million kg per year of oats, respectively.

5.1.2.6.3 Eastern Europe

Figure 5.19 shows crop production in Eastern European countries with starch or sugar content allowing production of ethanol. Ukraine produces 28.5 billion kg of corn annually to lead the region. Ukraine actually leads in several crop categories. No wonder they call this country the "breadbasket" of Eastern Europe. Romania, the second largest corn producer in this region of Europe, provides 11.4 billion kg of corn annually. Albania produces the least amount of corn (300 million kg annually). Clearly, corn is a major crop for the countries of this region.

Ukraine produces 22.8 billion kg of wheat. Poland is second at 9.5 billion kg of wheat annually. Bulgaria produces 5.38 billion kg of wheat annually. The smallest producer of wheat is Slovakia at 1.57 billon kg annually.

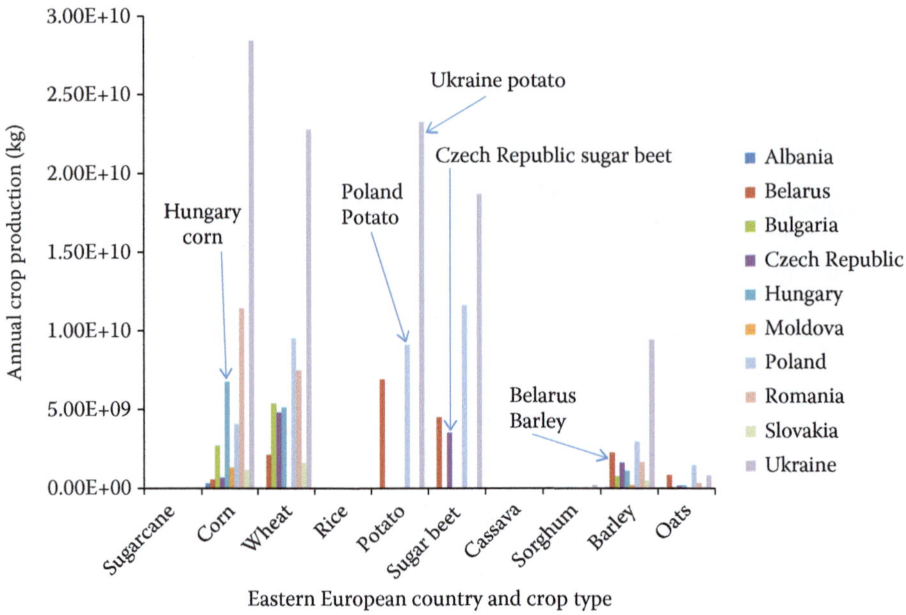

FIGURE 5.19 Production of starch- and sugar-containing crops in Eastern European countries.

Ukraine leads the production of potato in this region at 23.3 billion kg annually. Poland is next at 9.09 billion kg of potato annually. The remaining country listed for potato is Belarus at 6.09 billion kg annually. From personal experience, potato is a very challenging crop to grow since it is a root crop subject to many diseases except for early generations of seed potatoes for this tuber root crop. Seeds from the potato berry can be used initially to prevent disease formation to a large extent (Goldstein, W.E. 1988–1994).

Ukraine leads the production of sugar beet in this region at 18.7 billion kg annually. The Czech Republic is the lowest producer of sugar beet in this region on this chart at 3.5 billion kg annually.

Sorghum is a lesser produced crop in this region. Ukraine has the most production of sorghum in this region at 180 million kg annually. Romania is next at 49.9 million kg of sorghum annually. The lowest producer of sorghum in this region is Albania at 9.66 million kg annually.

Ukraine leads the production of barley in this region at 9.45 billion kg annually. Moldova is the lowest producer of barley in this region at 200 million kg annually.

Poland leads oats production in this region at 1.44 billion kg annually. Hungary and the Czech Republic are the lowest producers of oats in this region at 140 million kg annually each.

5.1.2.6.4 Northern Europe

As shown in Figure 5.20, the United Kingdom is the leading producer of four crops that contain starch or sugar and could be directed to ethanol production.

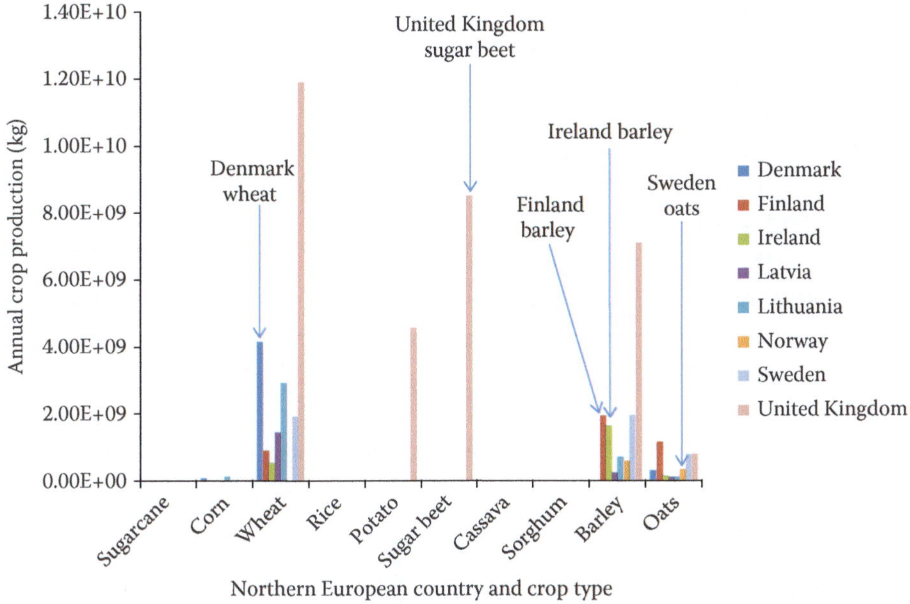

FIGURE 5.20 Production of starch- or sugar-containing crops in Northern European countries.

Corn is produced to a much lesser extent in this region than, for example, in Eastern Europe. Lithuania produces 121 million kg annually while Denmark produces 76 million kg annually, and Sweden produces 11 million kg annually.

The United Kingdom produces 11.9 billion kg of wheat annually. The second highest producer is Denmark at 4.15 billion kg annually. The lowest wheat producer in this region is Latvia at 1.44 billion kg annually.

The United Kingdom annually produces 8.5 billion kg of potato and 4.553 billion kg of sugar beet. It is surprising that other countries are not identified in this region that produce significant levels of these two crop products.

Several countries produce barley with the United Kingdom leading at 7.092 billion kg annually. Finland and Sweden produce barley at about the same quantity, 1.94 billion kg annually. The lowest producer of barley noted is Latvia at 233 million kg annually.

Finland produces oats at 1.16 billion kg annually to lead this region of Europe in this category. Latvia produces the least amount of oats noted for this region at 100 million kg annually.

5.1.2.7 Indian Ocean

As shown in Figure 5.21, there are six countries identified with the potential to produce crops that can provide feedstock for ethanol manufacture from the starch in the crop. In the first graph in this figure, Madagascar produces 366 million kg of corn annually. Madagascar produces 4 billion kg and Sri Lanka, 3.8 billion kg, of rice annually. Madagascar produces 3.11 billion kg per year of cassava, Sri Lanka, 303 million kg per year, and Comoros, 70 million kg per year.

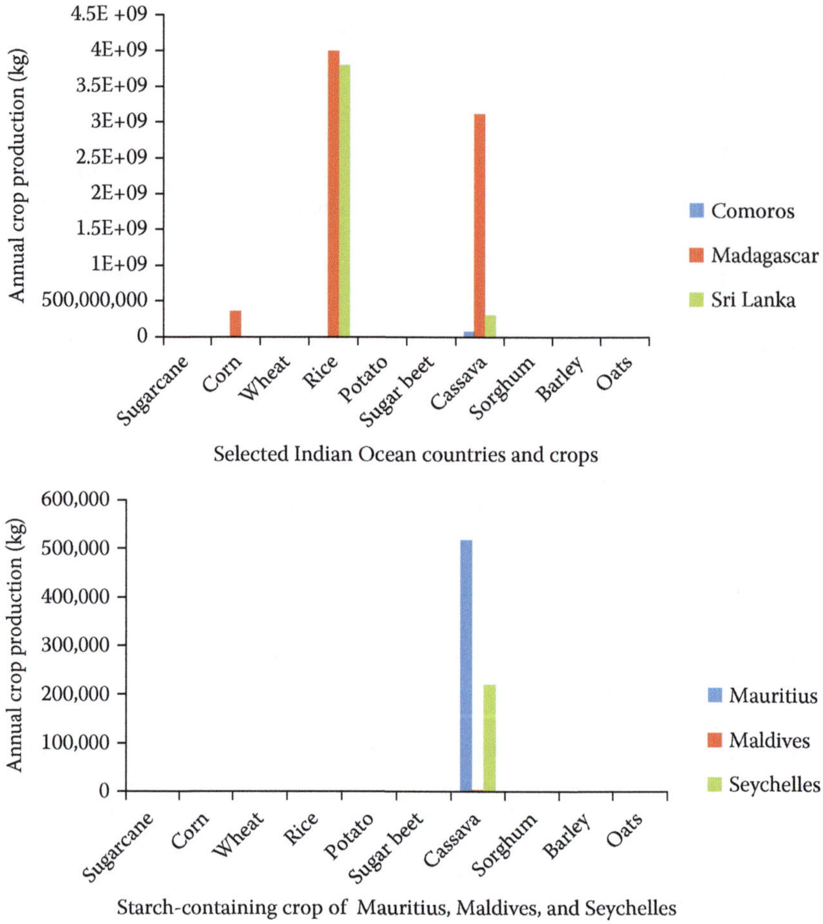

FIGURE 5.21 Starch-containing crops in countries located in the Indian Ocean.

In the second graph on Figure 5.21, Mauritius produces 517,000 kg of cassava annually. Seychelles produces 220,000 kg per year of Cassava, and the Maldives, 4,000 kg per year of cassava.

For these locales in the Indian Ocean, it would be interesting to determine how much more land could be applied to increase production and if arrangements could be made to use this land to grow crops to produce ethanol, which would bring in a much higher revenue than the crop and conceivably be worth the cost to make this come about.

5.1.2.8 North America

Figure 5.22 shows the crop production that may provide (or has provided) feedstock for ethanol manufacture for Canada, Mexico, and the United States. Figure 5.23 shows this same crop production for Canada and Mexico on a different scale for purposes of added clarity.

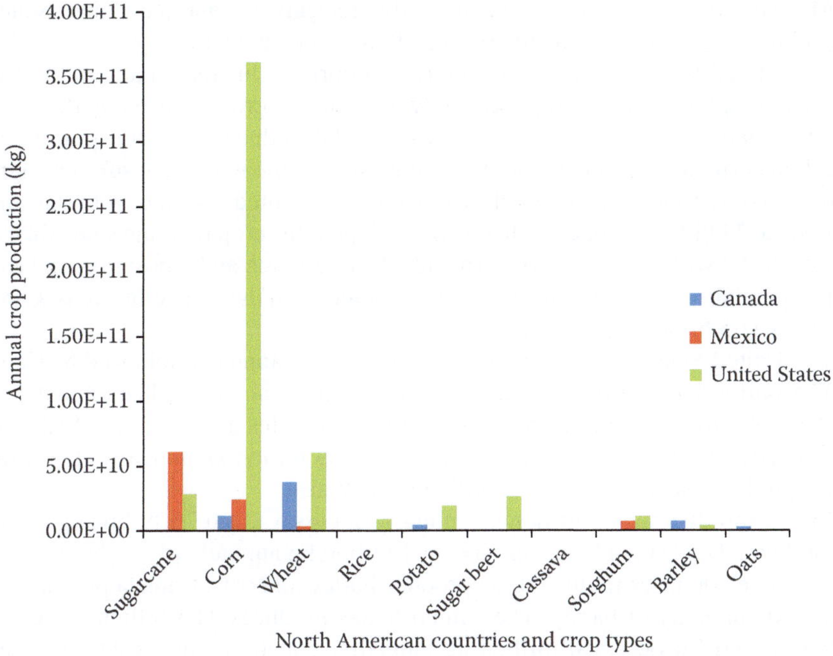

FIGURE 5.22 Crop production of starch and sugar sources for ethanol in North America.

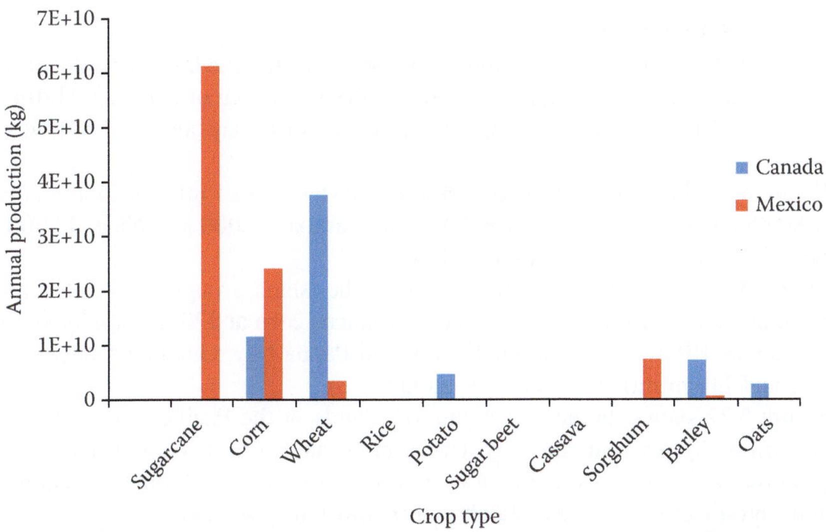

FIGURE 5.23 Annual production for starch- and sugar-bearing crops for Canada and Mexico.

Mexico produces 6.12 billion kg annually of sugarcane, and the United States is second in this category in this region at 2.79 billion kg annually.

The United States is the major producer of corn in this region and in the world at 361 billion kg annually. Approximately 40% of this crop is currently directed to ethanol manufacture. The corn industry itself and the ethanol industry spinoff that is based on corn has been in operation for quite some time with the result that significant technology and commercial development has resulted. Mexico, with the ability to produce 24 billion kg of corn, has also developed this crop for quite some time and houses the CIMMYT facility for corn and wheat research and repository—a facility that is quite famous for its contribution to agriculture in these specific areas. Canada produces 11.5 billion kg per year of corn.

The United States produces 60 billion kg of wheat annually, followed by Canada at 37.5 billion kg, and then Mexico at 2.4 billion kg annually. The United States produces 9 billion kg of rice annually. The United States also produces 19.2 billion kg annually of potato, followed by Canada at 4.59 billion kg annually. The United States produces sugar beet at 26.2 billion kg annually.

Mexico produces 24.3 million kg annually of cassava, and 7.3 billion kg per year of sorghum. The United States produces 11 billion kg annually of sorghum.

All three countries in this region produce barley and oats. Canada produces 7.12 billion kg annually of barley. The United States produces 3.85 billion kg per year of barley, and Mexico, 550 million kg annually. Canada produces 2.69 billion kg annually of oats. The United States produces 929 million kg annually of oats, and Mexico, 70 million kg annually.

From all the countries in North America, there is substantial productive capability in the crop areas covered so far in this analysis. The countries do cooperate and perhaps the productive capability is a positive result of that cooperation.

5.1.2.9 Pacific Ocean

Figure 5.24 shows starch-containing crops for Australia, the largest producer in the Pacific Ocean. Australia annually produces 400 million kg of corn, 22.9 billion kg of wheat, 2 billion kg of sorghum, 7.95 billion kg of barley, and 1.05 billion kg of oats.

Figure 5.25 shows annual cassava production (less than 1 million kg annually for each South Pacific island group) for American Samoa (85,000 kg), Niue (42,000 kg), Samoa (415,000 kg), and Singapore (5,000 kg).

Figure 5.26 shows the crop production for the island groups of Fiji, Papua New Guinea, and New Zealand. New Zealand produces corn at 220 million kg annually and barley at 410 million kg annually. Fiji and Papua New Guinea produce 94 million kg and 140 million kg of cassava annually.

Figure 5.27 shows the next grouping of islands in the Pacific Ocean. These are in the million kg per year category. The crop of interest in this case is cassava. The largest producer is the Federated States of Micronesia at 9.5 million kg annually. The smallest producer is the Cook Islands at 1.03 million kg annually.

Given the expanse of the Pacific Ocean, exploring cooperation in the area of cassava production directed toward ethanol and not interfering with food areas or ambiance seems warranted.

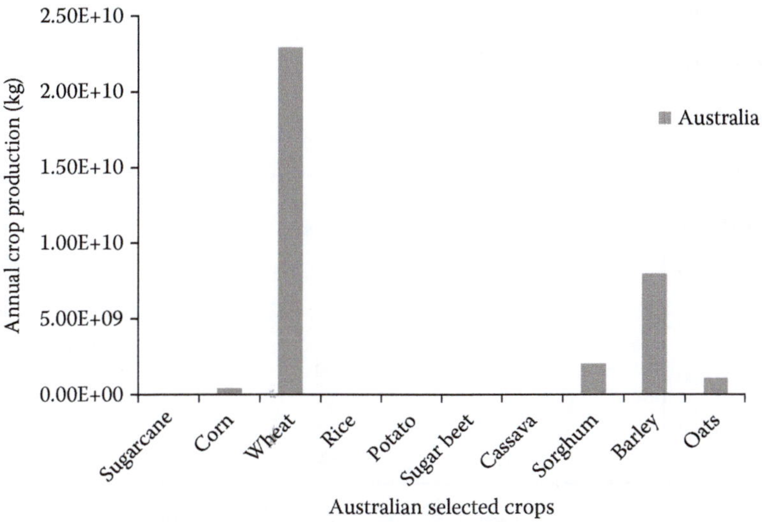

FIGURE 5.24 Starch-containing crop production in Australia.

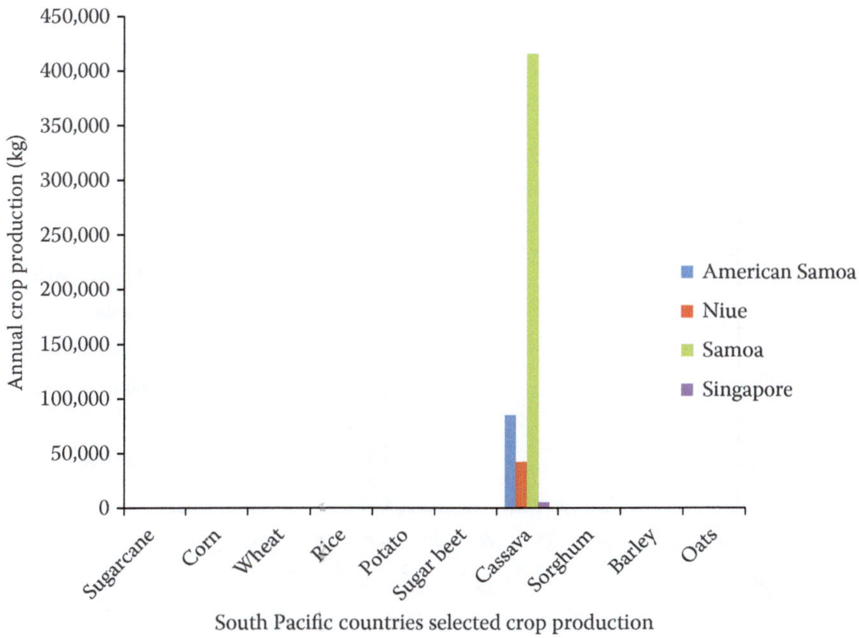

FIGURE 5.25 Starch-containing crop production in the South Pacific below 1 million kg.

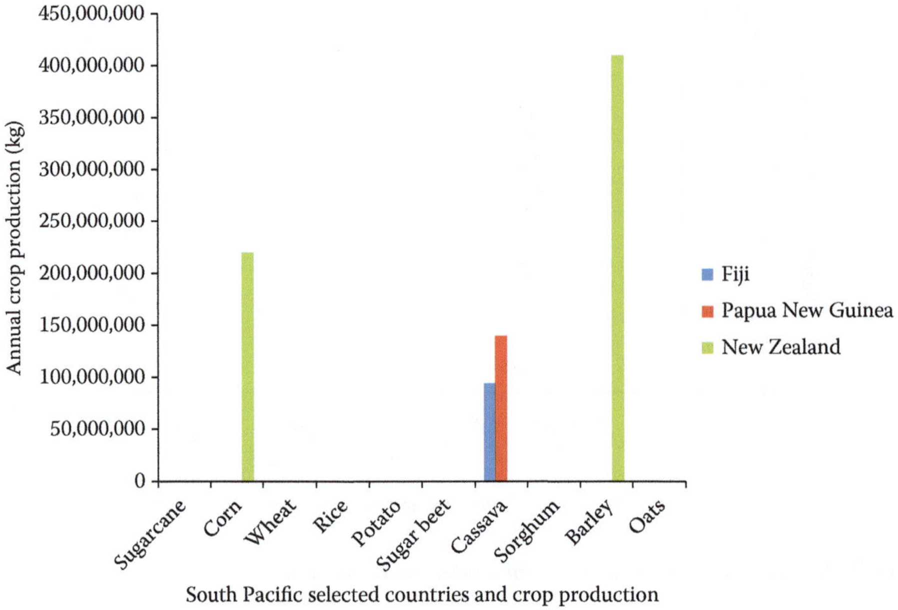

FIGURE 5.26 Starch-containing multicrop production in the South Pacific.

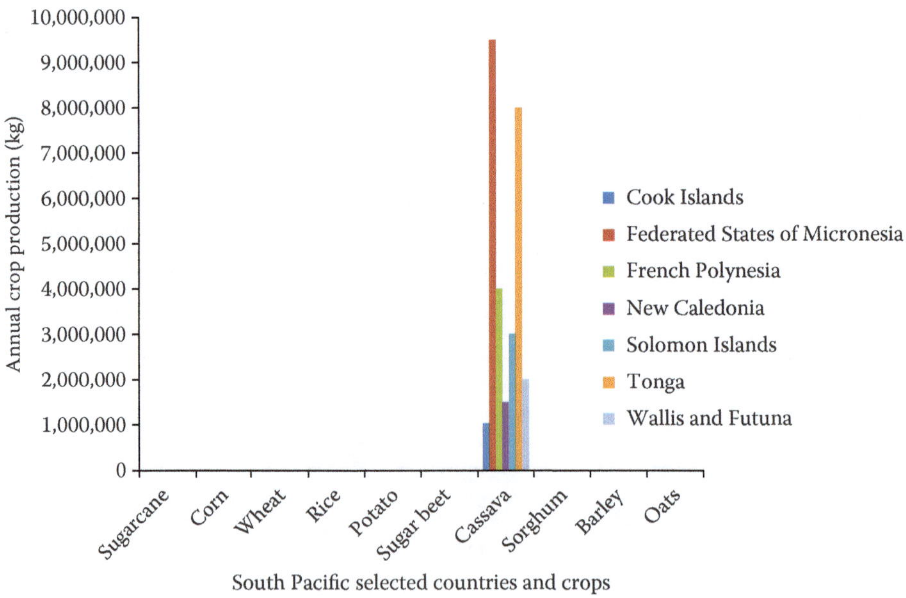

FIGURE 5.27 Starch-containing crops in the South Pacific in the million kg category.

5.1.2.10 South America

The countries in this region are presented using several charts. The first chart, Figure 5.28, shows Brazil, a country which is the most significant producer of sugarcane (739 billion kg annually), an important food source and raw material source for producing ethanol. Brazil also produces large amounts of other crops that could be used as feedstock for ethanol if that were desired. These are (in billions of kg annually) corn (75), wheat (5.7), rice (11.5), potato (3.73), cassava (21.5), sorghum (2), barley (0.33), and oats (0.4).

The second chart, Figure 5.29, shows a chart for two producers of sugar- and starch-containing crops, Argentina and Colombia. This chart indicates that Colombia produces 34.9 billion kg annually of sugarcane, 1.75 billion kg annually of corn, 2.23 billion kg annually of cassava, 200 million kg annually of sorghum, and 20 million kg annually of barley. Argentina produces 24 billion kg annually of corn, 8 billion kg annually of wheat, 184 million kg annually of cassava, 4 billion kg annually of sorghum, 3 billion kg annually of barley, and 450 million kg annually of oats. Separately (or together if they cooperate), these two countries can be substantial providers of ethanol as well food crops.

The next chart (Figure 5.30) shows starch-containing crops in eight South American countries. There are six prominent corn-producing countries in this set. These are Paraguay (3.08 billion kg annually), Peru (1.7 billion kg annually), Chile (1.4 billion kg annually), Ecuador (1.22 billion kg annually), Bolivia (725 million kg annually), and Uruguay (450 million kg annually). Guyana produces 5 million kg of corn annually.

As shown in Figure 5.30, Peru produces 4.47 billion kg of potato annually. The four countries, as shown, prominent in cassava are Bolivia (288 million kg annually), Ecuador (45.8 million kg annually), Paraguay (2.8 billion kg annually) and Peru

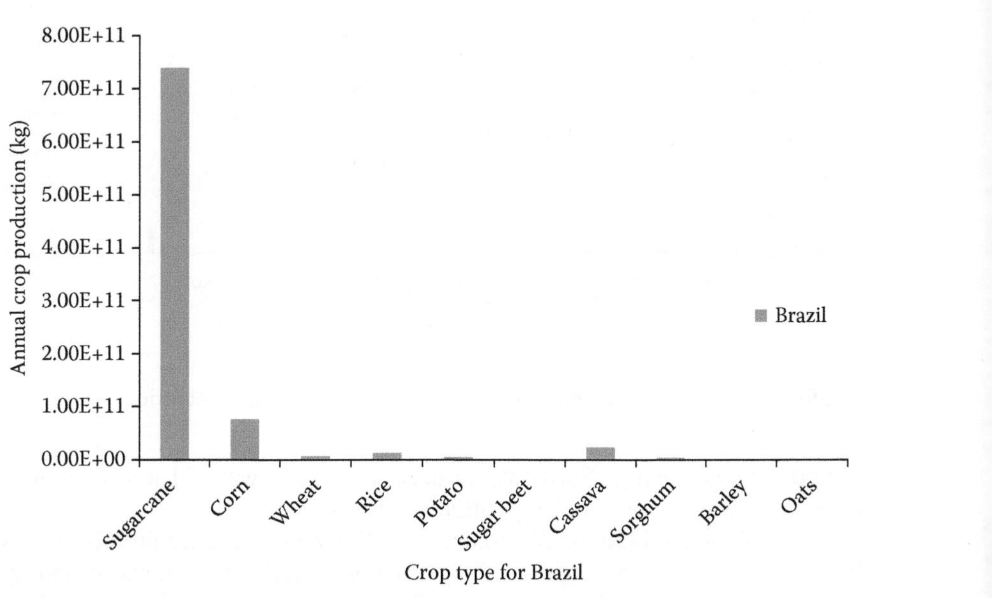

FIGURE 5.28 Annual crop production for Brazil.

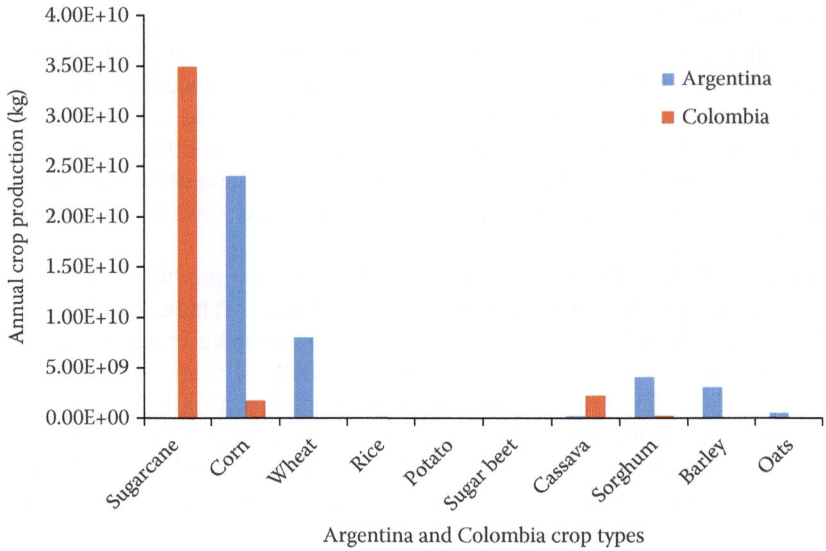

FIGURE 5.29 Annual crop production for Argentina and Colombia.

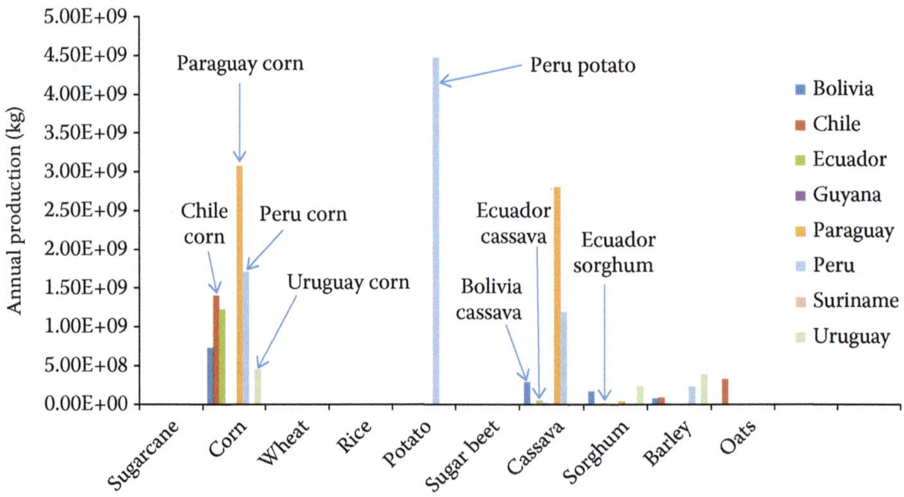

FIGURE 5.30 Starch-containing crops in select South American countries.

(1.18 billion kg annually). Suriname produces cassava as well (7.1 mm kg annually). Guyana produces cassava at 7.65 million kg annually.

As also shown in Figure 5.30, countries in this region that are prominent in sorghum are Bolivia (160 mm kg annually), Ecuador (10 million kg annually), Paraguay (41 million kg annually), Uruguay (240 million kg annually), and Peru (1 million kg annually).

Figure 5.30 shows that the countries prominent in barley (millions of kg annually) are Bolivia (70), Ecuador (15), Peru (230), and Uruguay (390). As shown in this figure, Chile produces 330 million kg annually of oats.

5.1.3 Value of Ethanol to a Country Based on Feedstock Potential, Pricing and Use in Automobiles

This assembled data described in Section 5.1.2 are helpful to judge a country's capability to produce ethanol and/or provide their crops to gain financial return. The worthiness of this economically is best evaluated by the businesses' (and a nation's) desire and culture to determine if developing ethanol makes sense for them. The primary driver for this is economics since the parties will likely find a way to accomplish what is needed if the financial return is sufficiently positive.

The value of this is a function of many aspects. One aspect (assuming a biological fermentation-based route) is the scale of the project, the size of the crop produced in the country, and the potential to considerably increase the amount so as to not interfere with the food supply. This is dependent on the parties' capability to increase crop yield and the availability of land to grow additional quantities. If the facility to make the ethanol is to be located in the particular country, then the return must be sufficient to justify the capital investment involved in the ethanol production facility and to also reasonably expect that continual operation can be conducted so that sales revenue will exceed costs.

The case for each country has to be evaluated on its own. For some, it will be better to join with others, perhaps to supply one main ethanol facility.

Let us examine some of the basic economics of this. First, it is necessary to examine the basic driving force for ethanol development. For the present, and some time to come, ethanol will be blended with gasoline to be used as an automotive fuel. There are alternatives to gasoline-powered automobiles and maybe not even using ethanol at all. However, a case is being made in this book for long-term ethanol use.

Consider the use of automotive fuel globally to judge the potential magnitude of the value of the ethanol business. To begin with, the world is presently approaching the production of nearly 100 million barrels of oil per day (Fleming, C. 2015). From each barrel, approximately 19 gallons of gasoline are produced (U.S. Dept. of Energy. Energy Information Administration. 2015). Therefore, the amount of gasoline produced daily is approximately 1.9 billion gallons. In metric units, this is approximately 7.2 billion liters of gasoline per day. According to present practice, the oil is not banked into inventory to any extent to cover a catastrophic event of severe shortage (Wikipedia. 2015a). Basically, only a few months are maintained as inventory. Avoiding maintaining a sizeable inventory of petroleum of (say) 2 years instead of just 2 months or so is a questionable practice. However, that is the way this is presently done. Therefore, the 7.2 billion liters of gasoline produced daily is consumed, mostly to power automobiles.

Present practices call for the replacement of 10% by volume gasoline with ethanol for use as a car fuel. The ethanol industry would like this percentage to increase. However, let us assume that the 10% replacement is fixed for the near term. Given this were to occur globally as a maximal potential figure, then that will involve in the vicinity of 720 million liters per day of ethanol.

Presently, the United States produces about 20% of this figure (*Ethanol Producer Magazine.* 2015). In the United States in 2014, approximately 60 billion liters of ethanol was produced, essentially all using corn as a feed stock. On a daily basis, this is approximately 0.2 billion liters per day. The other major world producer of ethanol at present is Brazil. They presently produce a little less than 50% of that produced in the United States, primarily from sugarcane (Renewable Fuels Association. 2015). Therefore, the present world production of ethanol is about 0.3 billion liters or 300 million liters per day. Therefore, the present production is less than half of the maximum the world might use (using the 10% ethanol substitution for gasoline criterion). If the percentage of ethanol by volume in the ethanol/gasoline blend goes up say to 15% instead of 10%, then the ethanol usage per day would exceed 1 billion liters. Therefore, using this very rough calculation, and the prospects in discussion, there is a future need for the product in that respect.

In the United States, at present, the retail cost of gasoline at the gas station pump is approximately $3/gal or $0.8/liter. Assume for purposes of this analysis (prior to examining ethanol economics in detail) that the value of the ethanol substituted into the gasoline is the same as that of gasoline. Therefore, the retail price of the ethanol blended into the gasoline is also $0.8/liter.

5.1.3.1 Corn Feedstock Value

Parties incur costs and gain profit to obtain the ethanol that goes into the fuel mixture. This starts with providing the crop that contributes the starch or sucrose that eventually is converted to ethanol. In the United States, corn that is used for ethanol currently is priced at approximately $4 per bushel (*Ethanol Producer Magazine.* 2016). The current yield of ethanol from corn is quoted to be 2.8 gallons of ethanol per bushel of corn (Renewable Fuels Association. 2015). This figure is based on using a well-functioning process that converts whole ears of corn to ethanol using an enzyme-based digestion, yeast-based fermentation process, and a product recovery process that is based on using distillation to isolate the ethanol. Using metric units, the amount of ethanol recovered is 10.6 liters of ethanol per bushel of corn.

Dividing $4/bushel by 10.6 liters per bushel results in $0.38 per liter. This represents the primary raw material cost that goes into preparing the ethanol. Costs involving other raw materials (such as enzymes, yeast, and specialty chemicals), capital charges for depreciation, maintenance, taxes, and insurance, labor, utility costs, and overhead are not yet factored into this (such costs are covered later in this book in Chapter 7). The by-product credit that will occur (e.g., production and sale of distiller's dry grains, DDGS) also has to be factored in from standpoint of addition to income or reduction of costs.

Before these additional costs and the by-product credit are included, it is clear (using corn in a U.S. process) that there is a margin available (the difference between $0.8 and $0.38/L) to produce the ethanol that will go into the gasoline–ethanol blend. This difference in this example is not, however, substantial. Therefore, a careful analysis of each situation (crop, country, and organization) is required to assess what needs to be done, and the potential value to each party considering getting into this business. If a sufficient margin can be shown, then considerable value may result for a company and its host country. The raw materials covered in this section relevant to each country listed in the charts enter into this analysis.

5.1.3.2 Potato Feedstock Value

Table 5.1 indicates the raw material type, two major countries producing ethanol, and approximate overall value as a useful starting point. Lignocellulose entities are not considered in this table, only present starch (and sugar)-bearing crops that are also food sources. One can see corn is listed with the $0.38/liter of ethanol raw material cost as shown. For another material, potato, a difficulty arises. Potato is also grown as a food source, and really is not yet a source for ethanol. Potato has a substantial water content that is part of its very nature and likely an important component of development for its characteristic desirable flavor as a food source. The raw potato contains 75% moisture content while corn contains about 12% in the product ears (Blau, D.A. et al. 2001; Wikipedia. 2013c).

When corn is provided for ethanol (presently mostly as whole corn for processing) what is primarily of interest for ethanol production is the starch content which is approximately 73% by weight in the original ears of corn. In regular white whole potatoes, the starch content is much less, 15%. This is a big difference since more potato has to be used to provide an amount of starch equivalent to that provided in a given weight of corn.

Given ethanol yields from a kilogram of potato starch are in effect the same as from a kilogram of cornstarch, the difference has to be made up by the potatoes being made available at a lower relative cost than corn. From what the writer can presently gauge, this is not going to happen without some other factors coining into play.

Examining Table 5.1, corn at $4/bushel, with 56 lbs per bushel is therefore valued at about $0.07 per lb or $0.16/kg. Since it is 73% starch, then its value per lb of starch is about $0.10/lb (or $0.22/kg) of starch in the corn. Potato is valued at $0.08/lb or $0.18/kg, in effect the same as corn. However, on a per kg of starch basis, it is valued at $0.52/lb of starch or $1.14/kg. Since the yield of ethanol from a given weight amount of starch is likely the same for corn and potato (about 0.26 liters per lb of starch), then the cost of the ethanol from potato on the basis of the raw material alone is about $2/liter compared to the $0.38/liter for corn. These prices are F.O.B. (Free-on-Board) prices. Unless the potato can be produced at much less a cost in a country and the reduced cost is reflected in its F.O.B. price, then whole raw potato is not an option as a raw material feedstock for ethanol.

Examining the price of dried potato chips (that will have the cost of drying added to the price), no price advantages result as the chips just cost more without any gain in relative cost reduction for the amount of water removed. High solids potatoes of the type needed do not seem to be a product of commerce and perhaps not technically feasible for this crop in the range needed. Therefore, a country that has potato as its best option has to look at this carefully since potato does not seem to be practical from a cost standpoint as compared to corn unless the costs to produce the crop can be reduced compared to other places and that is translated into selling price.

TABLE 5.1
Crop Costs by Country and Raw Material to Manufacture Ethanol

Crop	Country	F.O.B. Price RM ($/bushel)	Liters Ethanol/RM Quantity (L/bushel)	RM Cost ($/L)	Gasoline ($/L)	Price/lb Starch ($/lb)	Liters Ethanol/lb Starch Basis (lb/bushel)	RM Cost ($/L)	By-Product DDG	By-Product	By-Product	Starch Fraction	lb/ bushel
Corn	United States	4	10.6	0.377358	0.8	0.099206349	0.262896825	0.377				0.72	56
Sugarcane	Brazil	$/kg cane sugar	L/kg cane sugar										
		0.525	0.63	0.833333									
Sugarcane	Brazil	$/kg cane juice	L/kg cane juice solids						DDG	Molasses	Vinasse		
Juice		To ethanol plant	Should be same as above										
		0.2625	0.63	0.416667									
		Save evap, etc. $											
Potato	United States	$/lb of potato	$/lb of starch				L/lb starch						
		0.08	0.519480519	1.975962			0.2629					0.154	
Chips		0.682	0.817745803	3.110482			0.2629					0.834	
Rice ($/kg)													
Brown rice		0.4	0.571428571	0.988631		L/kg starch	0.578					0.7	
Rice ($/kg)													
Paddy rice		0.2	0.285714286	0.494315		L/kg starch	0.578		DDG	Rice chaff	Rice fiber-bran	0.7	

(Continued)

TABLE 5.1 (*Continued*)
Crop Costs by Country and Raw Material to Manufacture Ethanol

Crop	Country	F.O.B. Price RM ($/bushel)	Liters Ethanol/RM Quantity (L/bushel)	RM Cost ($/L)	Gasoline ($/L)	Price/lb Starch ($/lb)	Liters Ethanol/lb Starch Basis (lb/bushel)	RM Cost ($/L)	By-Product DDG	By-Product	By-Product	Starch Fraction	lb/ bushel
Sugar beet		$/kg of beet	$/kg of sugar in the beet										
Subsidized		0.055	0.305555556	0.528643		L/kg sugar	0.578					0.18	
Market		0.011	0.061111111	0.105729		L/kg sugar	0.578					0.18	
Use lower nonsubsidized price for beet													
Problem with regulations in Europe													
Which require subsidy, maybe that can change													
Farmers need to get their money from stake in ethanol production													
Elsewhere as well													

(*Continued*)

TABLE 5.1 (*Continued*)
Crop Costs by Country and Raw Material to Manufacture Ethanol

Crop	Country	F.O.B. Price RM ($/bushel)	Liters Ethanol/RM Quantity (L/bushel)	RM Cost ($/L)	Gasoline ($/L)	Price/lb Starch ($/lb)	Liters Ethanol/lb Starch Basis (lb/bushel)	RM Cost ($/L)	By-Product DDG	By-Product	By-Product	Starch Fraction	lb/ bushel
Wheat													
Whole grain		0.4	0.571428571	0.988631		L/kg starch	0.578					0.7	
Spread													
Varies no													
Pattern													
per kg													
Wheat flour		0.15	0.214285714	0.370737		L/kg starch	0.578					0.7	
Cassava													
Raw root ($/kg)		0.05	0.133333333	0.230681		L/kg starch	0.578					0.375	
Chips ($/kg)		0.17	0.197674419	0.341997		L/kg starch	0.578					0.86	
Flour ($/kg)		0.35	0.406976744	0.704112		L/kg starch	0.578					0.86	
Sorghum													
Grain ($/kg)		0.225	0.247252747	0.427773		L/kg starch	0.578					0.91	
						L/kg starch							
Barley													
Grain ($/kg)		0.18	0.2	0.444444		L/kg starch	0.45					0.9	
						L/kg starch							
Oats													
Grain ($/kg)		0.425	0.472222222	1.049383		L/kg starch	0.45					0.9	

The selling price has to be examined in terms of exchange rates set and adjusted by international banking adjustments, any arbitrage (ability to gain advantages through financial trading based on price differences between countries) that is possible involving different export products, and the complex vagaries involve with this (Baillie, R. and Bollerslev, T. 1989; Kuepper, J. 2016). The right person or power broker may be able to help a country trying to get into the ethanol business. There may be a statistical spread in pricing between exporting and importing countries that may be helpful if the basic agricultural production cost can be reduced sufficiently in a particular local. The price reduction will likely be a function of technological advances in a country that can also benefit price reduction. This is not to be negative. However, it is useful to check this out carefully so one knows what is possible.

5.1.3.3 Sugarcane Feedstock Value

Sugarcane is an important raw material source for ethanol as already demonstrated in Brazil (USDA. 2015). In this case, the cane sugar juice water extracted from the original cane sugar stalks is separated from bagasse (residual biomass after elution of the cane sugar—see Figure 5.31) and is sent to ethanol plants in Brazil for processing. Molasses (from subsequent evaporation–crystallization to make solid and purer cane sugar) adds a few additional gallons of ethanol per ton of molasses when molasses is added to ethanol fermentation feedstock as a supplement. Therefore, the costly crystallization and drying step is avoided in regard to the ethanol fermentation from crude liquid cane sugar supply to the ethanol fermentation. Based on the literature, this is estimated to save about 50% of the processing cost estimated for dry cane sugar that is exported (USDA. 2006). Optimization of the evaporation step has been studied (Encinas, A.V. et al. 2007). The resulting cost of \$0.26/kg of cane sugar divided by the quantity of ethanol produced in liters of ethanol per kg of cane sugar solids (0.63) results in a cost of raw material to produce ethanol of \$0.42/liter of ethanol (this is in proximity to the figure for corn above). Clearly, cane sugar is another viable alternative as already demonstrated by Brazil. Furthermore, by-products of bagasse, molasses, and vinasse (solids from the ethanol fermentation) provide opportunities for development to secure by-product credits and cash flow (even if these advantages are not presently taken into account through proactive work). Several countries already produce cane sugar as is covered in the charts and table already presented. Most of such possibilities involving by-product credits in use of cane sugar to make ethanol have not been followed up on commercially as yet (at least as far as the writer can tell).

5.1.3.4 Rice Feedstock Value

Table 5.1 shows figures for rice. Rice is processed by taking the original unmilled seed stock and milling it to eliminate chaff. The chaff which is lignocellulose and related material present more by-product processing possibilities. The first milling results in brown rice (long grain type) and next milling eliminates the hull and bran (also a potentially useful by-product) to give white rice (short grain type). The rice is dried as it will otherwise likely rot.

As indicated, the cost for milled rice whether white or brown likely results in a raw material ethanol cost of near \$1/liter of ethanol. If one can supply the paddy rice

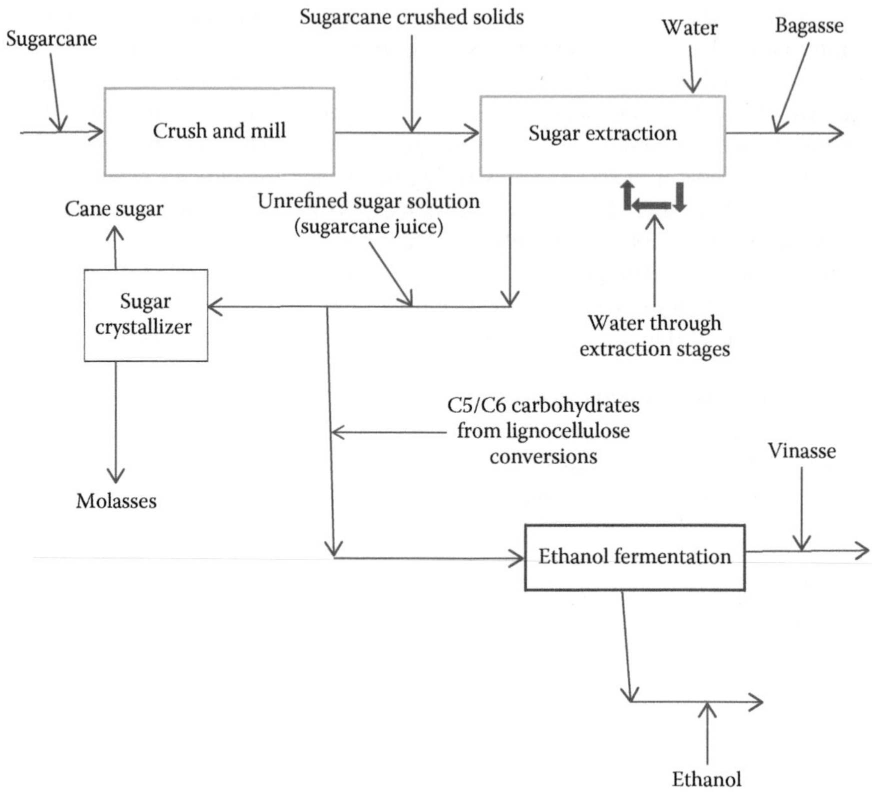

FIGURE 5.31 Sugarcane extraction process for ethanol fermentation feedstock.

(type before milling) and maybe get to processing before any drying is required (or maybe do dry it to an extent so it does not rot), then perhaps the cost drops in half or so and it is in the neighborhood of the corn and cane sugar options.

With some work and investment, this is a viable option for ethanol. This is speculation since an F.O.B. commercial offering for paddy rice does not seem to exist. What one would do is send the wet paddy rice directly to the local ethanol plant and suitably preserve it properly in transit and in storage before it is processed to ethanol.

Rice will have several by-product credits to develop that include foliage, and the chaff, fiber-bran, and waste yeast biomass. The entities should be properly separated so that specialized markets and uses can be developed for each of them such as nutritious and safe animal feed. As noted, some of the waste products may be developed as additional feedstock for ethanol based on the lignocellulose-related route discussed further. This represents considerable opportunities for interested parties.

5.1.3.5 Sugar Beet Feedstock Value

Sugar beet is a crop where the roots contain the sugar. It is produced globally with varieties tailored to particular areas so sugar beet can be grown in Europe, India,

Russia, and Brazil as indicated in the afore-presented charts in this chapter. It is not grown in other parts of Asia. The root contains much water and about 18% sugar.

Table 5.1 shows the figures for sugar beet. Sugar beet is a subsidized product in many countries, for example, in Europe. At the subsidized price, the raw material cost in terms of ethanol in $/liter is too close to the price at the gasoline pump, at least in the United States. In Europe, since gasoline there costs twice that in the United States, there is still a differential. The nonsubsidized price is much lower so as shown, the raw material cost for ethanol is about $0.10 per liter, more manageable for the ethanol as well as the gasoline by the thinking in this book chapter. Ethanol is formulated into a blend with gasoline at $0.8 per liter per the base case in this chapter.

Since laws in Europe require subsidization, that may be a problem unless the subsidy for gasoline remains as well thereby maintaining a large differential in price between the retail cost in the gasoline blend and the raw material cost expressed in terms of $/liter of ethanol. In other locales, use of nonsubsidized beet sugar will result in lower more competitive feedstock costs to make ethanol from beet sugar.

If the farmers get a share of the ethanol revenue then loss of the subsidy in the case of ethanol should not matter to them as long as it is sustained. The net result is that use of sugar beet (with the proviso of dealing with the subsidy perhaps) is preliminarily viable for ethanol.

The remains from sugar beet processing already go to animal feed. The by-products from sugar beet processing should add revenue once developed and approved. Molasses produced in refining of the sugar from beet sugar can be provided back to the ethanol fermentation as a feed stock. By-products that contain lignocellulose, hemicellulose, and pectin can be converted to feed stocks using appropriate enzymes if desired or else these can go to animal feed which is the present practice. This can be developed as a fine alternative opportunity to produce ethanol. In Europe, this will require developing a special understanding with government and modifying laws. It is unclear if this is practical.

5.1.3.6 Wheat Feedstock Value

Wheat is a grain that is an important food source. Its properties that vary with variety (particularly its functional gluten protein) lead to important baking products such as bread and pasta. The gluten-laden flour is an important product. The gluten has to be isolated by washing wheat grain flour out of it both to derive the value of the gluten plus prevent it "gumming" up process equipment.

Its use for making ethanol is price sensitive with the gluten-rich product being an important product or by-product depending on where the gluten is isolated from the starch. The fiber and related material represent an important additional by-product that can generate cash flow. The prices indicated for wheat flour with gluten removed range from $150 to $400/metric ton. At the higher price end, assuming comparable ethanol yield as from corn (in this case, this is isolated starch), the raw material cost is near $1/liter as noted in Table 5.1. This is likely unacceptable even with a by-product of wheat gluten sales. At the lower price of $0.15/kg, the price of the ethanol is in the range of that of corn. Again, if the farmers can share in the added value of ethanol sales, getting a lower price is a reasonable business proposition. Therefore, wheat is certainly a viable candidate to be ethanol feedstock.

5.1.3.7 Cassava Feedstock Value

As covered in the previous sections on regions, countries, and crops, cassava (or manioc, or tapioca) is a root crop prevalent in many areas of the world that produce nothing else that is a major crop. Therefore, this represents an opportunity for the smaller, developing country with maybe a greater need for some augmentative income to help their people.

The cassava plants (e.g., leaves and root) have a significant concentration of glucocyanides. These must be eliminated before consumption by humans or animals. The native peoples know to do this. Similarly, this must be done in the processing of cassava to be used as ethanol. One of the glycosides is limnarin wherein the glucose entity is cleaved from the cyanide compound releasing the cyanide, for example, as hydrogen cyanide. Perhaps the plant has this as an effective measure of protection. Also, the roots when separated from the rest of the plant turn black due to an oxidative process which is undesirable and may interfere with processing to ethanol as might the cyanides. Native concentrations of limnarinase help facilitate release of the cyanide and additions of oxidase may prevent the blackening. In an industrial process, therefore, it may be desirable to treat the root stock with enzymes and/or allow residence in a bioreactor that has water fed to the root stock so the stock is flushed, the poison eliminated, and the blackness also prevented. This (or a similar treatment) should be effected. Such treatment is likely already done in preparing the dried cassava chips and also in processing to cassava flour.

As shown in Table 5.1, there are three cases considered that include using the original raw root sans the vegetative plant material, using dried cassava chips, and using more highly refined cassava flour. The predicted approximate cost for raw roots in terms of raw material component in ethanol is $0.23/liter. However, this is a price where it is not yet diced or sliced nor treated to remove cyanide content or content causing black spots. The raw material cost for chips is predicted to be $0.34/liter (near that for corn, sugarcane, paddy rice, and wheat flour). The cassava flour is predicted to be $0.70/liter ethanol which is too high from this approximation. Cassava chips seem to be the most practical and lowest raw material cost possible for making ethanol within the cassava array of options, obviously subject to further analysis on processing steps noted. The material will allow numerous by-product formations to help reduce costs from base plant materials and also that from the alcohol fermentation.

Cassava starch readily converts to glucose and small oligosaccharides using the dual enzyme conversion process practiced for ethanol. The writer personally was involved in developing starch conversions from many materials and applying these to a submerged liquid fungal fermentation to form citric acid. Processes were developed to convert starches from this and similar materials for process development and worldwide manufacturing purposes (Goldstein, W.E. 1978–2008).

Cassava chips seem to be the best choice and it seems that is what is being done currently. Many factors can change due to technology advance so ranking in terms of this first cost prediction can certainly vary. One variable is of course the arbitrage that occurs in trading as well as opportunities to use technical skills to reduce cost in any location anywhere. Some locations such as in Africa have more problems with plant disease that affect yields. These are challenges to be addressed by talented people in the country affected and the parties in other countries with whom they choose to partner in order to improve themselves and help their partner at the same time.

5.1.3.8 Sorghum Feedstock Value

Sorghum is a grain with nutritional aspects similar to corn. However, it has to be processed to be digestible as the protein encases the starch. The isolated sorghum starch reportedly retrogrades very readily during jet cooking so the process of dual enzyme digestion of the sorghum grain will need process development as would captive isolation of the starch from the grain (Sullins, R.D. and Rooney, L.W. 1975; Claver, I.P. et al. 2010; Souilah, R. et al. 2014).

Tannins (in sorghum) affect nutritional aspects so this would be a factor in animal feed with which one has to deal. This material has already been used to make ethanol so presumably some of these challenges have been (or are being) addressed. The material's starch content in the grain is comparable to corn and sorghum also a relatively low moisture content of 9%. It yields a predicted raw material cost which seems to be on the high side at $0.43/liter. So even though there is experience with this crop for ethanol (it is also prominent in making a sour mash beer that uses malting to release nitrogen, a lactic acid fermentation that adds sourness, and then an alcohol fermentation). Carrying out work on sorghum to industrial ethanol and cost reduction should be a fruitful activity.

5.1.3.9 Barley Feedstock Value

Like sorghum (and wheat to an extent), barley is a source grain for brewing alcoholic beverages. This experience does translate to an extent to industrial ethanol processes produced through yeast fermentation. However, the grain itself is a complex entity with components naturally relevant to brewing beer. The grain is rich in carbohydrates and proteins and glucan-related entities that add to making the rich flavor of beer. The starch content is relatively high at about 66% and the moisture content is what is expected of a grain, nearly 10%.

As seen in Table 5.1, the price of ethanol from barley intended for an industrial blend is in the neighborhood of others, near $0.44/liter. So this is feasible even if on the high side in price. However, this assumes only a slightly reduced yield based on what is observed with corn which may be a stretch in assumptions since barley use in ethanol manufacture is really oriented toward a craft creating a tasty brew where yield is not really an objective. The entities in barley would make raising yield a challenge.

This is likely a research project that should probably not be undertaken unless other feedstock alternatives are not available for a particular country. The literature bears this out and hulled barley is not suitable for industrial ethanol while dehulled ethanol is a possibility (Ingledew, W.M. et. al. 1995; Teunissen, P. 2007). However, why would one bother with this if other feedstock sources are available? The chart and crop sources for a country as covered in this book allow for a first basis selection for crops if one investigates the production of ethanol intended for gasoline blends.

5.1.3.10 Oats Feedstock Value

Oats are a very rich nutritious grain, suited for food. It could be converted to ethanol. However, as shown in Table 5.1, it is too expensive for this application with a raw material cost contributing approximately $1/liter to the cost of ethanol if it were pursued for this application. The content of fiber and fat is high and this would create difficulties in fermentation except if one was trying to create an alcoholic beverage for consumption. Its products may be valuable as a supplement for feed to enhance nutrition. It is too costly to be considered as a feedstock for industrial ethanol unless there are other mitigating factors.

5.1.3.11 Soybean Feedstock Value

Soybean is not an appropriate feedstock for ethanol. It is also a high-protein, nutritious food suitable for humans and for animals perhaps as a supplement in the latter's feed. Its carbohydrate content consists of unusual polymers that sometimes may have to be removed to avoid antinutritional effects.

5.1.3.12 Preliminary Comparison of Feedstocks for Ethanol

From the analyses in Section 5.1.3, we have an understanding of which feedstocks should best be considered for research work and commercial development. Based on approximate commercial value as shown in Table 5.1, we can see that the possible feedstocks, approximately equally ranked that are most appropriate are: corn, sugarcane, sorghum, paddy rice, beet sugar, barley, wheat flour, and cassava chips. The assumptions inherent in these choices have been covered in the sections within Section 5.1.3.

The prices show some that are much more favorable than others from this table. However, these are approximations based on F.O.B. quotes, assumptions on yield matching that of corn (with corrections for starch or sugar content) and assumptions on which are more practical based on individual crop characteristics as described already.

5.1.3.13 Taking Advantage of Opportunities to Gain Favorable Financial Positions in Feedstocks for Ethanol

All prices are subject to prevailing supply and demand and the arbitrage (trading and negotiation possibly involving several countries) that occurs. So, if captive production is practiced, one can control the prices of their preferred feedstock commodity to match costs with the understanding that economic gross margins occur when ethanol is made and sold, and when by-product credits occur. The lesson from this is that there are multiple opportunities worldwide if parties take advantage of them.

One opportunity is to become skilled at trading and entering into contracts to purchase and sell crops at advantageous prices to position one-self to obtain lower-cost raw materials for ethanol processing besides gaining financially in the process. This area requires much expertise, knowledge of the field, mathematical and analytical skills, and the judgment to be able to act correctly plus the courage to do so, and to have the knowhow to hedge and protect purchases. If one does not possess such skills, then they should contract to retain them from the outside (perhaps entering into an encompassing business arrangement and partnership that is international in scope).

There are parties worldwide skilled at this that perhaps will partner in order to gain profit for themselves. They have to be sought out. Companies in the U.S. corn industry are particularly skilled at this as well as being highly competent producers and technologically very knowledgeable. One also has to be careful to avoid charlatans if one engages in trading, arbitrage, and future options activity since mistakes in judgment in this can result in business and economic disasters.

Whether or not one can participate in the futures business, there is another opportunity. This opportunity involves engaging in captive production of the raw material, maybe avoiding the mark-up as one proceeds to F.O.B. status (ready to ship without incurring the shipping cost). This also assumes one will not be subject to subsidies and tariffs on products imported to countries that will automatically raise the cost of a feedstock to (for example) protect the native agricultural industry. If such subsidies are apparent, then in order to proceed, a party has to partner with others to change the law only in the case of ethanol (if that is possible). Some crops affected by such laws have been noted already in previous sections on crops.

In the case of corn, for example, corn is currently priced (currently meaning as of August 2015) at \$3.80 per bushel or \$0.36/liter (assuming one bushel is converted to 2.8 gallons of ethanol and since each gallon is equivalent to 3.8 liters) (*Ethanol Producer Magazine.* 2015). Ethanol is currently quoted as being sold at \$1.72/gallon or \$0.45 per liter. From the analysis in Section 5.1.3.1, this is below the price of ethanol (\$3/gal or \$0.8/liter) when incorporated in gasoline expressed in gasoline blend retail prices. The difference reflects the markup in price from \$0.44/liter to \$0.8/liter. One can see that \$0.44/liter provides only a small margin separating the cost of corn from its value when incorporated in a fuel blend with gasoline. A drop in the price of gasoline also plays havoc in the ethanol production industry since this erodes profit margins. Use of corn futures purchasing and captive provision of the corn will likely result in a corn raw material cost that is much lower than \$3.80/bushel. This tactic and the strategy that goes with it are likely practiced. Therefore, futures participation and captive production of corn represent important routes to consider for a party anywhere wishing to get into the ethanol business. What applies to corn applies to all relevant crops anywhere in an analogous manner.

DDGS, produced as a by-product of ethanol manufacture, are selling at \$0.17/kg (*Ethanol Producer Magazine.* 2016). This is substantial and an important component to achieving an acceptable economic margin for an ethanol production facility. The sale of DDGS can defray raw material costs to an important degree. This by-product must be maintained and the technology advanced for anyone in this industry. All crops noted in Table 5.1 have or can have by-product credits in the production of ethanol. Each case is different and research is required to insure that this credit is realized since it will likely make a big difference in a party successfully achieving an adequate economic return for their participation in the ethanol business. Research can also improve the impact of this product on animal feed and may serve the basis to create new foods for human consumption in a world where shortages occur and some foods are coming into question for health reasons.

5.1.4 Applying Crops and Developing Processes Directed to Ethanol Production

There is a unique scenario for each crop that is to be used as feedstock for ethanol. This properly starts with corn since this is the current eminent source for provision of ethanol. The case continues for sugarcane as a source crop focusing on Brazil as the second leading source. The cases continue for other countries where these two crops offer possibilities, and where other crop raw materials are possible since they are sources-in-development for ethanol. This assumes this development will occur globally so that countries not proficient or capable in corn or sugarcane can get involved based on expertise in the crops more indigent to their country, and can become involved for their country's benefit in different opportunistic ways. Examining the corn industry and its well-developed practices should be helpful in developing feedstocks for ethanol in any country based on corn or other crops noted.

5.1.4.1 Corn Processing and Impact on Knowledge Development in Other Crops

For corn, many varieties have been developed over time. There is a rich history stemming back to sources in South America and to research origins in centers like CIMMYT in Mexico (CIMMYT. 2016). Many varieties were introduced into the United States from these origins. Of course, the Native Americans planted corn in the United States that served as an additional starting point.

American genetics and breeding skills have been highly developed over time as they have in other world locales. It is clear that varieties are adapted to particular areas. A corn that is bred and grown in the United States may not perform the same in another locale such as in China. It has to be adapted to the locale. A famous breeder and developer of corn in the world whom I was once fortunate to be linked with professionally (Elmer Johnson, PhD) often noted that the resulting corn is a function of genetics and the environment (Goldstein, W.E. 1988–1994).

Genetics refers to the native genes present in the plant and how they are expressed to control corn traits such as growth rate, starch content, etc. The manner of expression is controlled by the environment. The environment is where the corn is grown (e.g., the locale in the United States). The factors in expression due to the environment include water, weather, pollination, soil and characteristics, fauna, and factors in the corn intrinsic to genetic expression, etc. (Conventional Plant Breeding. 2004; National Research Council. 2004).

The corn industry in the United States settled on #2 Yellow Dent corn as the source raw material for ethanol. This is also a major source for animal feed so this choice was very logical. As noted, its yield has been increased with time (Butzen, S. and Smith, S. 2011). It is not clear that another variety might not be better for ethanol produced in the United States by locale. Another variety might not be chosen for development because it is imagined that that spending the money to do so may not be worthwhile, and it has been decided for practical business and scientific reasons that this variety is good enough for the application. It is likely a similar story and rationale would be repeated for any crop anywhere.

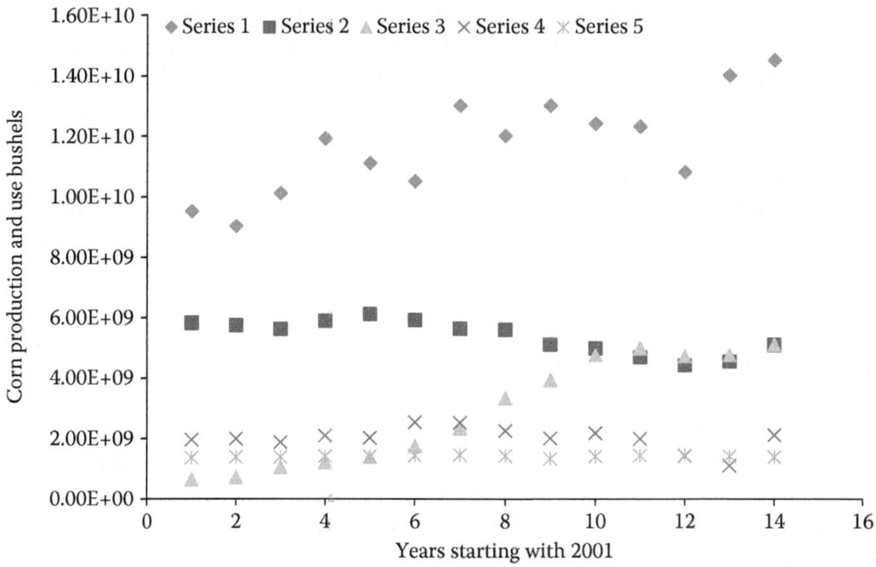

FIGURE 5.32 Whole corn production and its use in United States or exported. Series 1: total corn production, Series 2: whole corn used for U.S. animal feed, Series 3: whole corn used for ethanol, Series 4: whole corn and its products exported, Series 5: whole corn and its product for human food in the United States.

There may be practical reasons not to seek a new variety within a crop such as expenditure in agricultural R&D. However, the question needs to be addressed if we are seeking improvements since maybe there is a better hybrid variety for ethanol derived from corn. For example, this can involve one with a higher starch content with equivalent agronomical characteristics and with an equal opportunity for by-product sales (Bode, J.W. 2013; Wikipedia. 2013a; Fernandez-Cornejo, J. et al. 2014).

There are limitations on increasing acreage for corn in the United States. We know that yield is increasing every year. If we want to produce more ethanol we need more corn than we have presently at least by current data. Figure 5.32 shows the corn production per year and the corn applied to ethanol per year (USDA-NASS. 2015a,b; Foreign Agricultural Service. 2016). With essentially fixed acreage for corn and other crops and the development of U.S. society and infrastructure, where land expansion may be limited, we need greater yield increases. Therefore, new technology and methods have to be provided for the American farmer.

5.1.4.2 Role of Genetic Engineering in Crop Advancement

What techniques are available to increase the corn obtained per acre per year? It is not clear that if ethanol derived from corn is to proceed to grow significantly in the United States that advances in classical breeding genetics will be enough or certainly apply quickly enough to help the industry move forward. Modern practice involves genetically modifying the corn strictly to provide resistance to herbicides and pesticides so more of these compounds can be applied to protect crops and improve yields. It is not clear to everyone that modern genetic modification to improve plant protection is totally safe (Whitacre, P.T. 2008; Fernandez-Cornejo, J. et al. 2014). Some say this has not been thought through properly, and that safety issues exist in using genetically engineered corps developed for the plant protection area.

The fact that genetically engineered crop protection is allowed to improve a variety of characteristics while use of genetically modified yeast in fermentation is not as yet allowed is inconsistent (Schill, S.R. 2014). Why is one considered safe practice and the other not as yet safe practice?

Crop protection advances through genetic engineering have been allowed. However, commercial introduction of genetic engineering-related advances for crops (other than plant protection) or new yeast technology, both of which may help the ethanol industry, have not been allowed by regulatory authorities.

It makes one think there is the possibility of risk avoidance on the part of regulatory authorities. If this is a safe practice, proceed to advance it. If the practice is not safe, then let this be known so measures can be taken that are proper. To proceed partially is strange and sounds like there is confusion and indecision about what to do. It is logical that genetics is used for molecular diagnostics to track markers for improvement.

The thinking in this area is not clear to this writer. However, the need may be great due to weed and pest infestation so maybe it was felt there was no choice in the matter. It is necessary that information and tests be developed and applied to show that genetic engineering of yeast and plants (nonplant protection) is not dangerous to humans or to other life forms on which we depend that include feed animals and bees that pollinate our crops. Genetic engineering of crops and yeast can provide widespread benefit if shown to be safe as well as effective.

5.1.4.3 Responses to the Demand for Crops

The demand in the world for crops that are used for ethanol as well as food will increase significantly with time. Indeed, the Food and Aricultural Organization (FAO) projects this in their analyses (Conforti, P. 2011). If the United States does not provide sufficient yield increase, other nations will use their land and be competitive to the United States in ethanol. This could put us in the same situation of being "behind the proverbial eight ball" as we were in the 1970s due the oil embargo imposed on the United States for political reasons. If anything could have led to a war, events in that period could have well done so. So, the crops of choice will be used in all countries. If we cannot plant enough corn to keep up with growth in the supply of ethanol, and insure that the costs and pricing is competitive, we will gravitate to lignocellulose sources from crops that may provide solutions. What this step will do is to provide other challenges (Lee, H.V. et al. 2014; University of York. 2014).

The crops currently of choice for ethanol in different countries are listed in Table 5A.1 in Appendix 5A plus this has been covered through a number of charts by region and country in this chapter. The lignocellulose from the waste from such crops will be the second-generation crop. Any of these crops are a candidate for ethanol production as long as they have sufficient starch or saccharide content. The crop waste is also a source since the lignocellulose can be degraded so that the carbon sources are available for the yeast to use to proliferate and produce ethanol. As an example, the content of bagasse, a waste by-product of sugarcane processing, is cellulose (45%–55%), hemicellulose (20%–25%), with noncarbohydrate content being lignin (18%–24%), ash (1%–4%), and waxes (<1%) (Sanchez, A.P. 2016).

Cellulose and hemicellulose (beta-linked carbohydrate chains) can be potentially feedstocks for ethanol, though perhaps not as available as starches (alpha-linked carbohydrates) from crops like corn and sucrose from crops like sugarcane. The gap between use of lignocellulose, sugar, and corn will be closed as enzyme systems are developed and applied to convert cellulose and xylose sources in lignocellulose to ethanol (Gao, D. et al. 2011).

The crops vary, so processes will be developed (or have been developed) for each of them. What is interesting is to see what has been done with the DDGS (i.e., Distillers Dry Grains consisting of biomass and feed waste discharge from the ethanol fermentation) so far in different countries for different crops as a way to gauge development due to the formation of the important by-product or preferably co-product. Also, those countries that have the capability for larger amounts of crops across many crop types may be viewed to be more formidable as competitors given the advantages pointed out in this book come to pass and develop further. Let us look at these two subject areas (the case for DDGS by crop and the accumulative capability in total starch from many crops).

Figure 5.33 shows the crop types (as well example major producing country) that have potential for ethanol globally based on crop starch or sucrose content. The

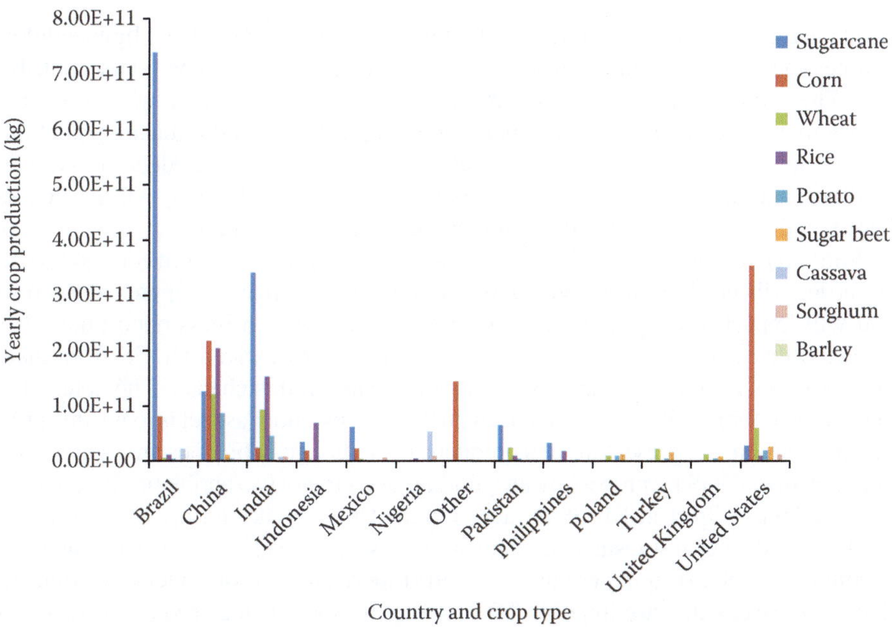

FIGURE 5.33 Yearly crop production as a function of country and crop type.

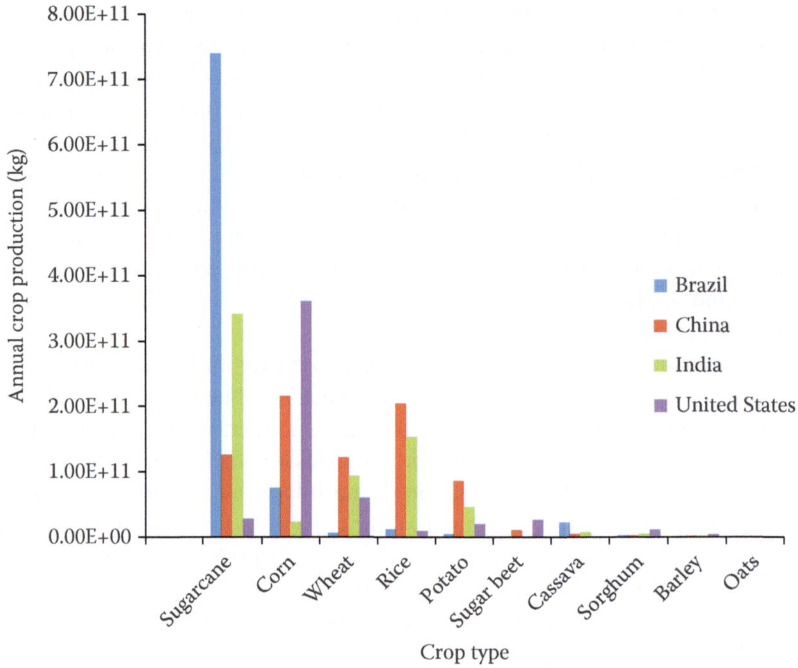

FIGURE 5.34 Annual crop production for major producers of starch- or sucrose-containing entities.

plants in these examples all have potential for producing ethanol from lignocellulose sources that are provided by these plants. Figure 5.34 shows some major contributors. The United States and Brazil currently lead the field. However, many countries also demonstrate potential in this area. For example, China and India are potentially major contributors. With the existing value of ethanol, and the evident growth possibilities, and justification of value as pointed out in this book, many will participate based on their agricultural and technology development potential.

Similarly, just as the United States produces DDGS primarily from corn now, other countries will produce their own equivalent DDGS once they are approved to do so, and will import from United States at first to build their business opportunity. It is expected that many will produce the equivalent of corn-derived DDGS from many crops sources available to them as described already in this chapter. The potential is clearly there for DDGS and an advanced DDGS. An essential aspect is to insure antibiotic residues are not present in such products. Even though antibiotics are cleared from being in any DDGS by process means, finding ways to not use antibiotics in fermentation at all will help clear DDGS produced in the United States for use overseas extensively in feed since test results on many animals show its value and cost advantages.

Import problems may perhaps result in import and use of DDGS by countries due to embargos that are imposed for different reasons such as protection of native industries. To qualify such products for use, the absence of microbial contamination to safe levels will have to be demonstrated. Microbes are effectively eliminated from

DDGS in processing but microbes can contaminate DDGS during shipping of this ethanol fermentation by-product. This contamination can be controlled by proper sanitary practices and by using beneficial microbes similar to those derived from fermentation processes and presently in food supplies and as a means to promote health protection and benefits.

5.1.4.4 Optimal Corn Feedstock and Corn Industry Supply for Ethanol

There are many corn varieties (Corn Refiners Association. 2013). Perhaps it is desirable to seek a new variety that might, for example, produce corn at higher yield per unit time to satisfy ethanol needs. These varieties can vary by locale since a particular corn plant may grow best and most successfully in a given locale. This is probably true for most any plant. Some corn varieties are bred and grown for human consumption. They have particularly positive organoleptic characteristics (that means they taste good). A very sweet corn with particular flavor notes might be developed for the food market. That is not necessarily something that would be targeted in addition to an ethanol market because the characteristic needs of these markets differ. The corn might be targeted if it was so ample there was plenty for both markets, so one has to keep that in mind.

The other side of the issue is that the corns that may be optimal for processing, for example, to canned goods or foods or for making corn starch and sweeteners may not be best for corn directed to ethanol. Some corn involves special varieties that are targeted to specific starches for producing carbohydrate structures that have desired characteristics in foods or specialty materials. Examples of these include high amylose starches which have properties for use in forming plastic-like films and waxy starches which may be applications to impart special processing and traits in formulated foods (Nagle, B.J. et al. 1999; Bode, J.W. 2013).

So, it may be appropriate to target a corn variety to ethanol that may be easier to process, and where the feed by-product is good for animal nutrition. The latter is not necessarily automatically happening since animals can have taste preferences. The writer remembers an incident where animals taught both his colleagues and him a lesson. A group the writer indirectly managed was testing a new crop modified by food microbes for animal feed. This product was very beneficial to the particular animal's health, weight gain, and feed efficiency (i.e., where the amount consumed by the animal provided a good weight gain per unit of feed consumed). The animals' health could also be judged by the qualities of their stools. In one experiment, the animal had a choice to select the experimental crop or a control. To access either, they had to nudge a button to release feed. They quickly learned to select the experimental crop and ignore the control. Maybe they had a sixth sense about the experimental crop, or maybe it tasted better to them. Besides noting the need to revise our experiments so we had a decent control, we learned something about animal feeding habits and behavior.

Maybe it would be better to isolate a corn variety that would have processing and fermentative characteristics that are optimal besides providing excellent corn oil and animal feed. It has to have acceptable growing characteristics as well that include water usage, crop rotation requirements to keep the soil fresh, and also the ability to control crop pests in combination with other methods used such as natural or chemical crop protection (Conforti, P. 2011).

If the market for ethanol suddenly went down, the farmers would have to have contracts protecting them in regard to commitments to purchase this crop and cover all the preparation required for it. If that is the case, then that land may not be available for growing corn for other applications.

Each application has its own specifications and requirements. Switching from one to the other application is not so easy. Given this targeting of corn crops and varieties, then land is accordingly allocated. Also, if the money obtained in one area versus the other is quite different, then that is one basis for the farmer to go in a particular direction. This can affect prices certainly since the field is very complex as can be judged from these comments. In a free enterprise system, the farmers will do what they have to in order to be successful. If there are market growth opportunities, the farmers will plant to meet those opportunities.

Therefore, as the corn refining industry notes, the farmers can likely meet needs for corn for human and animal consumption and for ethanol with resources available, since that is their business and what they are trained to do (Corn Refiners Association Annual Report. 2010). Probably then, the model might best be for the farmer to make selections for field planting targeted to an area or application, keeping in mind how to be able to switch if the events call for that. It does not make sense to this writer to be concerned about land shortages and higher corn costs since the farmers will manage this and supply both markets as long as the demand and the financial return are apparent.

Table 5.2 shows the industry figures for corn production and allocation. This table shows acreage, corn production, ethanol production, and amounts for each product of the corn plant by years (USDA. 2006; USDA-NASS. 2015a,b; Corn Refiners Association Annual Reports. 2010; Corn Refiners Association. 2013). Corn grown in the United States in 2014 amounted to 14.5 billion bushels. Apparently, by industry figures, 35% of this is currently applied to ethanol (Renewable Fuels Association. 2015). The industry projects that industrial processes are yielding 2.8 gal of ethanol per bushel (or 10.6 liters per bushel). This then yields a projected gallons of ethanol produced country-wide which compares to the projected figure the industry reports. The difference is the ethanol that is produced from noncorn sources.

In the United States, currently, effectively all ethanol produced is derived from use of corn as a raw material. As seen in Table 5A.1 in Appendix 5A, corn will not necessarily be the preferred source for ethanol in the rest of the world. What is common is that all of these sources contain starch or sucrose to be processed to ethanol. However, in the writer's experience, not all these raw materials can be processed equivalently. For example, wheat starch is associated with very functional gluten that has to be separated in processing first to avoid it gumming up process equipment. Corn gluten does not have to separately and can be processed along with the starch. The corn gluten simply ends up in animal feed. However, that simplicity in processing does not necessarily translate to equivalent quality animal feed. Each source has to be studied on its own merits. This is what occurs worldwide for starch-bearing crops. For crops based on lignocelluloses, there are also differences among the entities that have to be studied and determined.

Let us see what the literature data show. Data from the corn industry and the USDA are helpful. From Table 5.2, for the United States, we can see how much

TABLE 5.2
Ethanol and Corn Industry Statistics

Year	RFA Report Ethanol gal/year × 10^9	Ethanol Density (lb/gal) 6.836 Ethanol lbs/year × 10^{10}	Corn for Ethanol bushels/year × 10^9	Ethanol gal/bushel 2.8 lb Corn/year × 10^{11}	lb/bushel 56	Annual Yield 2 USDA Data Corn bushel/year × 10^{10}	Ethanol Data Over USDA Annual Corn bushels Fraction for Ethanol bushel/bushel × 10^{-1}	USDA bushel/acre	USDA Data Acres Planted Corn × 10^7	USDA Data Calculated Acres Harvested Corn × 10^7
2001	1.80	1.23	0.642	0.360		0.951	0.675	138.2		6.88
2002	2.00	1.37	0.715	0.400		0.901	0.793	130		6.93
2003	2.91	1.99	1.04	0.581		1.01	1.03	142.2		7.09
2004	3.35	2.29	1.20	0.670		1.19	1.01	160	8.05	7.44
2005	3.90	2.67	1.39	0.780		1.11	1.26	149	8.30	7.42
2006	4.88	3.34	1.74	0.976		1.05	1.66	149.5	7.85	7.02
2007	6.52	4.46	2.33	1.30		1.30	1.79	150.5	9.35	8.64
2008	9.31	6.36	3.33	1.86		1.20	2.77	152.5	8.58	7.87
2009	10.9	7.48	3.91	2.19		1.30	3.00	164	8.63	7.94
2010	13.3	9.09	4.75	2.66		1.24	3.83	152	8.85	8.16
2011	13.9	9.52	4.98	2.79		1.23	4.03	147.5	9.20	8.36
2012	13.2	9.04	4.72	2.64		1.08	4.37	122	9.75	8.85
2013	13.3	9.09	4.75	2.66		1.40	3.40	158.5	9.55	8.80
2014	14.3	9.78	5.11	2.86		1.45	3.52	171	9.05	8.48

(*Continued*)

TABLE 5.2 (*Continued*)
Ethanol and Corn Industry Statistics

Year Frac Starch	Corn Refinery Products (lb × 10^{10}) 0.717	Corn Refinery Products USDA (bushels × 10^8)	Corn Refinery Products (bushels × 10^8) 1.094[a] 56	Starch Products USDA Data (bushels × 10^8)	Starch Products (lb × 10^9) Moist Factor 1.16	Export Products (lb × 10^9)	Export Products (bushels × 10^7) Calculated as Corn
2001	3.27	7.73	7.45	2.49	5.90	1.78	3.19
2002	3.31	7.82	7.53	2.49	5.99	1.23	2.19
2003	3.27	7.76	7.45	2.58	6.10		
2004	3.29	7.80	7.48	2.77	6.59		
2005	3.26	7.73	7.41	2.83	6.70	2.06	3.69
2006	3.39	8.04	7.72	2.79	6.63		
2007	3.25	8.06	7.39	2.76	6.47	2.94	5.24
2008	3.12	7.85	7.10	2.61	6.08	2.96	5.29
2009	2.99	7.42	6.81	2.31	5.51	2.79	4.99
2010	3.00	7.88	6.82	2.54	6.03	5.03	8.98
2011	3.00	8.08	6.84	2.57	6.04	5.40	9.64
2012	2.98	8.17	6.79	2.53	5.81	5.38	9.61
2013	2.98	7.91	6.78	2.49	5.84	4.49	8.01
2014		7.99		2.09			

(*Continued*)

TABLE 5.2 (*Continued*)
Ethanol and Corn Industry Statistics

Year	Corn Oil Products ($lb \times 10^9$) Fraction of Oil	Corn Oil Products Calculated Fraction of Syrup + Starch $\times 10^{-2}$	Corn Oil Products Calculated Fraction of Syrup Only $\times 10^{-2}$	Corn Feed Glut and Oilmeal ($lb \times 10^{10}$)	Corn Feed Glut and Oilmeal Plus Glut Meal ($lb \times 10^{10}$)	Corn Glut Meal ($lb \times 10^9$) Fraction of the Sum in lbs	Corn Feed Glut, Oilmeal, Glut Meal, Sum Ratio Syrup and Starch $\times 10^{-1}$
2001	1.12	2.60	2.95	1.04	1.30	2.65	2.93
2002	1.21	2.80	3.18	1.01	1.28	2.70	2.89
2003							
2004							
2005	1.08	2.50	2.89	1.00	1.25	2.47	2.79
2006							
2007	1.21	2.78	3.19	1.07	1.31	2.39	2.86
2008	1.20	2.87	3.28	1.04	1.27	2.25	2.87
2009	1.11	2.81	3.19	0.986	1.19	2.08	2.86
2010	1.37	3.33	3.81	1.07	1.29	2.16	2.86
2011	1.55	3.71	4.24	1.10	1.34	2.39	2.91
2012	1.12	2.78	3.18	1.04	1.26	2.21	2.83
2013	1.02	2.47	2.82	1.15	1.38	2.35	3.08
2014							

(*Continued*)

TABLE 5.2 (*Continued*)
Ethanol and Corn Industry Statistics

Year	Corn Direct to Animal Feed bushels × 10^9 2011 Corn Refiners Association	Corn Direct to Animal Feed bushels × 10^9 USDA	Alcohol Beverage and Manufacturing Cereals and Foods and Seed bushels × 10^8 USDA	Whole Corn Exported bushels × 10^9 USDA Stats	Whole Corn Export Corn Meal Syrup Starch Express bushel × 10^9 USDA	Total Calculated Corn bushels × 10^{10} Used
2001	5.85	5.82	3.41	1.88	1.96	1.08
2002	5.55	5.74	3.44	1.89	1.98	1.08
2003	5.78	5.61	3.45	1.70	1.87	1.09
2004	6.14	5.89	3.48	1.78	2.08	1.16
2005	6.12	6.12	3.50	2.26	2.01	1.20
2006	5.55	5.91	3.52	2.26	2.53	1.27
2007	5.86	5.63	3.70	2.25	2.51	1.30
2008	5.18	5.61	3.69	2.12	2.25	1.36
2009	5.13	5.11	3.56	1.87	2.00	1.33
2010	5.00	4.98	3.58	1.99	2.17	1.43
2011	4.70	4.68	3.63	1.80	1.98	1.41
2012	4.31	4.42	3.67	1.24	1.42	1.31
2013	5.03	4.55	3.73	0.944	1.10	1.28
2014		5.10	3.72	1.95	2.11	1.47

(*Continued*)

TABLE 5.2 (*Continued*)
Ethanol and Corn Industry Statistics

Year	Total Calculated Corn bushels × 10^{10} Acres × bush/ac USDA Estimate	Corn bush Used/Calculated Corn Yield bushel/bushel	Calculated Corn Yield-Corn Used bushels × 10^{9}	Calculated Corn Yield -Corn Used/Corn Yield bushel/bushel × 10^{-1}	Ale Beverage, CRA Data Cereals, etc. Plus Syrup and Starch bushels × 10^{9} CRA Data	Ale Beverage Cereals, etc. Plus Syrup and Starch bushels × 10^{9} USDA Data
2001	0.951	1.14	−1.29	−1.36	1.23	1.36
2002	0.901	1.20	−1.83	−2.03	1.25	1.38
2003	1.01	1.08	−0.843	−0.836	1.24	1.38
2004	1.19	0.977	0.275	0.231	1.26	1.40
2005	1.11	1.08	−0.933	−0.0844	1.26	1.41
2006	1.05	1.21	−2.21	−2.10	1.29	1.44
2007	1.30	1.00	−0.0103	−0.00789	1.27	1.45
2008	1.20	1.14	−1.64	−1.37	1.23	1.42
2009	1.30	1.02	−0.295	−0.0226	1.17	1.33
2010	1.24	1.16	−1.94	−1.57	1.19	1.40
2011	1.23	1.15	−1.80	−1.46	1.20	1.43
2012	1.08	1.21	−2.27	−2.10	1.19	1.44
2013	1.40	0.921	1.11	0.793	1.20	1.41
2014			−13.7			1.38

[a] Moisture factor.

corn is produced per year, how much goes to ethanol, how much to animal feed directly as corn, how much to export, and how much to human food products. By-products from corn processing and ethanol (using as much as possible from the corn kernel) are also to supply animal feed (as shown in Table 5.2). Some of the corn is processed to food ingredients such as sweeteners and some to food, either canned or fresh corn as noted in this table. This table provides a material balance on the corn industry from corn refining association and USDA data. This same sort of information should be tabulated for each crop in the world contemplated for ethanol use since that can define the commercial economic opportunity for such crops anywhere. Really, the Science of Ethanol begins with the "Science of Corn" as a baseline.

Besides information on corn yield per acre changes and also ethanol production, Table 5.2 shows the products coming out of corn refineries and the number of bushels of corn equivalent directed to such products as well as to animal feed or to export.

The ethanol segment derived from corn has been growing fast in the last 10 years (see Figure 5.35) though the growth rate has flattened lately. So there is concern that it will grow so fast that it will cause corn and food prices to rise due to shortage. Some say this is a myth. If we look at Table 5.2, we can see specifically how much corn goes to ethanol and how this has been increasing. If one looks at USDA data for all crops and for corn, the total amount of acreage planted is constant year to year. So, if more corn is needed, given land that we use is about fixed, then we need to get this from yield increases. Companies that should know about corn and technology to increase yield (DuPont Pioneer) and the USDA, due to their advanced and significant expertise, point out that yield is increasing about 1.5 bushels of corn per acre per year and that this is ample (Butzen, S. and Smith, S. 2011). Other sources such as the USDA indicate a trend approximating 1.8–2.0 bushels of corn per acre per year (USDA-NASS. 2015b). Looking at Table 5.2, comparing corn used for ethanol and

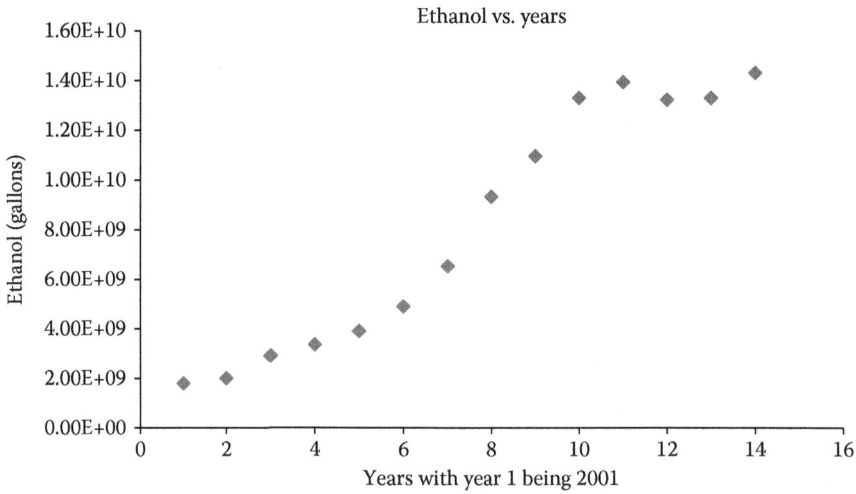

FIGURE 5.35 Ethanol production in the United States as a function of time.

the effect of increased yield on the corn supply to supply the increased corn going to ethanol, it seems that for now, using past years as a reference point, the yield increase is not quite enough. Between 2001 and 2014, the corn yield in bushels per acre per year has advanced between 1.2% and 1.5% per year. The use of corn to produce ethanol was at 7% of the crop in 2001 and reached 35% in 2014. Given this continues, if whole corn were to be the sole source of feedstock for ethanol, greater yield increases will be required if ethanol demand continues to rise.

If the corn yield rise is not practical, then alternative sources are needed. This then justifies the need for lignocellulose sources in parallel to corn sources. In some years, if more corn is produced, it is not stored or diverted to produce ethanol nor is it held in inventory since that is too costly a tact and impractical, as the corn will not keep unless preserved. So, putting the lignocellulose source aside for the moment, it appears that the yield in bushels per acre must go up maybe to 3 bushels per acre per year to compensate for ethanol increase projections. This has to be done through typical classical genetics or plant breeding and continually improved agronomy practices. This is not to say genetic engineering cannot do the job at some point. As noted, it appears regulatory parties are too leery of that in food crops for now except for plant protection against insect, microbial, and plant pests such as weeds. However, even the plant protection application causes concerns. So, it is expected that the yield of corn will go up higher per year to help contribute to what is needed to produce the supply for ethanol needs. Furthermore, this will be augmented by lignocellulose sources for ethanol coming on stream in a win-win situation for both sources. For this to occur, several issues must be addressed. One being that the yield from all sources must be sufficient so that pressure causing prices to rise does not occur. Consumers will not accept overt and obvious price increases.

5.1.4.5 Isolation of Corn Plant Components in Processing

The components of the corn plant are identified in Figure 5.36 (White, P.J. and Johnson, L.A. 2003; Ericksen, A. 2006; DDG Users Handbook. 2012). Components that contain carbohydrate compounds in portions of the corn plant are amenable to be raw materials for the organism (usually yeast) to convert them to ethanol. For example, the stalks and leaves of the corn plant are suitable to be converted to ethanol as long as there is a means to convert the cellulose and hemicellulose portions of the lignocellulose to simple carbohydrates that the yeast can utilize. Lignocellulose is a structure that combines complex carbohydrates (polymers) of cellulose and hemicellulose, with another complex chemical structure called lignin. These compounds reinforce the turgor (osmotic strength of the plant) and are maintained by a multitude of hydrogen bonds. Hydrogen bonds share hydrogen, but are not chemically reacted together (Noller, C.R. 1958). This bonding means that in principle, the cellulose, hemicellulose, and lignin can theoretically be cleanly separated if the right and economical means can be found to do so.

Looking at the corn kernel in Figure 5.36, the pericarp contains bran and fiber, which also contain cellulose compounds. The pericarp structure also contains protein (polymers of amino acids). Though some nitrogen from a protein is needed by the yeast, this is in small amounts. So this portion of the corn plant is normally a component of other products such as those intended for animal feed. However,

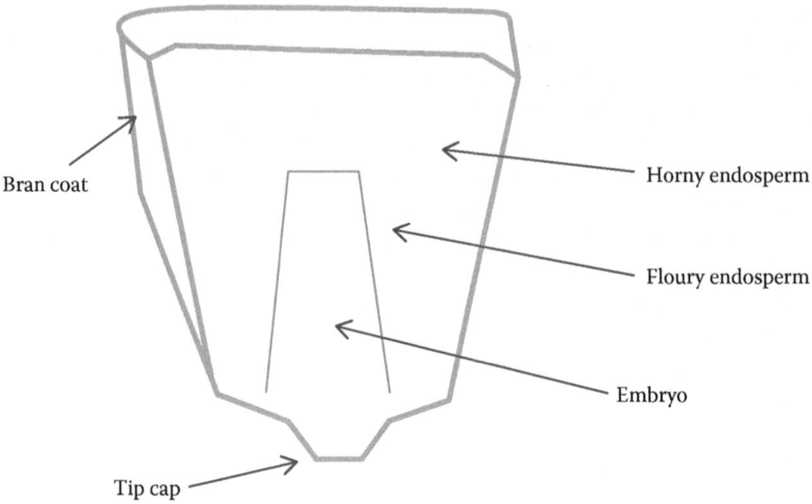

FIGURE 5.36 Corn kernel and constituents.

depending on the processing, this portion of the plant can also be used to make ethanol.

The portion of the corn kernel called the endosperm is primarily composed of starches and gluten (a protein/carbohydrate combination). The starches are convertible to simple carbohydrates that can be used to make ethanol. The glutens typically become a component of animal feed, unless of course something else can be made from them. The starches are also sources (following conversion) of many sweetener products based on glucose, since the starches are polymers of alpha-D-glucose (Shapley, P. 2012).

The area of the kernel referred to as the germ is the source of corn oil, an important product. The surface entrapping the oil is a cellulosic structure that is potentially convertible to carbohydrates that can form one of the bases of the raw material to supply the ethanol fermentation process. This source can also be a component of other products such as those for animal feed.

It can be seen that this area is quite involved whether one is trying to make ethanol or other products such as feeds or else food components such as sweeteners, specialty food starches, and chemically modified starches for other industrial applications.

So, if one could ideally do so, should these components be cleanly separated apart to supply what they are best suited to supply? That may be an important quality if having a pure material would lead to added-value product applications. Advanced enzyme systems, for example, may offer a way to accomplish this separation by selectively digesting areas that separate one fraction from another.

As a practical matter, in corn wet milling, these fractions are cleanly separated as best possible in normal processing, so products have evolved based on what is possible (at least so far) to be produced (Ericksen, A. 2006). For some corn dry milling processes, used predominantly for ethanol, some product separation is conducted to make specialty products such as corn grits, corn flour, and pregelatinized starches for some food preparations (Weller, C. et al. 1984; Kwiatkowski, J.R. et al. 2006; Renewable Fuels Association. 2015).

The supply of entities as food for people or for animals should not be passed over lightly. Following the adage that we are what we eat, food can be designed and contain additives to promote and protect health. This is undiscovered territory. If one thinks this way, then a process designed to make the best and least costly ethanol can be designed to make extremely valuable co-products that could, for example, target the animal feed market (maybe a "super DDGS" that is fortified to optimize animal health and meat quality). With further future development, the products fed to animals can be processed and formulated to become human foods. This can follow the soy protein model but taken steps further to create markets that attract consumers based on appearance, taste, and health attributes (Marketsandmarkets.com. 2015). The study of the feed market and analysis of quality is already practiced to a significant and sophisticated degree by the corn refining industry (Ericksen, A. 2006).

Typically, probably for cost reasons, most ethanol derived from corn presently involves use of a dry milling process, where whole ground corn is milled and ground in a dry state and is converted via enzymes to glucose and related polymers of glucose. So from this process, the fuel component, ethanol, and another complex of animal feed components are produced. In 90% of the facilities used to make ethanol in the United States, whole ground corn, is used to produce ethanol, recycling some of the resulting biomass, and sending the rest for animal feed (Renewable Fuels Association. 2015). The typical process (following milling of the whole ground corn) to produce ethanol is as shown in Figure 4.2, where the carbohydrates derived from the milled dry corn are provided for conversion of the carbohydrates to ethanol (Yang, R.D. et al. 1982).

This straightforward approach has great value and has spawned much research and product development. It would have even greater value if the by-product biomass, a complex of many entities, could be formed into a special and valuable entity or entities. The feed product from the ethanol facility differs in many respects from that from the corn refining facility (used to produce corn foodstuffs based on wet or dry milling of the corn). However, the by-product feed from the ethanol facility is equally important with similar nutritional and health-giving possibilities as that from the corn refining facility. Indeed (see Figure 5.37), the amount of by-product feed from the ethanol facility has grown to be far in excess of that provided from U.S. corn processing facilities (Table 5.2; AgMRC. 2015).

Thus, one reason for further development is to achieve greater understanding of the nutritional and health needs of animals receiving this as feed, and use this as a basis to see what might be done with such substances to create nutritional and tasty foodstuffs for humans. The animals have tailored needs, for example, for digestion, a matter that is a subject of historical and present research. The knowledge extends to better understanding how to improve nutrition and the health needs of humans. Further research is expected to turn up some product possibilities of interest that were previously unknown or predicted, both for animals and humans.

With all of the attention to the role of foods in health, understanding how proteins are separated from corn, and seeing if the corn gluten can have value in food, is

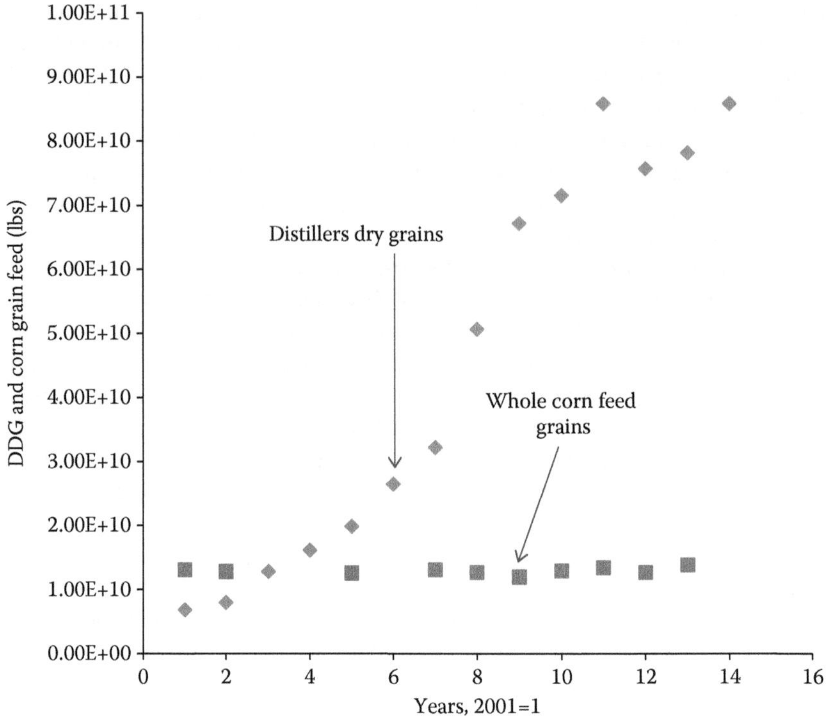

FIGURE 5.37 Distillers dry grains and corn animal feed grain as a function of time.

useful both for animal feed and human nutrition (Schroeder, J.W. 1997). This is, of course, far from a new topic, but it should be revisited for corn and other crop entities being considered for ethanol feedstock. The nature of the separation and what is needed for products is important since it is a valued component of understanding how to better the economics for use of ethanol. An improved and more valuable by-product of the ethanol process can create greater revenue since more can be charged for it and thereby improve the economics of ethanol. Perhaps gluten functionality can be altered so that it is more useful in foodstuffs, possibly examining elimination of gluten allergic reactions that occur it seems for gluten derived from another grain, wheat (Shewry, P.R. 2009).

The wet milling process involves soaking the corn kernels in a hydrogen sulfite solution for softening to facilitate separation of corn components for processing (Ericksen A. 2006). Fractions are separated out by centrifugal means, either by motor-driven centrifuges or by inverted conically shaped hydroclones (International Starch Institute. 2014). The isolated starch is converted in some facilities and used for ethanol manufacture. The dry milling process avoids any such soaking.

Each corn processing method serves its own purpose and product areas. What may be significant for either wet milling or dry milling is the way the fractions are separated and the nature of the barriers to separation and use of fractions. This is worth some discussion since it bears on the use of materials for ethanol fermentation and to form products from materials not specifically part of the ethanol fermentation. The lessons here also provide background for developing processes for other crops. The lessons learned in the corn industry for decades can readily apply to other crops so they can become better sources of ethanol.

5.1.4.6 How Does Diversion of Corn for Ethanol Affect the Supply and Pricing of Corn for Human Food and Animal Feed Supply?

Looking at industry historical data can provide some insights. There are some critical questions to address that are current hot topics. One question involves the concern that if so much corn is diverted to making ethanol, will that lead to a shortage of corn for human and animal and higher prices that go with such shortages? What forces come into play to understand this? This leads us to the field of agricultural economics and supply. This is certainly a relevant topic since it is important to avoid shortages and cost spikes.

To address this, we start with understanding the process further and how products and by-products are formed. Whole corn is either used as is or broken down and processed. Whole corn direct uses include that supplied to make ethanol, fed to animals in the United States, or used in corn processing. In corn processing, it may be used to make syrups, such as High Fructose Corn Syrup, glucose or other syrups, or to make starch or starch variations. The components of corn are all isolated at least to an extent to make animal feed, or to make corn oil, or applied to other applications where the components are used. Some of the corn fiber is sent on to make ethanol. From Table 5.2, we can see that fractions of feed and oil products on a yearly historical basis roughly correspond to the composition in the corn kernel. This should be intuitively expected. This is to be expected as this industry is very efficient at selling all components of the corn plant (and minimizing waste that is simply discarded at a cost). The amount of corn oil is less than would be expected based on plant kernel content perhaps since some of the oil remains in animal feed products.

A mass balance on the corn kernel can be made involving bushels of corn supplied from growing and then used. Figure 5.38 shows that the amount for ethanol has grown till it matches that for animal feed. The lower curves show the amounts for export and for human food have remained fairly constant over time. Figure 5.38 shows that the amount of corn produced is proximate to the total amount used (as a check of the values in Figure 5.38 and Table 5.2) and has been growing with time to supply needs. This chart is derived from industry and government data (USDA-NASS. 2015a,b; Corn Refiners Association Annual Report. 2010).

As indicated in Table 5.2, corn yield has increased to help sustain growth in bushels produced. If acreage is not changed, which is the case year-to-year, then the yield has to increase to sustain growth. Based on the history of the corn industry in regard to only food and feed supply there is no reason to believe that this cannot occur (Butzen, S. and Smith, S. 2011).

However, putting great increases in ethanol in the picture changes matters and adds another important variable. Based on Table 5.2, we can see that the corn acreage has increased in recent years. However, it is basically constant as a fraction of total acreage in the United States. Also the acreage allotted to different crops is fairly constant year to year as all crops are valuable. The total acreage also seems to be

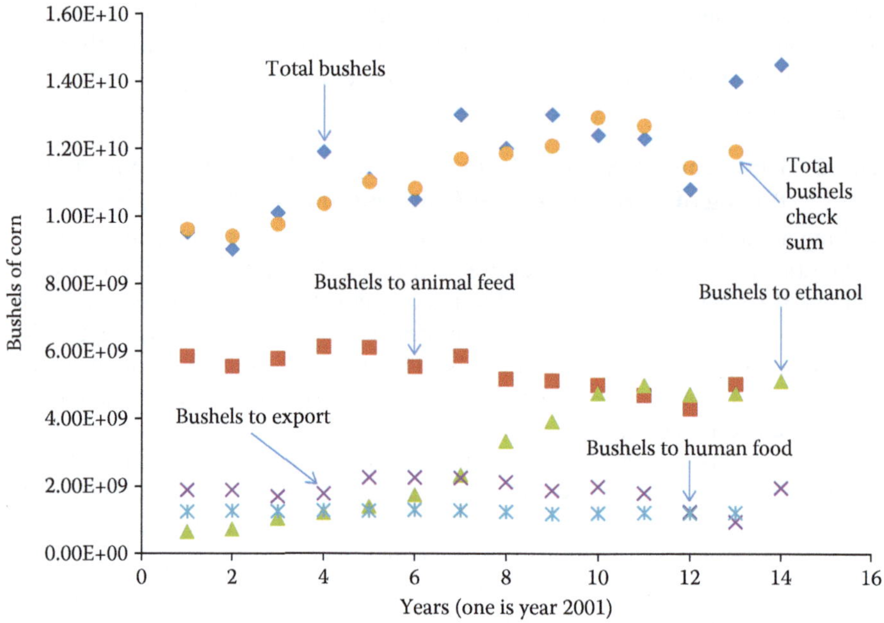

FIGURE 5.38 Whole corn production and use.

fixed in the United States. The growth of land use for nonfarming purposes pushes parties to raise yields to accommodate increased needs such as needing more corn due to growth of the ethanol market (where the ethanol is derived by using corn as a raw material). There is pressure on the corn crop for supply and yield, since it can only increase so much due to breeding based on traditional genetics. Plant protection is the basis for application of genetic engineering, and the latter is viewed suspiciously by those who are not certain of the safety of the technology. There is not enough proof that we understand safety sufficiently to avoid creating problems for ourselves through genetic engineering of the corn plant or other plants (Key, S. et al. 2008). A key action is development of better diagnostics for the presence and containment of transformants in the plant, and improvement of multispecies safety testing to help provide greater assurance. However, nothing is absolute, so there can never be perfect assurance. The probability of a safety problem due to transformed plants has to be reduced to a level where concern is minimal.

It can also be said that in some places, plant disease is so rampant that worries about negative effects of genetic engineering become of secondary importance since one needs any means to increase the food supply to avoid widespread starvation. Given these factors, and that corn applied to ethanol has increased so it is currently 35% of the corn crop, it does appear that animal feed from distiller dry grains from the ethanol fermentation of corn has become the dominant source of animal feed, displacing much of the animal feed from corn wet milling fractions. It is not clear that the animal feed from DDGS is a universally accepted feed for animals replacing that from corn wet milling. In addition there is cost pressure on ethanol obtained from corn milling. If lignocellulose can succeed, then that provides (perhaps) a lower-cost feedstock for ethanol fermentation plus allows more of the corn fractions from wet milling to be used to supply animal feed needs.

These factors leave a considerable opportunity then for growth of lignocellulose for supply with the expectation to this writer being that corn and lignocellulose in many forms will both be the basis of development of the ethanol market, the animal feed market, the human food market, and also other new product areas. The value of the ethanol market will be based on sound science and economics as covered in this book based on several expected initiatives.

An important aspect of this matter is by-products. Biomass is produced in corn processing and the components of the corn kernel are separated and sold to various markets, both animal feed and human food. Lignocellulose sources also supply animal feed as a by-product of fermentation. How the by-product from corn feed compares to that from lignocellulose feedstock to the ethanol fermentation still remains to be determined. Lignocellulose itself is not a source of human food at present nor for some animals that cannot digest cellulose plus the lignin content in lignocellulose is a negative factor in animal feed. However, who is say this will not change in the future based on research. The writer believes this will change significantly because there is demand for such change. The opportunities due to food scarcity are tremendous. Also, the opportunity for healthier and safer foods is unparalleled for the food industry. It all depends on the quality of the research that is conducted and also how well such research is implemented. The writer sees many opportunities to be possible to create food and feed opportunities from the ethanol fermentation (and use of multicrops) beyond those that exist at present.

The feed to animals from the corn plant is well developed in the United States and elsewhere and is well tested (Ericksen, A. 2006). It competes with forage crops, particularly for animals that can use (consume and convert cellulose in their digestive processes). Opportunities still exist to improve animal feed and thereby improve the meat of animals for human consumption. The balance between corn produced and what it is used for is shown in Table 5.2. Ethanol use is certainly a fast-growing segment of corn use as noted.

A huge proportion of the corn used is directly fed to animals. The by-product animal feed is a substantial and important business. Developing and maintaining its quality is obviously important since humans consume animal products. The cost of supply must be favorable and it must be safe to consume. The cost is favorable. There are too many reports of bacteria shedding and meat contamination which can end up in the human food supply. This needs to be directly addressed in research. There are ways to accomplish solving contamination issues using sanitation practices, including protective microbiology.

The new entry to animal feed since ethanol became prominent is feed from ethanol plants. As seen from Figure 5.37, the area is growing tremendously and is also resulting in considerable export. This feed is also sold along with feed from corn components and it will be sold from lignocellulose-based feed stocks. The impact of each of these sources on animal health and human health is still being developed. The management of these to animal feed supplies will have to be carefully done. Defining nutrition and safety are paramount in this.

One can see from examining Figure 5.37 that the growth of DDG from ethanol production has grown dramatically both domestically and in export (Makkar, H.P.S. 2012). Certainly, the ethanol exports which are also growing also result in more DDG exported than expected for the amount of ethanol shipped outside the United States. The DDG produced domestically is finding its way to export markets ever more significantly. For this to continue successfully, matters of nutrition, health, and safety become considerations of international scope that have to be dealt with properly to keep these foreign markets viable. Domestic DDG is being exported as perhaps a way to open new markets and relieve pressure on domestic markets which are already supplied by components of the corn plant independent of ethanol production.

Ethanol is the important target product. DDGS is also important since if it sold and favorably upgraded in terms of how it affects meat production and quality, it will act as a reduction in the cost of producing ethanol to benefit the producer and also in the end, customers that burn ethanol in their automobiles. If the health aspect is followed up on, then the value of the by- product is enhanced, it is worth more, and it also adds to improving animal and human health. This can be done in a way that can help lead to minimization of the use of antibiotics in animal husbandry and feed.

It is clear from work on DDGS that there are hints or weak signals across the board for enhancing this by-product and regular grain products whether sourced from grains like corn or from lignocellulose. The microbial contamination question (which may start with animals and be passed on to humans) can be addressed at the animal stage using DDGS by fortifying it, using enzymes, and adding entities to strengthen the immune system of animals and ward off infections (DDG Users Handbook. 2012). Where fortification is lacking, these can be added by use of enzymes to enrich the additives or mixing in additional additives. One this author is thinking about in particular is microalgae due to lipid content, for example (Pienkos, P.T. et al. 2011).

This fortification can be at the corn process DDG stage (Figure 4.2, where yeast biomass exits the process), or at the equivalent lignocellulose DDG stage since the lignocellulose may lack what is necessary in nutrition to take care of the animal. The matter of adding cultures to help the animals is another augmentative step. If these things are done, the product will get better, better protect the animals (and therefore, the public) and be worth more so it will command a higher value because it has these attributes.

The grains can be from almost anything that has carbohydrate. If something is a barrier to carbohydrate accessibility such is the case with some, then add an enzyme biocatalyst to make it more accessible. It is necessary to get to the point where all such animal feeds are equal except for the factor of economy of scale in regard to the raw material source. If more of a source is available to use, then the cost is naturally lower since the amount available to process to product is greater (economies of scale). However, that can be approached by combined operation where this is possible as noted in this chapter.

5.1.4.7 Lignocellulose Derived from the Corn Plant as an Plant Raw Material Resource

Lignocellulose from the waste of the corn plant is being developed as a raw material source for ethanol in the United States to (in effect) parallel ethanol derived from corn starch. This effort can serve as model as well for converting lignocellulose from the waste of other crops that have potential to produce ethanol. This development represents an economic trade-off of sorts between looking for a lower feedstock cost to make ethanol using lignocellulose sources that will help lower costs and perhaps giving up a valuable by-product credit to a degree since lignin related waste may not be as suitable for animal feed. This is the negative proverbial "glass half-full" argument.

On the contrary, it is possible that one can gain both lower feedstock costs and a replacement by-product credit through research. One could in principle find uses for the waste streams from using lignocellulose if not for animal feed then for something else.

One could fractionate the lignin ahead of the conversion to ethanol and make products of greater value than simply using this as a fuel (Lake, M.A. 2010). One could, for example, find a practical and low-cost way to pretreat cellulose so it can be enzymatically degraded more readily (Dadi, A.P. et al. 2006). The fact that use of ionic liquids may not yet be a commercial reality is beside the point. One conducts research to make discoveries to advance the field. You cannot always predict from where the discoveries will be found. The really opportunity lies in exploitation of advanced enzymology.

Certainly, there will still be the biomass waste that is yeast-laden. Furthermore, if one is integrated, then development of the DDG and related waste from use of corn should be pursued along with what is being done with corn stover (corn stovers include everything harvestable in the corn plant except for the corn kernels) (Kumar, R. et al. 2009).

The corn plant has cellulosic entities in it (leaves, tassels, the stalks, and also the components of the pericarp) that can provide raw material sources for making ethanol. These are the celluloses and hemicelluloses. These are "bound" through hydrogen bonding to polyphenolic structures termed lignins. Hydrogen bonds are not permanent chemical bonds and can be disrupted. The question, perhaps, is how to do this cleanly, that is, one needs to introduce an alternative entity for hydrogen bonding, that can disrupt the hydrogen bonds. Lignins can be potentially separated from the cellulose using enzymes such as those found in the guts of termites (so-called lignases, e.g., see Scharf, M. and Sethi, A. 2013). However, for this to happen, the enzymes have to reach the lignin–cellulose boundaries and get within them. The method that seems to make sense is one involving a physical force input such as ultrasonics (if made practical) and one that is mild chemically and preferably recyclable (Zhang, Y.-H.P. et al. 2007; Lake, M.A. 2010). Liquid salts come to mind, and are being explored, though one has to ignore the fact that these are undeveloped and far removed from being part of a process as yet (Dadi, A.P. et al. 2006). Many methods have and are being looked at to manage this separation (Harmsen, P.F.H. et al. 2010). None is an embraced solution as yet, meaning the economics to manage this are not yet acceptable. It would be better perhaps to separate the lignin as an entity that can be cleanly processed to something else, perhaps to a fuel as a last resort (only if something more valuable cannot be done with it).

Many materials have lignocelluloses in them. Many plants abound in cellulose and hemicellulose as well as pectins to hold them together and manage osmotic stresses (Jeffries, T.W. 1994). Some of these such as those that produce fruit have much less lignin associated with them (Meredith, C.E. and Doran-Peterson, J. 2012). Those such as trees include lignins to manage the protection much better since these are basic wall support mechanisms that offer protection against stresses.

Lignocelluloses are certainly found in wastes of all kinds such as sawdust, leaves, paper trash, bark of trees, the cell walls of plant cells, and therefore plants (Union of Concerned Scientists. 2012). Converting them in a favorable manner is part of the objective. Being able to collect them in a practical manner is a second part of the puzzle—for example, converting them to pellets for ease of handling and use (Portz, T. 2016). The real driving force is to discover value-added properties of isolated compounds to drive research to success. There is work going on this topic as well (Galletti, A.M.R. and Antonetti, C. 2011; PLET. 2016). This is absolutely critical, since discoveries in this area can significantly change the economics of ethanol production for processes involving lignocelluloses as a primary source.

This is analogous to using parts of the corn kernel and DDGS now derived from corn as a major source. Besides deriving ethanol from corn DDG, it is contemplated to derive value from the DDG produced by use of lignocellulose to make ethanol. The economics change because a new market and profit from a by-product can be viewed as a cost reduction in producing ethanol as noted in the previous section on corn. As discussed, it is valuable to have parallel sources from corn and lignocellulose sources to make ethanol since that opens up research, new directions, and opportunities to gain economic advantages for ethanol (to match advantages from research discussed in Chapter 3 to make ethanol valuable in mixed fuel combustion in vehicle engines). In theory, this can result in pressure to reduce ethanol prices if this makes sense. Once the materials are cleanly separated, then enzyme technology can developed (as it has been to this point) to breakdown components further leading to better ways to convert key components to ethanol in fermentation or else to valuable by-products.

One can look at the matter of constituent separation by comparing the corn kernel (for #2 dent corn) to the constituent composition among lignocellulose entities. The corn kernel, on a dry basis, has about 80% starch and sugars, maybe 4% cellulose and lignin, and 6% pentosans (xylans, sources of xylose) which are five carbon sugars rather than six carbon sugars such as glucose (Ericksen, A. 2006). Lignocellulose (such as the stover or leaves in the corn plant) has 40% carbohydrates based on six carbon sugars (this is about half or less of the six carbon based carbohydrates in the corn kernel), about 25% pentosans (about 3–4 times that in corn), and about 3 times the lignin (Zhang, Y.-H.P. et al. 2007). Therefore, the corn kernel carbohydrate is relatively richer in the six carbon carbohydrate (glucose). This does not necessarily detract from the lignocellulose source since there may be more of it in a locale, it may be less costly to obtain, yeasts can be derived to convert xylose, etc.

However, it does mean the separation challenge is different and there may be more by-products with which to deal. This is also acceptable as long as the by-products can be made valuable for animal feed and other uses. This is critical to economic success again for reasons discussed. If one uses lignocellulose sources to make ethanol, then the by-products must be developed through research, and a market established for selling such by-products in the manner done for the corn kernel.

This is not new as development of lignocellulose-based technology to produce ethanol is of wide interest and the subject of considerable investment (DuPont Industrial Biosciences. 2016).

In an important example in this regard, the by-product feed market for corn is highly developed and much research is ongoing to improve this for feeding animals and improving their health and well-being since that makes for a better and safer protein source. The feeds from ethanol fermentation must be shown to be as efficacious and safe as others. Since there are questions, the author believes, in regard to the safety of meat for humans due to infections in animals that may be traced to their feed supply, this opportunity applies to both lignocelluloses and corn sources for ethanol. Challenges in this area can be addressed through proper research and process development (including by-product treatment and the addition of protective substances such as beneficial microbes to help preserve the by-product and make it more nutritious and healthy).

5.1.4.8 Establishment of Processes for Corn-Based Lignocellulose Sources of Ethanol That Help Where Changes in Statutes Affect Pricing

The corn plant has foliage (and the remainder from harvest termed corn stover) that can converted to ethanol where the foliage is rich in lignocellulose. The same plant residue applies for other crops and expands the opportunity.

The ethanol industry growth has been enhanced by federal government programs that help provide capital and lower taxes for this fledgling industry (Renewable Fuels Association. 2014a,b). Furthermore, the fuel blending industry received subsidies in the form of tax credits (up to $0.60 per gal) over time so that they would be more receptive to accepting ethanol for blending as they were in effect paid to accept ethanol in the gasoline/ethanol blend. This of course resulted in higher prices at the gas pump.

This helped the ethanol industry's capital and operating cost structure, and also promoted the use of ethanol by the fuel blending industry (Post, W. 2013). The financial benefits to fuel blenders expired in 2012 (Schnepf, R. 2013). Legislation was enacted to renew incentives until 2017 for ethanol producers using noncorn starch-based fuels (Muuzii. 2015).

There is a federal mandate for formulation of ethanol with gasoline that was enacted into law in 2007. Increased ethanol production and formulation into fuel blends was mandated. Tax credits were provided for this as assistance to the developing ethanol industry. Strict accounting is required for those participating in this (Schnepf, R. and Yucobucci, B.D. 2013). The mandate involved increased production and use of ethanol for first-generation developments (corn starch-based) and second-generation developments (lignocellulose and other bases). Also, there are environmental requirements. The mandate required participation by fuel blenders subject to significant fines for failure to participate properly. This may have become more distasteful in some respects to blenders with the tax credit for ethanol blending ceasing in 2012 (Congressional Budget Office).

The EPA proposed standards for production and blending going into 2017 (USEPA. 2015). There is more extensive legislation being proposed to extend tax benefits and the mandate for ethanol to be blended into gasoline (Voegele, E.

2016). Basically, this area is in a state of flux and some uncertainty. A primary reason for this is that the production of ethanol from noncorn starch sources has not developed according to plans. It seems it was forgotten (or ignored) that research results implemented into production were necessary and that research results have unknowns and uncertain timetables. It does sound like the implementation and "forced feeding of ethanol development" was handled in a nonproductive manner. That is unfortunate since this is an important area like all areas of energy innovation and improvement.

The incentive tax credits to gasoline ethanol blenders to incorporate ethanol perhaps should have been continued in a fashion to help the ethanol industry is to continue to develop. Furthermore, the capital and tax incentives afforded bioethanol producers that were extended to 2017 need to be extended long term also to help support this fledging industry. One cannot make commitments to 2022 and then all of sudden shorten the commitment to 5 years prior to that. Short-term seemingly temporary support puts a lot of cost pressure on the ethanol producers since their costs have increased and their revenues may have been subject to decreases depending on the response of the gasoline refining/blending industries and the ability of ethanol producers to secure capital. In other words, if this industry is to be supported, then it should be done as a long-term commitment so research, process development, and manufacturing growth can continue. Doing this in a tepid and insufficient manner is a waste of time and money.

One consequence of this situation is development of lignocellulose sources with the parties participating in this hopeful that this initiative will help their cost situation. This work is essential so that lignocellulose as a source can be developed to augment the corn industry to provide new sources of raw materials for ethanol. One prominent new entry into ethanol production by biological means using lignocellulose is DuPont (to be DowDupont subject to regulatory approval). DuPont is (or has been) using enzyme technology patented by Genencor (DuPont. 2013; Genencor. 2011). Novo Industries, is also a prominent enzyme supplier to the corn processing and ethanol industries (Novozymes. 2013).

The input of information from the corn refining industry to the ethanol industry as to steps to take and general guidance is quite important to entrepreneurs in ethanol. The hope is that the cost to produce ethanol will drop with the use of lignocellulose sources. Due to a myriad of complex reasons, perhaps this will not happen so easily. It is the writer's view that this will happen with creative research in how the ethanol is made from many standpoints and also how it used in automobile combustion as covered in this book.

The mass balances for lignocellulose sources to products and by-products need to be assessed as well. However, all of that data are not as accessible as they are from the corn industry.

It is most logical to have lignocellulose processing near corn fields or other fields and near corn processing and corn processing ethanol plants as well since corn stover waste and similar entities become sources of feedstocks for ethanol manufacture. This is what DuPont, for example, has done (Feehery, W.F. 2015). This is a natural way to use synergy in sourcing a variety of raw materials. The corn processing plants can also process corn waste as an onsite alternative to corn itself. This makes for cooperation, cost savings, and added profits. It should also lead to advances in knowledge and technology.

5.1.4.9 Animal Feed Sources

Animal feed streams are produced from corn wet milling. These provide another revenue stream. An additional source of animal feed to promote nutrition and health is derived from ethanol production. This is a combination of feed components from the grain (feedstock source) that is fermented to produce ethanol resulting also in yeast biomass. The excess yeast biomass that is produced is sold as animal feed (see Figure 4.2). This source is called DDGS for Distillers Dry Grains and Solubles (DDG Users Handbook. 2012). This feed source is augmentative to that from corn or in competition with it. It is important to compare the corn and DDGS as feed sources, noting that the DDGS bears on the cost to produce ethanol as either a cost reduction or co-product profitability. By-product streams from corn wet milling or dry milling that are sold also reduce process costs. Therefore, these by-product streams will compete for sales revenue.

The amount of DDGS produced is approximately 30% of the corn processed to ethanol. As indicated in Figure 5.37, the amount of DDGS produced has grown in the United States from approximately 10 billion lbs (4.6 billion kg) to approximately 90 billion lbs (41 billion kg) from 2001 to 2014. At $0.17/kg, this represents income of approximately $0.8 billion becoming $7 billion by 2014 (price information from *Ethanol Producer Magazine*. 2016).

The waste materials coming out of the plants can be modified and adapted with other products so they have added value, including nourishment and resistance to pathogens by promoting health. Ways should be found to do this and obtain weight gain and feed efficiency and minimize use of antibiotics where not needed which is creating dangers in our society and also affects export revenues—in actuality or as a perception on the part of importing countries (US Grains Council. 2012. Chapter 6).

One of the challenges in regard to using waste (as animal feed) from lignocellulose supplied to ethanol plants is that the material is often not as nutritious as that from corn sources (US Grains Council. 2012). The way around this is to augment the waste with perhaps single components or maybe another source such as microalgae which is rich in nutrients and may provide what is not available in the lignocellulose waste from the ethanol plant (Committee on the Sustainable Development of Algal Biofuels. 2012). This may be acceptable from an economic standpoint (numbers have to be analyzed to evaluate this suggestion) since the waste lignocellulose material may be much lower in cost as compared to corn.

The upgrading of nutrition seems to be totally necessary to make the lignocellulose by-product source an acceptable feed to grow healthy animals that serves to satisfy needs for human food. It may be that the augmentation can include healthful bacterial sources for the animals and items protective for their health against infection. Enzymes can be incorporated to make the feedstock more nutritionally accessible to the animal. Some sources may actual serve to stimulate the animal's immune system to cut down on the formation of harmful organisms such as the wrong kind of *E. coli* (Segelken, H.R. 1998). The metabolic agent helping to counter *E. coli* and similar entities may also be in items such as microalgae since they are in lipid pathways. This is advocating proactive upgrading of feed for economic benefit.

Co-product credit and income are absolutely necessary in this business. Witness that if corn is \$4/bushel and a bushel is 56 lbs, then the value is \$0.07/lb or \$0.15/kg. The DDG derived from corn is priced under the corn cost so say it is \$0.035/lb or \$0.08/kg. Latest prices set the price at twice that or about \$0.17/kg (making the DDG price competitive with that for corn which sounds like an anomaly). However, this is not an anomaly since these markets move more according to trading and bartering, not just by the logic of cost structures.

Lignocellulose may eventually be at a cost that is 20% of corn as a guess. This makes lignocellulose raw material (e.g., corn stover) at a cost of \$0.03–\$0.04/kg. So if it is enriched in either the case of corn or lignocellulose supplied to the ethanol plant, there is room for the enriching additive from a cost standpoint if there are attributes that result matching these just suggested in preceding paragraphs.

These comments apply to other crop sources since almost anything that has a carbohydrate source in it can serve as a supply, including garbage, trash, and relatively unconverted waste. If a source is seemingly inadequate, it needs to be treated to make it adequate without making the costs absurd. The principles stated here apply to any waste. For some wastes, such as manure, these are better suited to make biogas which is methane-based and not appropriate as a source for ethanol for practical reasons.

Table 5A.1 in Appendix 5A shows a list of feedstock sources available in the United States and worldwide. DDGS (or its mimetic or equivalent) would be available from all of them if used to produce ethanol. However, work is required to make sure they are adequate for rumen and nonrumen animals (US Grains Council. 2012). Adequate means that the issues related to health and safety of supply are addressed.

Papers on economics and processes clearly point out the key parameters for ethanol sourced from corn and that from lignocellulose sources (see Chapter 7). For the latter, one first needs to separate the cellulose and hemicellulose from the lignin. Using a dilute acid treatment is one way. The author thinks doing this without chemicals in the long run may be better. Once separated as cleanly as possible, a mixture of cellulase and hemicellulase enzymes is used to break apart the cellulose into glucose and the hemicellulose into xylose and arabinans in a way that these are produced in highest yields (see Chapter 6). The yeast used in fermentation can be selected to have the ability to convert the xylans (pento sugars) that remain and as well as the glucose to ethanol. This is derived by traditional selection and mutation and screening and fermentation control. Genetic engineering in regard to the yeast fermentation is allowed in research, but not as yet commercially.

5.1.5 Sugarcane as a Feedstock for Ethanol

In Brazil, processing of sugarcane to obtain ethanol is accomplished as a national policy. India is also a major producer. They have improved yields of sugarcane significantly over many years. This yield improvement is significant including in comparison to improved yield in corn in the United States. This yield improvement is an accomplishment since sugarcane exhibits a variable polyploidy, number of chromosomes per genetic locus (Edmé, S.J. et al. 2005) This compares to humans that normally have 23 pairs of chromosomes in their genetic makeup in cells, and have a pair of chromosomes per genetic locus. Potatoes are a tetraploid typically with four chromosomes per genetic locus. From personal experience in leading a breeding group, cross-breeding to alter traits has to proceed over several crosses in several generations to get the genetic change to effect a trait to appear on all chromosomal strands so one can get a dominant trait (Goldstein, W.E. 1988–1994). Therefore, the accomplishments for sugarcane breeders and geneticists is commendable.

The two major centers for genetic breeding of sugarcane are in Florida in the United States and in India. Given this polyploidy, there should be concern about genetic drift where all of sudden a sugarcane plant this year is not expressing the same traits it did last year. It is usual to apply environment pressure and controls to insure that such drift does not occur. Use of modern diagnostic techniques should be mandatory to stay on top of this situation. From the standpoint of control of the biology and agronomy of such an important crop, this is absolutely mandatory. Oversight in this area is obviously important.

As shown in Figure 5.31, a process diagram to convert sugarcane to product sugar and product ethanol is presented. Features relevant to this writer's observations and recommendations are covered.

First, sugarcane enters the mill as feedstock This feedstock must be fresh and not older than 24 hours (rule-of-thumb) since harvest. Otherwise, the sugar will invert, meaning it hydrolyzes to glucose and fructose, changing the product and its characteristics so that it will not process properly to sugar (McLaren, J. 2009; Wikipedia. 2015b). It may process properly to be used for ethanol. So this is a point to be checked since perhaps there can be savings in allowing longer waiting times when the sugarcane is destined for ethanol manufacture. However, if there is too long a time interval, bacterial and fungal infestation may result.

As indicated in Figure 5.31, the sugarcane is first crushed and milled to a sufficient extent so that aqueous extraction of the sugar from the sugarcane can occur. The crushing and milling step is shown separately from the extraction step to separate the unit operations though perhaps combining crushing, milling, and sugar extraction in one step consisting of several stagewise operations may be better and perhaps easier to keep matters in a sanitary state.

The diagram shows a countercurrent extraction where water enters in one end and crushed sugarcane at the other. With sufficient residence time, and drainage time for the extracted biomass and then fluid recycle the fluid exits, the reactor as shown. The solids exiting pictorially at the other end are termed bagasse and can be directed to animal feed uses. The fortification of the bagasse stream in ways used and suggested for corn processing may make much sense and increase the dollar value of the bagasse by-product creating more lucrative marketing opportunities. The biomass has to be substantially delignified in order for it to be suitable as animal feed. This lignin can be isolated using modern enzyme systems as part of degrading xylans and cellulose since these latter entities can presumably be sent to reactors that reduce these to glucose and xylose to provide feedstock for the ethanol fermentation in combination or separate from sucrose streams.

A use can be found for the lignins from research as already noted. The bagasse solids contain pentyl (xylan) and hexyl cellulosic beta-linked carbohydrates which can be treated with the set of enzyme systems to convert them to xylose and glucose suitable for use by the appropriate yeast system. This is an improvement that can be effected (and perhaps already is) to enrich the fermentation feedstock directed to the ethanol fermentation as shown.

As indicated, the sugar solution effluent from aqueous extraction is directed to sugar stream evaporation and crystallization which can be multieffect to improve steam economies and energy efficiency. The waste liquid following evaporation–crystallization is termed molasses and may be sold as another by-product or directed also to the ethanol fermentation. The molasses may have to be treated to make it suitable for the yeast to ethanol fermentation. Ideally, this is all done in a way to minimize microbial contamination so the facility is kept clean. The crystallized sugar is dried, bagged, and sold.

The liquid sugar stream along with other streams as noted in sent to the ethanol fermentation. This fermentation and recovery process (along with the process for producing ethanol from corn) is analogously applicable to ethanol potentially produced from many crop sources.

The waste stream exiting from the ethanol fermentor in the case of the fermentation of sugar to ethanol basically consists of the biomass effluent that may be produced batchwise or else continuously. The biomass effluent is termed Vinasse for the sugar to ethanol fermentation. This biomass has to be treated to make it suitable for animal feed and likely enriched. This is worthwhile since it can lead to a more valuable product along with the bagsasse which presents an additional by-product opportunity.

It may be desirable, perhaps for new facilities dedicated to ethanol, that process research and development be conducted to see if the milled sugarcane could be directly sent to the ethanol fermentation and decent yields of ethanol and rates of production resulting. This would mean one waste stream that may be subsequently divided into fractions. This simplifies the process in one respect but unnecessarily complicates it if the fractions have to be separated. Perhaps, modern enzymes systems from suppliers can help effect this to advantages if they can facilitate the ease of such separations from lesser units of recovery equipment. It is suggested, as in the case of other ethanol feedstocks, this crop represents a wonderful opportunity for application of enzymes beyond what is presently being done.

5.1.6 Sorghum as Feedstock for Ethanol

It is assumed that the whole sorghum grain can be converted with a dual enzyme process similarly to that done for corn. Since in the case of sorghum, the starch retrogrades rapidly, the pH, temperature, and reactor residence time in liquefaction and saccharification will have to be optimized from process development (Sullins, R.D. and Rooney, L.W. 1975; Owuama, C.I. 1999; Souilah, R. et al. 2014). Since protein encases the starch, perhaps granules will have to be treated with protease and perhaps other enzymes including cellulase and pectinase to expose the starch to facilitate the starch conversion in liquefaction via thermostable amylase in a next stage reaction. This may be analogous to malting barley so in this case we would be malting sorghum. Malted barley goes through a process to activate enzymes to release starch from protein (Claver, I.P. et al. 2010). The enzymes naturally germinate in the barley during the process. If such enzymes are not present in sorghum, they can be exogenously added to the reaction.

The initial chaff from the plant will be isolated and converted via a cellulase family of enzymes so that the glucose from this can be supplied to the ethanol fermentation perhaps in a separate fermentor as the yeast fermentation may be different for use of converted lignocellulose feedstock. The unconverted chaff would represent a separate animal feed supply that will be optimize with additives to achieve an optimal feed balance for animals that would include natural agents to prevent the sorghum DDG from putrefying. Animal feed must be free of infectious microorganisms in all cases. The fermentor DDG waste would be blended with the treated chaff waste or else kept separate depending on animal feed testing results that would be done. Fermentation conditions for the yeast are expected to be invariant from other feedstocks unless there is evidence to the contrary.

5.1.7 Paddy Rice as a Feedstock for Ethanol

Paddy rice (the rice before any milling) is the suggested preferred feedstock for ethanol (Wikipedia. 2012; Ofongo, S.T. et al. 2008). Once harvested and freed of chaff (which can be processed as a lignocellulose), it must be quickly dried in an inexpensive manner so it does not rot (maybe passed through hydroclones under vacuum and low heat to strip water off and draw moisture away). It may be milled, maintaining a low water environment, and then subjected to a two-stage enzyme treatment (liquefaction using thermal stable amylase followed by saccharification with amyloglucosidase and sent to the ethanol fermentation system).

It should evident that the similarities among feedstocks and processing could allow one to design a facility that can accept a multitude of feedstocks indigenous to the region so that if one particular feedstock was more amply available at the best lowest cost, it might be used for ethanol. This multiprocessing may allow for cost reduction optimization for the facility in the selected locale.

As indicated, side streams allow for treatment of lignocellulose-related material for use and processing of fermentor by-product for sale as the treated equivalent of DDGS (once this is substantiated through research and commercial development). The DDGS-presumed equivalent would be treated and fortified, so it is an optimal feedstock for animal feed. At some point, as in other cases, research may result in testing of food materials (based on treating fermentor DDGS equivalent) so that once fortified, it can become a successful human food supplement. This was always the dream for single-cell protein and soybean products. Single-cell protein did not result in a successful product. Soy protein and other products of soy still represent a growing business. The soy protein could be in part used in the DDGS material for animal feed and maybe later for human feed once this is shown to be viable, sufficiently tasty, accepted, and also economically attractive.

5.1.8 Sugar Beet as a Feedstock for Ethanol

Sugar beet is a root crop with the sucrose basically contained in the root. It is processed analogously to processing sugarcane. The root can be isolated from chaff (chaff saved for processing of its cellulose and hemicellulose) and the root milled and then treated with enzymes to access the sucrose in milled stock. The isolation should be effected by treating with cellulase, hemicellulose, perhaps pectinase, and maybe protease to make the sucrose more accessible for fermentation to ethanol. The protein present in the root may interfere with beet processing if not removed (Parpinelo, G.P. et al. 2004). The alternative (for ethanol) is to use part of the sucrose from extraction of the process that results in sugar derived from beet and also molasses. It seems desirable to also evaluate recovering glucose and xylose from the lignocellulose in the chaff and use this source to make ethanol as well.

It may be unnecessary to use the sugar extracted from the beet sugar evaporation and concentration in the ethanol fermentation as a less refined stream may be preferably used. However, it may be that the facility could end up producing refined sugar as well as ethanol such that separate facilities are desired for beet sugar processing to refined sugar and to ethanol.

Note that the sucrose in sugar beet begins to be lost in a manner similar to that for sugarcane. It may be that fast processing at elevated moisture can minimize the loss (Jansen, A. nee Buschmeier. 2009; Lafta, A.M. and Fugate, K.K. 2009; FAO. 1999a,b). Enzyme reactions such as those involving invertase are responsible in-part for the loss (converting sucrose to glucose and fructose). The elevated moisture content may lower respiration which seems to trigger enzyme activity and degrade plant membranes.

Again with these options as indicated, by-products involving original plant chaff, sugar beet molasses, and DDGS equivalents from fermentation are possible. Having a multistream process offering flexibility seems to make the most sense for this feedstock as well.

5.1.9 Barley as a Feedstock for Ethanol

As discussed in Section 5.1.3.9, dehulled barley would be the choice if one were to develop a process to make industrial ethanol from barley. This is not a new development as this has already been proposed. Perhaps this is an auxiliary plant to one producing alcohol for brewing purposes. If that is the case, then one take the harvested hulled barley grain and remove the hull through a combination of enzymes and milling. Actually enzymes can be employed to dehull the barley and allow it to be amenable for fermentation. This involves some pretreatment to open up the grain so starch processing to glucose can be effected with hydrolysis enzymes. Use of beta glucanase (beta cellulase class enzymes) to remove cellulose and hemicellulose substances is recommended as this can avoid "gumming-up" problems and if done properly, increase the rate of fermentation. Yield depends on having adequate fermentation time to produce the ethanol. As discussed, this is eminently feasible and can draw on brewing experience as well as experience possessed by commercial enzyme companies such as Genencor and Novo. Apparently there is much interest in this and the feeling that the barley can be competitive to other sources.

5.1.10 Wheat Flour as a Feedstock for Ethanol

Wheat flour is suggested as this is expected to be the least expensive alternative to different substances derived from wheat grain. The chaff from the plant would be separated and converted to feedstock for ethanol unless something more valuable can be done with it. The wheat flour has wheat gluten which musts be separated for its value and since it will cause equipment plugging due to binding gluten to surfaces. The remaining material, wheat starch, can be processed using dual enzyme systems in preparation of providing a high glucose stream for the ethanol fermentation by yeast. Wheat is very rich in minerals and protein, so it is expected that the DDGS equivalent from the fermentation will be a competitive and rich source of nutrition even with the vital wheat protein removed prior to starch conversion. It may be desirable to fractionate the wheat grain as the corn industry does with corn as this may result in some new product opportunities. Remember, one must think this way as any valuable by-product will reduce the cost of ethanol no matter how the accounting is done.

There is experience making wheat beer by combining wheat with malted barley. This affords another by-product opportunity plus the chance to gain knowledge of enzyme usage for this grain. The gluten is very functional and will adhere to itself allowing washing out of the starch. This has been done in many ways including via a patented process. This is very amenable to centrifugal related separation once sufficient time for gluten self-binding and separation has occurred.

5.1.11 Cassava Chips as a Feedstock for Ethanol

This mode of providing cassava is recommended (already treated so that cyanide has been removed, though there should be less in the root than in leaves) and sliced chips provided (Oguntimein, G.B. 2006; Wikipedia. 2015c). Providing these steps in a captive process may be more cost effective with more assurance of maintaining quality. Quality control testing would be applied to assure absence of the cyanide and making sure chip quality meets specifications. The chips would be milled to specifications to be supplied for conversion to glucose and oligosaccharides using dual enzyme conversion.

The leaves and chaff would also be freed of cyanide and supplied for lignocellulose processing as indicated. Cassava is such an important food supply that it seems logical to set up or run a modified facility to meet needs of food from cassava as well as ethanol. The biomass waste resulting from fermentation may have ample protein in it for animal feed. This may even extend to a superior and rich food supply based on a type of single-cell protein if this can made suitably tasty and attractive and provide reasonable revenue for the cost. Again, these different avenues in effect can be considered to reduce the cost of ethanol by deriving and providing feedstuffs.

APPENDIX 5A

TABLE 5A.1
Potential Crop Sources for Ethanol by Country

	Crop Production (kg)									
Country	**Sugarcane**	**Corn**	**Wheat**	**Rice**	**Potato**	**Sugar Beet**	**Cassava**	**Sorghum**	**Barley**	**Oats**
Albania		3.00E+08						9.66E+06	7.00E+06	
Afghanistan		3.00E+08	5.20E+09						4.00E+08	
Algeria		1.00E+06	3.20E+09						1.30E+09	
American Samoa							8.50E+04			
Angola		1.69E+09					1.64E+10			
Anguilla							1.56E+07			
Antigua and Barbuda							1.15E+05			
Argentina		2.40E+10	8.00E+09				1.84E+08	4.00E+09	3.00E+09	4.50E+08
Armenia									1.60E+08	
Australia		4.00E+08	2.29E+10					2.00E+09	7.95E+09	1.05E+09
Austria		1.64E+09	1.54E+09			3.00E+09		1.02E+07	7.34E+08	1.00E+08
Azerbaijan		1.60E+08							6.00E+08	
Bahamas							9.00E+05			
Bangladesh				3.39E+10	8.21E+09				7.00E+06	
Barbados	3.00E+07						3.20E+05			
Belarus		5.70E+08	2.10E+09		6.91E+09	4.50E+09			2.25E+09	8.10E+08
Belgium		8.38E+08	1.64E+09		2.93E+09	5.00E+09			3.91E+08	3.40E+07
Belize							4.42E+05			
Benin		1.10E+09					3.70E+09	1.20E+08		

(Continued)

TABLE 5A.1 (*Continued*)
Potential Crop Sources for Ethanol by Country

	Crop Production (kg)									
Country	Sugarcane	Corn	Wheat	Rice	Potato	Sugar Beet	Cassava	Sorghum	Barley	Oats
Bhutan		7.00E+07							5.00E+06	
Bolivia		7.25E+08					2.88E+08	1.60E+08	7.00E+07	
Bosnia and Herzegovina		6.50E+08	2.65E+08						6.50E+07	
Brazil	7.39E+11	7.50E+10	5.70E+09	1.15E+10	3.73E+09		2.15E+10	2.00E+09	3.30E+08	4.00E+08
Botswana		3.50E+07						3.00E+07		
Brunei							3.00E+06			
Bulgaria		2.70E+09	5.38E+09					5.01E+06	7.18E+08	4.00E+07
Burkina Faso		1.42E+09					4.35E+06	1.84E+09		
Burma				3.30E+10			6.30E+08			
Burundi		1.50E+08					2.23E+09	3.20E+07		
Cambodia		1.08E+09		9.30E+09			8.00E+09			
Cameroon		1.60E+09					4.60E+09	1.15E+09		
Canada		1.15E+10	3.75E+10		4.59E+09				7.12E+09	2.68E+09
Cape Verde		1.00E+06					4.12E+06			
Cayman Islands						1.50E+04				
Central African Republic		9.50E+07					7.00E+08	2.50E+07		
Chad		3.50E+08					2.50E+08	9.00E+08		
Chile		1.40E+09							9.00E+07	3.30E+08
China	1.26E+11	2.16E+11	1.22E+11	2.04E+11	8.59E+10	1.07E+10	4.60E+09	2.70E+09	1.55E+09	5.40E+08

(*Continued*)

TABLE 5A.1 (*Continued*)
Potential Crop Sources for Ethanol by Country

Country	Crop Production (kg)									
	Sugarcane	Corn	Wheat	Rice	Potato	Sugar Beet	Cassava	Sorghum	Barley	Oats
Colombia	3.49E+10	1.75E+09					2.23E+09	2.00E+08	2.00E+07	
Comoros							7.00E+07			
Congo		1.30E+07						7.00E+06		
Cook Islands							1.03E+06			
Costa Rica		1.80E+07					1.45E+08			
Croatia		1.91E+09						1.39E+06	1.78E+08	
Cote d'Ivoire		6.50E+08					2.44E+09	5.00E+07		
Cuba		4.26E+08					5.14E+08			
Cyprus			1.80E+07						6.70E+07	2.70E+06
Czech Republic		6.75E+08	4.79E+09			3.50E+09			1.59E+09	1.40E+08
Democratic Republic of Congo							1.65E+10			
Denmark		7.60E+07	4.15E+09							3.00E+08
Dominica							1.20E+06			
Dominican Republic		4.50E+07					1.76E+08	4.00E+06		
Egypt		5.96E+09	9.50E+09	5.90E+09	4.50E+09	7.40E+09		7.57E+08	1.08E+08	
Ecuador		1.22E+09					4.58E+07	1.00E+07	1.50E+07	
El Salvador		7.69E+08					3.86E+07	1.50E+08		

(*Continued*)

TABLE 5A.1 (*Continued*)
Potential Crop Sources for Ethanol by Country

	Crop Production (kg)									
Country	Sugarcane	Corn	Wheat	Rice	Potato	Sugar Beet	Cassava	Sorghum	Barley	Oats
Equatorial Guinea							7.00E+07			
Eritrea		2.00E+07						8.00E+07	6.50E+07	
Estonia			4.06E+08						4.39E+08	7.00E+07
Ethiopia		6.50E+09	4.00E+09					4.00E+09	2.10E+09	
Federated States of Micronesia							9.50E+06			
Fiji							9.40E+07			
Finland			9.00E+08						1.94E+09	1.16E+09
France		1.51E+10	3.86E+10		7.00E+09	3.81E+10		2.79E+08	1.03E+10	4.50E+08
French Antilles	2.00E+07									
French Polynesia							4.00E+06			
Gabon							3.15E+08			
Gambia		8.00E+06					1.15E+07	9.00E+06		
Georgia		3.00E+08							3.00E+07	
Germany		4.39E+09	2.50E+10		6.34E+09	2.50E+10			1.03E+10	6.68E+08
Ghana		1.82E+09					1.60E+10	2.93E+08		
Greece		2.19E+09	4.70E+08					1.60E+06	3.53E+08	1.20E+08
Grenada							2.10E+05			
Guinea		6.52E+08					1.20E+09	4.50E+07		

(Continued)

TABLE 5A.1 (*Continued*)
Potential Crop Sources for Ethanol by Country

	Crop Production (kg)									
Country	**Sugarcane**	**Corn**	**Wheat**	**Rice**	**Potato**	**Sugar Beet**	**Cassava**	**Sorghum**	**Barley**	**Oats**
Guinea-Bissau		1.00E+07					2.30E+07	1.40E+07		
Guatemala		1.64E+09					2.00E+07	4.50E+07		
Guyana		5.00E+06					7.65E+06			
Haiti		2.53E+08					4.18E+08	9.00E+07		
Honduras		4.30E+08					2.40E+07	4.00E+07		
Hungary		6.73E+09	5.10E+09					5.01E+06	1.07E+09	1.40E+08
India	3.41E+11	2.25E+10	9.35E+10	1.53E+11	4.50E+10		7.24E+09	5.00E+09	1.83E+09	
Indonesia	3.37E+10	9.40E+09		6.90E+10						
Iran		1.25E+09	1.40E+10		5.40E+09	3.80E+09	2.39E+10	2.00E+07	2.60E+06	
Iraq		2.50E+08	3.30E+09					6.50E+07	1.00E+09	
Ireland			5.34E+08						1.63E+09	1.40E+08
Israel		3.10E+07	4.40E+08					2.25E+07	1.20E+07	5.00E+07
Italy		6.50E+09	3.24E+09			3.25E+09		3.20E+08	7.72E+08	3.20E+08
Jamaica	1.60E+08	2.00E+06					1.74E+07			
Japan		1.00E+06		1.07E+10		3.00E+09			1.70E+08	
Jordan		5.00E+06							1.50E+07	
Kazakhstan		6.64E+08	1.39E+10		3.13E+09				2.41E+09	1.40E+08
Kenya		2.65E+09			2.92E+09		1.11E+09	1.20E+08	7.50E+07	
Korea (North)		1.93E+09								
Korea (South)		7.80E+07		6.40E+09					1.31E+08	
Kyrgyzstan		4.50E+08							2.25E+08	

(Continued)

TABLE 5A.1 (*Continued*)
Potential Crop Sources for Ethanol by Country

	Crop Production (kg)									
Country	**Sugarcane**	**Corn**	**Wheat**	**Rice**	**Potato**	**Sugar Beet**	**Cassava**	**Sorghum**	**Barley**	**Oats**
Laos		1.15E+09					1.12E+09			
Latvia			1.44E+09						2.33E+08	1.00E+08
Lebanon		3.00E+06							3.30E+07	
Lesotho		8.60E+07						5.00E+06		
Liberia							5.20E+08			
Libya									1.01E+08	
Lithuania		1.21E+08	2.90E+09						6.82E+08	1.10E+08
Luxembourg		2.00E+06	9.10E+07						4.20E+07	6.00E+06
Macedonia		1.19E+08						6.01E+04	9.00E+07	
Madagascar		3.66E+08		4.00E+09			3.11E+09			
Malawi		3.93E+09			3.26E+09		4.81E+09			
Malaysia		5.60E+07					8.17E+07			
Maldives							4.00E+03			
Mali		1.60E+09					3.80E+07	1.30E+09		
Mauritius							5.17E+05			
Mauritania		1.50E+07						6.50E+07		
Mexico	6.12E+10	2.40E+10	3.40E+09				2.43E+07	7.30E+09	5.50E+08	7.00E+07
Moldova		1.30E+09							2.00E+08	
Morocco		2.00E+08	6.90E+09			2.75E+09		6.00E+06	1.70E+09	
Mozambique		1.90E+09					1.00E+10	2.25E+08		
Myanmar		1.75E+09		1.08E+10						

(Continued)

TABLE 5A.1 (*Continued*)
Potential Crop Sources for Ethanol by Country

	Crop Production (kg)									
Country	**Sugarcane**	**Corn**	**Wheat**	**Rice**	**Potato**	**Sugar Beet**	**Cassava**	**Sorghum**	**Barley**	**Oats**
Namibia		6.80E+07								
Nepal		2.30E+09		5.10E+09	2.58E+09				3.00E+07	
Netherlands		2.53E+08		1.34E+09	6.77E+09	6.00E+09			2.08E+08	6.00E+06
New Caledonia							1.50E+06			
New Zealand		2.20E+08							4.10E+08	
Nicaragua		4.21E+08					2.77E+08	1.10E+08		
Niue							4.20E+04			
Niger							1.50E+08	1.00E+09		
Nigeria		7.50E+09		4.80E+09			5.30E+10	6.30E+09		
Norway									5.77E+08	3.20E+08
Pakistan	6.38E+10	4.99E+09	2.42E+10	9.40E+09	4.10E+09			1.45E+08	2.65E+08	
Panama		8.50E+07					2.15E+07			
Papua New Guinea							1.40E+08			
Paraguay		3.08E+09					2.80E+09	4.10E+07		
Peru		1.70E+09			4.47E+09		1.18E+09	1.00E+06	2.30E+08	
Puerto Rico							7.50E+05			
Philippines	3.19E+10	7.98E+09		1.80E+10			2.36E+09			
Poland		4.04E+09	9.50E+09		9.09E+09	1.16E+10			2.92E+09	1.44E+09
Portugal		8.49E+08	8.50E+07						3.00E+07	3.60E+07

(*Continued*)

TABLE 5A.1 (*Continued*)
Potential Crop Sources for Ethanol by Country

	Crop Production (kg)									
Country	**Sugarcane**	**Corn**	**Wheat**	**Rice**	**Potato**	**Sugar Beet**	**Cassava**	**Sorghum**	**Barley**	**Oats**
Republic of the Congo							1.25E+09			
Romania		1.14E+10	7.45E+09					4.99E+07	1.65E+09	3.10E+08
Russia		1.13E+10	5.21E+10		2.95E+10	4.76E+10			2.00E+10	4.03E+09
Rwanda		4.80E+08					2.95E+09	1.45E+08		
Saint Lucia							1.25E+06			
Samoa							4.15E+05			
Sao Tome and Principe							1.43E+06			
Saint Vincent and the Grenadines							6.00E+05			
Saudi Arabia		8.00E+07						2.65E+08	1.50E+07	
Senegal		1.02E+08					1.54E+08	4.50E+07		
Serbia		7.70E+09	2.70E+09					1.00E+07	3.67E+08	
Seychelles							2.20E+05			
Sierra Leone		3.80E+07						3.00E+07		
Singapore							5.00E+03			
Slovakia		1.13E+09	1.57E+09						4.46E+08	4.00E+07
Slovenia		2.37E+08	1.39E+08						6.90E+07	5.00E+06
Somalia		9.50E+07					9.00E+07	1.40E+08		
South Africa		1.13E+10						1.58E+08	3.07E+08	

(*Continued*)

TABLE 5A.1 (*Continued*)
Potential Crop Sources for Ethanol by Country

	Crop Production (kg)									
Country	**Sugarcane**	**Corn**	**Wheat**	**Rice**	**Potato**	**Sugar Beet**	**Cassava**	**Sorghum**	**Barley**	**Oats**
Spain		4.85E+09	6.69E+09			4.00E+09		4.54E+07	1.01E+10	7.99E+08
Slovakia		1.40E+09						9.59E+05		
Solomon Islands							3.00E+06			
Sri Lanka							3.03E+08			
Sudan							1.75E+07	6.28E+09		
Suriname							7.08E+06			
Swaziland		1.00E+08						1.00E+06		
Sweden		1.10E+07	1.90E+09						1.94E+09	7.76E+08
Switzerland		1.46E+08	4.38E+08						1.85E+08	
Syria		8.00E+07	3.10E+09						3.50E+08	
Taiwan	6.50E+07	1.08E+08						4.00E+06		
Tajikistan		8.00E+07							1.50E+08	
Tanzania		5.00E+09					4.76E+09	8.40E+08		
Thailand	1.00E+11	4.85E+09		3.78E+10			3.02E+10	5.50E+07		
Timor-Leste							2.65E+07			
Togo		8.33E+08					9.03E+08	2.93E+08		
Tonga							8.00E+06			
Trinidad and Tobago		5.00E+06					3.19E+06			
Turkmenistan		5.00E+07							2.00E+07	

(*Continued*)

TABLE 5A.1 (*Continued*)
Potential Crop Sources for Ethanol by Country

	Crop Production (kg)									
Country	**Sugarcane**	**Corn**	**Wheat**	**Rice**	**Potato**	**Sugar Beet**	**Cassava**	**Sorghum**	**Barley**	**Oats**
Tunisia									9.00E+08	
Turkey		5.90E+09	2.21E+10		4.82E+09	1.61E+10			3.95E+09	2.60E+08
Uzbekistan		1.20E+08	6.80E+09						2.40E+08	
Uganda		2.75E+09					5.23E+09	3.17E+08		
Ukraine		2.85E+10	2.28E+10		2.33E+10	1.87E+10		1.80E+08	9.45E+09	7.90E+08
United Kingdom			1.19E+10		4.55E+09	8.50E+09			7.09E+09	7.84E+08
United States	2.79E+10	3.61E+11	6.00E+10	9.00E+09	1.92E+10	2.62E+10		1.10E+10	3.85E+09	9.29E+08
Uruguay		4.50E+08						2.40E+08	3.90E+08	
Venezuela		1.30E+09					5.58E+08	1.10E+08		
Vietnam		5.63E+09		4.37E+10			9.76E+09			
Wallis and Futuna							2.00E+06			
Yemen		8.00E+07						4.30E+08	3.50E+07	
Zambia		3.38E+09					1.07E+09	1.20E+07		
Zimbabwe		1.30E+09					2.30E+08	6.50E+07	6.00E+07	
	1.56E+12	9.84E+11	6.85E+11	6.80E+11	2.98E+11	2.49E+11	2.73E+11	6.38E+10	1.34E+11	2.09E+10

REFERENCES

AgMRC. April 10, 2015. *Estimated U.S. Dried Distillers Grains with Solubles (DDGS) Production and Use*. Agricultural Marketing Research Center, Iowa State University, Ames, IA.

Baillie, R. and Bollerslev, T. 1989. The message in variable exchange rates: A conditional variance tale. *American Statistical Association and the Journal of Business and Economic Statistics*. 7(3), 297–305.

Blau, D.A., Brouk, M.J., Drouillard, J., and Montgomery, S.P. February 2001. Corn Gluten Feed. Contribution No. 01-101-E from the Kansas Agricultural Experiment Station, Kansas State University.

Bode, J.W. 2013. *Corn Starch*, 12th Edition. Corn Refiners Association. http://corn.org/.

Bryan, T., Ed. 2016. *Ethanol Producer Magazine*, Grand Forks, ND 58203. Publisher: BBI International. http://www.ethanolproducer.com

Butzen, S. and Smith, S. 2011. *Corn Yield Gains due to Genetic and Management Improvement (Crop Insights)*. DuPont Pioneer, Ames, IA.

CIMMYT. 2016. http://www.cimmyt.org/.

Claver, I.P., Zhang, H., Li, Q., Zhu, K., and Zhou, H. 2010. Impact of the soak and the malt on the physicochemical properties of the sorghum starches. *International Journal of Molecular Sciences*. 11, 3002–3015.

Committee on the Sustainable Development of Algal Biofuels; Board on Agriculture and Natural Resources; Board on Energy and Environmental Systems; Division on Earth and Life Studies; Division on Engineering and Physical Sciences; National Research Council. 2012. *Sustainable Development of Algal Biofuels*. National Academies Press, National Research Council of the National Academy of Sciences. ISBN 978-0-309-26032-9.

Conforti, P. (ed.) 2011. Looking ahead in World Food and Agriculture Perspectives to 2050. Food and Agriculture Organization (FAO). Agricultural Development Economics Division, Economic and Social Development Dept. ISBN 978-92-5-106903-8.

Congressional Budget Office. https://www.cbo.gov/publication/45477.

Conventional Plant Breeding. December 2004. Pocket K No. 13. Global Knowledge Center on Crop Biotechnology. http://www.isaaa.org/kc.

Corn Refiners Association. 2013. *Corn Starch*. 12th Edition. http://corn.org/.

Corn Refiners Association Annual Report. 2010. *Telling the Positive Story of Corn*. Corn Refiners Association. Washington, D.C. http://corn.org/.

Dadi, A.P., Varamaso. S., and Schall, C.A. 2006. *Enhancememt of Cellulose Saccharification Kinetics Using an Ionic Liquid Pretreatment Step*. Wiley Interscience. http://www.interscience.wiley.com.

DDG Users Handbook. 2012. *A Guide to Distiller's Dried Grains with Solubles (DDGS), U.S. Grains Council*. 3rd Edition. http://www.grains.org/buyingselling/ddgs/ddgs-user-handbook.

DuPont. 2013. Accellerase® Trio™. http://accellerase.dupont.com/.

DuPont Industrial Biosciences. February 10, 2016. 5 minute guide and DuPont files lawsuit against EPA. http//www.biomassdigest.com.

Edmé, S.J., Comstock, J.C., Miller, J.D., and Tai, P.Y.P. Determination of DNA Content and Genome Size in Sugar Cane. USDA-ARS Sugarcane Field Station. Canal Point, FL. http://www.saa.ars.usda.gov.

Encinas, A.V., Nebra, S.A., Lozano, M., and Serra, L. September 2007. Design of evaporation systems and heater networks in sugar cane factories using a thermoeconomic optimization procedure. *International Journal of Thermodynamics*. 10(3), 97–105.

Ericksen, A. 2006. *Corn Wet Milled Feed Products*. 4th Edition. Corn Refiners Association, Washington, D.C. http://www.corn.org.

Ethanol Producer Magazine. February. 2016. Corn Futures Prices. BBI International. p. 21.
Factfish. 2013a. Cassava Production Quantity (tons). http://www.factfish.com/statistic/cassava
Factfish. 2013b. Sorghum Production Quantity (tons). http://www.factfish.com/statistic/sorghum.
FAO (Food and Agricultural Organization). 1999a. Sugar Beet Handbook.
FAO (Food and Agricultural Organization). 1999b. Sugar Beets/White Sugar. Agribusiness Handbooks. Vol. 4.
FAO (Food and Agricultural Organization) Stats. 2011. World Sugar Beet Production by Tonnes.
Feehery, W.F. October 30, 2015. DuPont Opens Commercial Scale Cellulosic Ethanol Plant in Iowa. http://www.dupont.com/.
Fernandez-Cornejo, J., Wechsler, S., Livingston, M., and Mitchell, L. February 2014. Genetically Engineered Crops in the United States. Economic Research Service, Economic Research Report No. 16. USDA.
Fleming, C. November, 2015. *Industry at a Glance*. World Oil, Gulf Publishing Company. p. 31.
Foreign Agricultural Service. January 2016. World Agricultural Production. Circular Series WAP 1-16. USDA.
Galletti, A.M.R. and Antonetti, C. September 19, 2011. Biomass Pretreatment: Separation of Cellulose, Hemicellulose, and Lignin. Exiting Technologies and Perspectives. UROBIOREF.
Gao, D., Uppugundla, N., Chundawat, S.P.S., Yu, X., Hermanson, S., Gowda, K., Brumm, P., Mead, D., Balan, V., and Dale, B.E. 2011. Hemicellulases and auxiliary enzymes for improved conversion lignocellulosic biomass to monosaccharides. *Biotechnology for Biofuels*. 4(5), http://www.biotechnologyforbiofuels.com/content/4/1/5.
Genencor. 2011. Accellerase® Trio™. A Division of Danisco.
Goldstein, W.E. 1978–1988. Personal Experience, Miles Laboratories Biotechnology Group, Division of Bayer, A.G., Elkhart, IN.
Goldstein, W.E. 1978–2008. *Personal Experience*. Miles Laboratories and Bayer, Inc., Elkhart, IN.
Goldstein, W.E. 1988–1994. *Personal Experience*. ESCAgenetics Corp., San Carlos, CA.
Harmsen, P.F.H., Huijgen, W.J.J., Bermúdez López, L.M., and Bakker, R.R.C. September, 2010. *Literature Review of Physical and Chemical Processes for Lignocellulose Biomass*. Report ECN-E-10-013. Wageningen University & Research Centre. The Netherlands.
Index Mundi. 2014a. *Agriculture*. http://www.indexmundi.com/Agriculture.
Index Mundi. 2014b. *Barley*. http://www.indexmundi.com/agriculture/?commodity=barley&graph=production.
Index Mundi. 2014c. *Sorghum*. http://www.indexmundi.com/agriculture/?commodity=sorghum&graph=production.
Index Mundi. 2014d. *Wheat*. http://www.indexmundi.com/agriculture/?commodity=wheat&graph=production.
Ingledew, W.M., Jones, A.M., Bhatty, R.S., and Rossnagel, B.G. 1995. Fuel alcohol production from hull-less barley. *Cereal Chemistry*. 72(2), 147–150.
International Starch Institute. 2014. Hydroclones Replacing Centrifuges. http://www.starch.dk/isi/profile/hc.asp.
ITTA (International Institute of Tropical Agriculture). 1997. Phenomenal Increase in Maize Production in West and Central Africa. http://www.worldbank.org/html/cgiar/newsletter/april97/8maize.html.
Jansen, A. nee Buschmeier. 2009. Modifying post-harvest sucrose loss in sugar beet: Assessment of transgenic approaches. A Dissertation. Ruperto-Carola University of Heidelberg, Germany.
Jeffries, T.W. 1994. Chapter 8. Biodegradation of lignin and hemicelluloses. C. Ratledge. (ed.), *Biochemistry of Microbial Degradation*, 233–277. Kluwer Academic Publishers, The Netherlands.

Key, S., Ma, J.K.-C., and Drake, P.M.W. 2008. Genetically modified plants and human health. *Journal of the Royal Society of Medicine*. 101, 290–298.

Kuepper, J. 2016. Trading the Odds with Arbitrage. http://www.investopedia.com/articles/trading/04/111004.asp.

Kumar, R., Mago, G., Balan, V., and Wyman, C. 2009. Physical and chemical characterization of corn stover and poplar solids resulting from pretreatment technologies. *Biosource Technology*. 100, 3948–3962.

Kwiatkowski, J.R., McAloon, A.J., Taylor, F., and Johnston, D.B. 2006. Modeling the process and costs of fuel ethanol production by the corn dry-grind process. *Industrial Crops and Products*. 23(2006), 288–296.

Lafta, A.M. and Fugate, K.K. 2009. Dehydration accelerates respiration in postharvest sugarbeet roots. *Postharvest Biology and Technology*. 54, 32–37.

Lake, M.A. August 4, 2010. Potential commercial uses for lignin. *Southeastern Bioenergy Conference*. Tifton, Georgia.

Lee, H.V., Hamid, S.B.A., and Zain, S.L. 2014. Conversion of lignocellulose biomass to nanocellulose: Structure and chemical process. *The Scientific World Journal*. Hindawi Publishing Corporation. http://dx.doi.org/10.1155/2014/631013.

Makkar, H.P.S. (ed.) 2012. *Biofuel Co-Products as Livestock Feed: Opportunities and Challenges*. Food and Agricultural Organization of the United Nations, Rome, Italy. p. 40.

Marketsandmarkets.com. August 2015. Soy Protein Ingredients Market by Type (Soy Protein Isolates, Soy Protein Concentrates & Soy Flours), Application (Bakery & Confectionery, Meat Alternatives, Functional Foods, Dairy Replacements & Infant Foods), & by Region. *Global Trends & Forecast to 2020*. Report Code: FB 2092.

McLaren, J. September 2009. Sugar Cane as a Feedstock for Biofuels. Report prepared by Strathkirn, Incl. for the NCGA (National Corn Growers Association). http://www.ncga.com/.

Meredith, C.E. and Doran-Peterson, J. 2012. Pectin-rich biomass as feedstock for fuel ethanol production. *Applied Microbiology and Biotechnology*. 95, 565–575. doi: 10.1007/s00253-012-4173-2.

Muuzii. December 16, 2015. Legislation aims to extend biofuel, bioenergy tax credits. http://www.muuzii.com/legislation-aims-to-extend-biofuel-bioenergy-tax-credits/.

Nagle, B.J., Emling, J.L., and Mason, W.R. September 21, 1999. Waxy maize starch derived from grain of a plant which is heterozygous for the sugary–2 allele. U.S. Patent 5954883A. National Starch and Chemical Investment Holding Corporation.

National Research Council. 2004. *Biological Confinement of Genetically Engineered Organisms*. National Academy Press, Washington, DC. ISBN: 0-309-52778-3.

Noller, C.R. 1958. *Textbook of Organic Chemistry*. 2nd Edition. W.B. Saunders Company, Philadelphia, PA. p. 59.

Novozymes. 2013. *Enzymes at Work*. Novozymes A/S, Denmark. http://www.novozymes.com/.

Ofongo, S.T., Kehraus, S., Iyayi, C.A. and Suedekun, K-H. 2008. Rice mill feed: An agro-industrial by-product with potential for rural development. *Conference on International Research on Food Security, Natural Resource Management, and Rural Development*.

Oguntimein, G.B. 2006. *Processing Cassava for Animal Feeds. Food and Agricultural Organization*. http//www.fao.org.

Owuama, C.I. 1999. Brewing beer with sorghum. *Journal of the Institute of Brewing*. 105(1), 23–34.

Parpinelo, G.P., Versari, A., and Riponi, C. 2004. Characterization of sugarbeet (*Beta vulgaris*, L. protein. *Journal of Sugar Beet Research*. 41(1 and 2), 40.

Pienkos, P.T., Laurenes, L., and Aden, A. 2011. Making biofuel from microalgae. *American Scientist*. 99, 474–481. http://www.americanscientist.org.

PLET. 2016. Pure Lignin Environmental Technology. PLET Lignin. Kelowna, B.C. V1Z-ZE6. http://purelignin.com.

Portz, T. (ed.) January 2016. Pellet exports pose no threat to southern forest or price changes. *Biomass Magazine*. 10(1), 20.

Post, W. October 15, 2013. The U.S. Corn to Ethanol Program. The Energy Collective. http://www.theenergycollective.com/willem-post/287061/us-corn-ethanol-program.

Renewable Fuels Association. 2014a. Energy Independence and Security Act. http://www.ethanolrfa.org/2014/12/the-rfs-celebrates-seven-years-of-success/.

Renewable Fuels Association. December 17, 2014b. The RFS Celebrates Seven Years of Success. www.ethanolrfa.org.

Renewable Fuels Association. 2015. *Pocket Guide to Ethanol*. Washington, D.C., http://www.EthanolRFA.org.

Sanchez, A.P. February 2016. Bagasse and blended biomass cogeneration advances in the Cuban sugarcane industry. *Power*. 160(2), 27. http://www.powermag.com.

Scharf, M. and Sethi, A. February 11, 2013. Novel lignases and aldo keto reductases for conversion of lignin-containing materials to fermentable products. US 20150184211 Patent Application Publication. Purdue Research Foundation. West Lafayette, Indiana.

Schill, S.R. March 13, 2014. Yearning for new yeasts. *Ethanol Producer Magazine*.

Schnepf, R. May 1, 2013. *Agricultural-Based Biofuels: Overview and Emerging Issues*. Congressional Research Service. 441282. 7-5700. http://www.crs.gov.

Schnepf, R. and Yucobucci, B.D. March 14, 2013. Renewable Fuel Standard (RFS): Overview and Issues. Congressional Research Service. 7-5000. R40155. http://www.crs.gov.

Schroeder, J.W. 1997. Corn Gluten Feed. North Dakota State University Extension. Report AS1127. http://www.ag.ndsu.edu. 2m-5-97.w-7-10.200-8-12.

Segelken, H.R. 1998. Simple change in cattle diets could cut *E. coli* infection. *Cornell Chronicle*. Cornell University.

Shapley, P. 2012. Cellulose and Lignocellulose. http://buane.chem.uiuc.edu/pshapley/genchem2/bi013.html.

Shewry, P.R. 2009. Darwin review—wheat. *Journal of Experimental Botany*. 60(6), 1537–1553. http://www.jxb.oxfordjournal.org.

Souilah, R., Djabali, D., Belhardi, B., Mokrane, Hl., Boudries, N., and Nadjemi, B. 2014. In vitro starch digestion in sorghum flour from Algerian cultivars. *Food Science and Nutrition*. 2(3), 251–259.

Strychar, R. 2010. *World Oat Production, Trade, and Usage. Oats: Chemical Technology*. 2nd Edition. AgCommodity Research Council. British Columbia, Chapter 1.

Sullins, R.D. and Rooney, L.W. 1975. *Light and Scanning Electron Microscopic Studies of Waxy and NonWaxy Endosperm Sorghum Varieties*. American Association of Cereal Chemists, St. Paul, MN.

Teunissen, P. April 24–25, 2007. Production of ethanol from barley and DDGS containing reduced beta-glucan and phytic acid. *3rd Bioethanol Technology Meeting*, Detmold, Germany. Genencor, A Danisco Division. Leiden, the Netherlands.

UN Development Report. 2015. Maps of the World. http://www.mapsof world.com.

Union of Concerned Scientists. 2012. Turning Agricultural Residues and Manure into Bioenergy. http://www.ucsusa.org.

University of York. September 25, 2014. Plant variants point the way to improved biofuel production. *Ethanol Producer Magazine*.

USDA. July 2006. The Economic Feasibility of Ethanol Produced in the United States.

USDA. May 2015. Sugar: World Markets and Trade. Foreign Agricultural Service.

USDA-NASS. January 12, 2015a. Corn for Grain Yield United States. http://www.usda.nass.gov/charts_and_maps/graphs/cornyld.pdf.

USDA-NASS. March 31, 2015b. Corn Acres United States. http://www.usda.nass.gov.

U.S. Dept. of Energy. Energy Information Administration. September 2015. Amount of Gasoline from a Barrel of Oil. http://www.eia.gov/tools/faqs.cfm?id=327&t=9.

USEPA. June 10, 2015. Final Renewable Fuel Standards for 2014, 2015, and 2016, and the Biomass-Based Diesel Volume for 2017. http://www.epa.gov/renewable-fuel-standard-proposal-for-renewable-fuel-stnadards.

US Grains Council. 2012. Comparison of different grain DDGS sources—Nutrient compositions with animal performance. In *A Guide to Distiller's Dried Grains with Solubles (DDGS)*, 3rd Edition. Chapter 6.

US Grains Council, Corn Refiners Association, World corn production and trade, 2015.

Voegele, E. January 6, 2016. Bill aims for long-term extension of renewable tax credits. *Ethanol Producer Magazine*. http://www.sanders.senate.gov/newsroom/press-releases/sanders-introduces-major-clean-energy-jobs-package.

Weller, C., Steinberg, M.P., and Rodda, E.D. 1984. *Fuel Ethanol from Raw Corn*. Biological Systems Engineering: Papers and Publications. University of Nebraska-Lincoln.

Whitacre, P.T., Rapporteur. 2008. Genetically engineered organisms, wildlife and habitat: A workshop summary. Planning Committee for the Workshop on Research to Improve the Evaluation of the Impact of Genetically Engineered Organisms on Terrestrial and Aquatic Wildlife and Habitat. National Research Council. ISBN: 0-309-12086-1.

White, P.J. and Johnson, L.A. (eds.) 2003. *Corn Chemistry and Technology*. American Association of Cereal Chemists. St. Paul, Minnesota.

Wikipedia. 2013a. *Genetically modified maize*. Wikimedia Foundation, Inc. https://en.wikipedia.org/wiki/geneticallymodifiedmaize. Last modified on August 10, 2015.

Wikipedia. 2013b. *Maize*. Wikimedia Foundation, Inc. https://en.wikipedia.org/wiki/maize. Last modified on July 3, 2015.

Wikipedia. 2013c. *Potato*. Wikimedia Foundation, Inc. https://en.wikipedia.org/wiki/potato. Last modified on June 30, 2015.

Wikipedia. 2012. *Rice*. Wikimedia Foundation, Inc. https://en.wikipedia.org/wiki/rice. Last modified on July 5, 2015.

Wikipedia. 2015a. *Strategic Petroleum Reserve*. Wikimedia Foundation, Inc. https://en.wikipedia.org/wiki/strategicpetroleumreserve. Last modified on September 12, 2016.

Wikipedia. 2015b. *Sugar Cane*. Wikimedia Foundation, Inc. https://en.wikipedia.org/wiki/sugarcane. Last modified on July 14, 2015.

Wikipedia. 2015c. *Cassava*. Wikimedia Foundation, Inc. https://en.wikipedia.org/wiki/cassava. Last modified on July 14, 2015.

Yang, R.D., Grow, D.A., and Goldstein, W.E. 1982. Pilot plant studies of ethanol production from whole ground corn, corn flour and starch. *Fuel Alcohol U.S.A.* 4(2), 13–18.

Zhang, Y.-H.P., Ding, S.-Y., Mielenz, J.R., Cui, J.-B., Elander, R.T., Laser, M., Himmel, M.E., McMillan, J.R., and Lynd, L.R. June 1, 2007. Fractionating recalcitrant lignocellulose at modest reaction conditions. *Biotechnology and Bioengineering*. 97(2), 214–223.

6 Biocatalyst Systems for Ethanol Manufacture

6.1 BIOCATALYST SYSTEMS AND PROCESSES TO ISOLATE AND CONVERT STARCH AND LIGNOCELLULOSE COMPONENTS

Figure 6.1 shows a generalized process schematic for preparing a feedstock for ethanol (with corn as the example). It is desired to complete this conversion while isolating, saving, and treating by-products so that they are most suitable for purposes such as for animal feed and other materials. It is intended that enzymes be used to the maximal extent (those available and those to be developed). This process will have corn as the example since this is the main entity or source of ethanol currently in the United States. It can serve as a basis model for other starch and by-product sources besides corn. The steps and enzymes used (and in a particular order) may differ. However, the principles still apply.

The crop is fed to a separator, which isolates the grain source from the plant chaff (crop plant material other than that which is starch bearing). As indicated, the chaff is sent to a separate system where enzyme systems will convert the chaff to a glucose stream that can be sent to ethanol fermentation and will also provide another by-product stream. The first step in the chaff processing is to apply conditions that involve grinding and sonication in the presence of enzymes in which the environment is analogous to an ionic liquid but where the working fluid is not an ionic liquid per se. Ionic liquids facilitate separation of components such as lignin (Harmsen, P.F.H. et al. 2010; Frölander, A. and Rødsrud, G. 2011; Galletti, A.M.R. et al. 2011). However, the ionic liquid option requires treatment, purification of the ionic liquid, and recycling of the ionic liquid for reuse. This is something that would preferably be avoided for the present since that will involve a cost and step that will inhibit the adoption of this method in many locales. It is necessary to define a solution condition that will achieve the same separations desired, similar to that which is got with an ionic liquid, and allow the enzymes to function optimally for separation and conversion. The "ionic liquid-like environment" will be such as to cause lignin to separate and be isolated. Lignin is not desired in animal feed as it may be unhealthy for the animals and a hindrance. Lignin as a by-product will be processed to something else that is valuable based on research (Lake, M.A. 2010). After the lignin is separated, the lignocellulose stream (sans the lignin) is carried forward for further conversion.

An enzyme mixture ("cocktail") is added (as indicated in Figure 6.1) to the lignocellulose to digest the lignin-less material to soluble carbohydrate in what is assumed to be a single reaction step as shown. The enzyme cocktail will have to be adjusted for each feedstock material considered based on testing. A stream exiting this reactor is sent to the ethanol fermentation area and a second stream (the remaining waste) is sent on as a possible by-product.

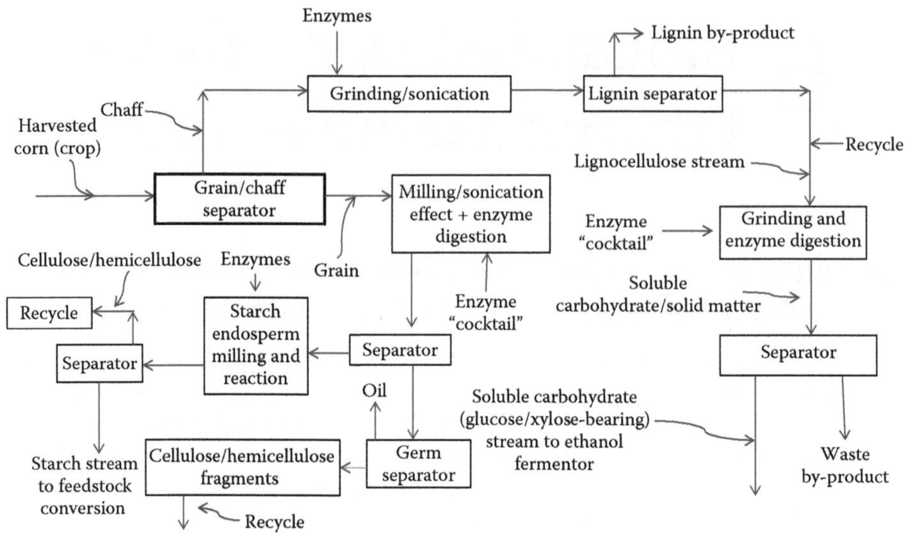

FIGURE 6.1 Corn (or crop) processing for ethanol feedstock.

The second exit stream from the harvester grain/chaff separator consists of the target grain entity containing the starch. In the case of corn, this is the corn kernel. The corn kernel has many components as shown in Figure 5.36. The kernel shell (the pericarp as well as the tip) entraps the starch and other entities. Using sonication and a set of enzymes, the pericarp and cap are taken off, and a combination of enzymes (cellulase, hemicellulase, protease, and pectinase) facilitate digestion while the stream is subjected to agitation in a bioreactor augmented by sonication and milling to facilitate separation.

Since the last step involves beta-glucosidase to convert disaccharides into glucose, this may be separate since the cellobiose disaccharide inhibits the endocellulase reactions (Novozymes 2013). It is possible that further work may alleviate this inhibition. Also, it may be helpful to derive more thermostable enzymes to allow cellulose degradation in all stages to proceed more rapidly and perhaps facilitate reducing reactor size and associated costs.

The endosperm areas containing the starch are freed and the germ component containing the oil is separated. Using hydroclones or a related entity, the germ-containing oil and the cellulose and hemicellulose fragments are separated as shown. This is important as the oil contained in the germ is a valuable by-product. Keeping it intact will allow the use of proven oil-processing extraction schemes. The cellulose and hemicellulose materials are separated from the germ containing the oil facilitated by protease, cellulose, and hemicellulase as also shown.

The oil fraction is isolated as a product and the hemicellulose/cellulose fraction is recycled to the conversion area where the lignocellulose is processed. The oil-laden material may be milled and the corn oil extracted. If a particular material has no oil at this point, this step may be modified or omitted.

The endosperm carrying the starch is treated again with cellulase/hemicellulase and protease so that the starch is more accessible to the starch-degrading enzymes and milling is applied to reduce particle size and facilitate reactions. A little enzyme only is needed to enable this reaction and make the starch accessible. The cellulose/protein fraction treated by enzymes may be recycled for digestion in the lignocellulose area or else kept within the starch stream since a certain amount of protein nitrogen is useful for the yeast fermentation (Martínez-Moreno et al. 2012). The starch stream is sent to starch liquefaction where thermostable amylase liquefies and digests the starch-forming oligosaccharides. These oligosaccharides are converted to mainly glucose, a stream in a reactor that is fed amyloglucosidase.

At this point, there are several streams converging on the ethanol fermentation area. It is possible that all these can be combined and blended for use unless studies show that these streams need to be kept separate.

6.2 BIOCATALYSTS SYSTEMS FOR ISOLATION AND CONVERSION OF STARCH AND CELLULOSE

Enzymes are commercially available to convert starch into oligosaccharides and glucose as feedstock for ethanol fermentations. Enzymes are likely needed to help free starch from where it is contained in the plant source. The array of enzymes commercially available by use, properties, and source of supply is presented in Table 6.1 (Genencor 2011; DuPont 2013; Novozymes 2013). The range of enzymes and their function is huge. If combinations of enzymes that may act synergistically is considered, the number of enzyme formulations expands dramatically.

If an enzyme hydrolyzes a bond in the interior of a polycarbohydrate molecule, it is referred to as an "endo" form. If it hydrolyzes specifically a bond just after the bond at the nonreducing end (or at the reducing end), it is referred to as an "exo" form. Polycarbohydrates in the α-1-4 configuration are found in starch (Corn Refiners Association 2013).

The polycarbohydrates in the α-1-4 configuration form bonds between carbon atoms (through oxygen as the intermediary atom) that are on the same side of the plane of a glucose molecule in the glucose chain comprising starch (see Figure 6.2). Those in the β-1-4 configuration are beta-glucans or beta-celluloses prevalent in lignocelluloses (Noller, C.R. 1958; Shapley, P. 2012). They form bonds between carbon atoms (through oxygen as the intermediary atom) that are on opposite sides of the plane of a glucose molecule in the glucose chain comprising cellulose (see Figure 6.3). This emphasizes that the glycosidic bonds to oxygen differ in the spatial arrangement between alpha and beta configurations.

The terms glucans and celluloses may be used interchangeably. However, beta-glucans are composed of a host of structures and cellulose is composed of a specific structure used in making paper.

As indicated in Table 6.1, and noted elsewhere in this book and in references, enzymes may be used in combination to achieve a task. The Accellerase enzyme of DuPont and Genencor provides just such a combination for degrading lignocellulose. The Avantec, Spirizyme Achieve, and Olexa combination from Novozymes provides another combination for degrading corn mash. The offering from DuPont/Genencor

TABLE 6.1
Enzymes Commercially Available by Function and Properties

Enzyme	pH	Temp	Function	Source	Trade Name
Alpha amylase			Endo glucosidase to degrade α 1–4 bonds in carbohydrate polymers to oligosaccharides		
Acid alpha amylase			Low pH endo glucosidase to degrade α 1–4 bonds in carbohydrate polymers to oligosaccharides	Novozymes	Liquozyme LpH
Amyloglucosidase			α 1–4 exoglycosidase to clip glucose off the end of the polycarbohydrate and endoglycosidase to degrade α 1–4 and α 1–6 bonds in oligosaccharides in the interior of the polycarbohydrate	Novozymes	ExtendA blend
Debranching enzyme			Cleaves α 1–6 bonds as an endoglycosidase to clip glucose off at branch points in the polycarbohydrate	Novozymes	Promozyme
Pectinase			Carbohydrase to degrade bond in pectin linkages		
Alkaline protease			Cleaves hydroxy amide bonds in polypeptides at basic pH		
Neutral protease			Cleaves hydroxy amide bonds in polypeptides at neutral pH		
Acid protease			Cleaves hydroxy amide bonds in polypeptides at acid pH		
Beta glucanase			Cleaves β 1–4 bonds in carbohydrate polymers (covers range of celluloses)		

(Continued)

TABLE 6.1 (*Continued*)
Enzymes Commercially Available by Function and Properties

Enzyme	pH	Temp	Function	Source	Trade Name
			Endo for cleaving within beta glucan (general beta cellulose structure)		
Cellobiohydrolase CBH1			Exo for cleaving at the reducing end of the beta glucan structure	Novozymes	
Cellobiohydrolase CBH2			Exo for cleaving at the nonreducing end of the beta glucan structure	Novozymes	
Beta cellulase			Cleaves β 1–4 bonds in carbohydrate polymers (specific for cellulose based on assay)		
			Covered within the general beta glucanase category unless otherwise specified		
Cellobiase (β-glucosidase)	5–6	50°C	Cleaves β 1–4 bonds in cellobiose (disaccharide)	DuPont Genencor	Accellerase BB
Lignases					
Lignocelluloseenzyme mixture	4.3/5.3	50°C	β Glucanase, cellulase, xylanase proprietary mixture	DuPont Genencor	Accellerase
Corn mash enzyme mixture			α Amylase, cellulose, amyloglucosidase, protease	Novo	Enzyme comb

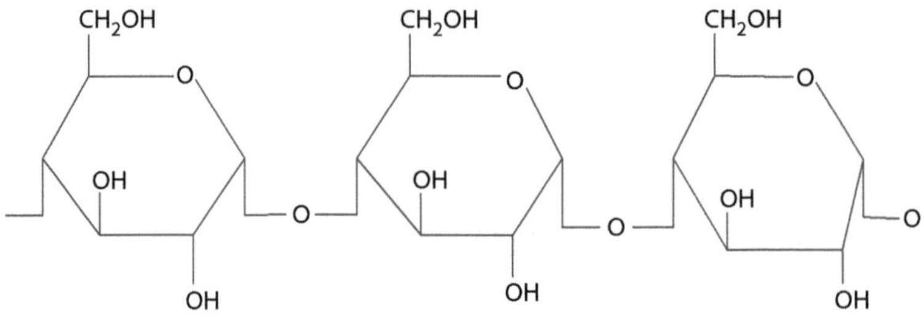

FIGURE 6.2 α 1–4 glycosidic bond linkage—amylose starch (individual hydrogen atoms to complete tetrabond to carbon implicit and not shown).

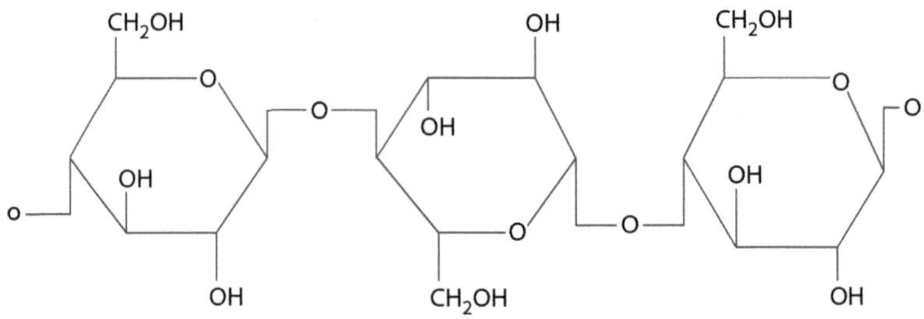

FIGURE 6.3 β 1–4 glycosidic bond linkage—cellulose (individual hydrogen atoms to complete tetrabond to carbon implicit and not shown).

has endo- and exo-glucanase activity as well as xylanase activity. Therefore, it can break up beta-glucans (cellulose if you like) by attacking the interior and exterior portions as well as break up polymers of five-carbon sugars (the hemicelluloses). The glucanase activity is valuable in degrading lignocellulose beta-chains and also perhaps in separating starch from lignocellulose coatings.

It is possible that additional enzymes are needed as indicated in Table 6.1 depending on the crop/carbohydrate source. The example combination from Novozymes is reported to have alpha-amylase, cellulase, amyloglucosidase, and protease activity. Enzyme combinations are selected to achieve a task such as selective and targeted digestion in the best circumstances. The selection should be based on breaking bonds between lignin, hemicellulose, cellulose, and protein, starch, and glucuronic acid structures as pertinent examples. It is best to start with the basic structure that is cleanly taken apart and then individual components would be broken down to be sent to the ethanol fermentation. This has to be validated experimentally since there is not any (as yet) proven theoretical way to achieve such a targeted response.

When enzymes are used together, enhanced activity effects are obtained, for example, in carbohydrate digestion or separating a carbohydrate from something to which it bonded such as a protein or different kind of carbohydrate structure. The opportunities in the area are basically undiscovered but likely very beneficial and will be lucrative if pursued through research and process development.

The term "reducing end" refers to the carbohydrate molecule at the end, where the carbohydrate can "open up" into a hemiacetal (aldehyde) form. The carbohydrate is capable of donating hydrogen (accompanies the electron that is donated) and so it is a reducing agent. The end without the donating hydrogen is the nonreducing end where the aldehyde is not located. As indicated in Table 6.1, enzymes cleave off beta-linked disaccharides (termed cellobiose) from the reducing and nonreducing ends of a cellulose molecule. The disaccharide is subsequently cleaved to glucose using a beta-glucosidase or "cellobiase." DuPont/Genencor term this enzyme as their product Acclerase BB as indicated in Table 6.1.

There is no commercial "lignase" though there is an application pending with the patent system about a mixture to degrade lignin and maybe isolate it (Scharf, M. and Sethi, A. 2013). It is possible that isolating lignin might result in a separate possible by-product or a degraded lignin may also be a valuable by-product.

6.3 BIOCATALYST SYSTEMS TO ACT ON LIGNOCELLULOSE AND ITS COMPONENTS

The enzyme systems to convert corn starch into glucose are already well developed commercially. The ones for converting lignocellulose have been developed but perhaps not fully proven as yet for commercial use (though as noted in Table 6.1, DuPont/Genencor have product offerings). Assuming that lignin can be separated and perhaps used for other coproducts, we are left with degrading and converting the remainder of lignocellulose (cellulose and hemicellulose). If the lignin can be conveniently separated, then it is used as a raw material for another product. If this is a failure (hopefully not), then it is processed to be burned for energy, which might not be very lucrative.

There has been enormous progress in identifying enzymes for cellulose and hemicellulose separation and conversion (Novozymes 2013). Different forms of cellulose have to be broken (crystalline and not so crystalline) and the entities broken cleanly and well to provide glucose for the ethanol fermentation. Novozymes indicates that their cellulase enzymes disrupt amorphous regions thereby opening up more refractory crystalline regions for disruption and digestion.

First, in cellulose conversions, if the enzyme activity results in the splitting of a glucose dimer from an entity that is more insoluble than soluble, then this reaction is termed cellobiohydrolase activity (a really very complex reaction). Second, if the enzyme acts on the interior of the molecule breaking off sections of oligoscaccharides or small oligomers (small polymers of glucose linked sections), then this reaction is an endoglycolytic activity termed beta-glucanase activity. This is on the beta-glucose-linked chain (hydroxyls in a specific direction around a primary or anomeric carbon). Third, if this activity is on a more soluble entity such as a disaccharide such as cellobiose, then this is called a beta-glucosidase activity. This might be best "packaged" as a preferred biocatalyst or a family of catalysts that can rapidly degrade cellulose in all forms. The commercially offered product will undoubtedly be improved. Two major enzyme suppliers are Novozymes and Genencor, and perhaps DuPont as well, depending on their strategic interests. This field requires substantial technology to be successful technically and commercially and this might be a barrier for entry into this field.

6.4 ETHANOL PROCESS SYNTHESIS FOR COST ANALYSIS

The biocatalyst system used is a synergy of glucose and growing and active yeast to form a more optimal reactive environment (Novozymes 2013). The hemicellulose reacts with specific hemicellulases (again a package of biocatalysts) to produce the (pento) five-carbon membered sugars, such as the class of xylose and arabinose. These can possibly be produced directly in the fermentor with the yeast in an optimal reactive arrangement. The modern yeast is developed to convert both the pentoses and hexoses (glucose) to ethanol (Zhang, W. and Geng, A. 2012). This is only useful if the reaction can reach its goal faster and at higher yield than if the conversions of cellulose and hemicellulose are kept in separate supply reactors. Keeping the reactions separate and managing substrates (feedstock) to be optimal for the ethanol fermentation seems quite logical and practical. Some of this may involve deriving new organisms and formulations that allow biocatalytic particles to be formed and used without undue mass transfer penalties.

Currently, genetic engineering of yeast for ethanol or the use of genetically engineered organisms is not approved for commercial use (Schill, S.R. 2015). This is neither good nor bad. Perhaps, we do not know enough about this entity to allow use in such a large public domain with safety. Someday, this will be possible when we know enough. Modern techniques and culture forms are studied in research. The use of genetics for testing is practiced successfully, leading to advances in the field of genetics.

Very efficient enzymatic processes are used to process corn using thermostable enzymes. For ethanol, this involves dry milling of the whole corn since that is the least expensive practice. A better way is to separate the ingredients ahead of time so that they can be used in coproducts and in applications that involve feed and also materials of value. The remaining component of the corn, the starch, can be converted with the dual enzyme process and introduced to the ethanol fermentation. This can be done in one stage if the yeast can grow and produce ethanol under the same conditions used for conversion to feedstock that is provided to make glucose and oligomers of glucose (Kunduru, M.R. 1994; Snoep, J.L. et al. 2009; Rubio-Arroyo, M.F. et al. 2011). Such a one-stage process involves common high temperatures and matching pH conditions. Perhaps, advances will be through brewing research and studies (Bisson, L. 2001; Curry, J. 2009). The conditions and process can be developed to relieve inhibitory reactions and also lead to the production of biomass that is both recycled and is the basis of a higher grade feed product to produce ethanol.

Similarly, for lignocellulose, the lignin will be cleanly separated for optimal use, the hemicellulose will be optimally split and used for xylose and related pentocarbohydrates, which will be converted to ethanol. The beta-cellulose will be split by a dual enzyme process or close to that, with the possibility of the yeast being thermophilic and operating at the same temperature and pH or near to that used in the carbohydrate hydrolysis. This is assuming that provision of glucose when formed in fast reactions will supply the yeast to form ethanol in a noninhibitory manner and also produce a biomass that is again optimal for use in healthy feed and maybe other applications. This situation will allow for best economics that can be managed prior to approval of recombinant organisms.

The association of enzyme with the cellulose biomass (carbohydrate portions) requires the enzyme imbedding in the cellulose or being in proximity. The enzyme reacts with the substrate and the substrate is further degraded and subject to mass transfer limitations. As these are relieved due to breakdown, mixing effectiveness, or temperature elevation causing motion, the diffusive access to the cellulose and the lack of interference in back reactions are facilitated.

The objective is to ensure reaction kinetics, where limitations on maximum reactivity are much alleviated. Furthermore, yeast accessibility to glucose and the relief from ethanol inhibition are also a function of mass transfer since the yeast likely agglomerates or forms small pellets, which are broken up. Maximization of yeast production of ethanol can be best achieved if factors in carbohydrate conversion and ethanol fermentations can be affected at the same time in the same place. This will not occur unless the yeast and enzymes operate at the same pH and temperature. Until then, the reactions to hydrolyze the cellulose and hemicellulose occur in one bioreactor and the yeast fermentation in another bioreactor.

In principle, this methodology is depicted in Figure 6.1. Looking at the basics at the front end of the process, the first step is the clean separation of lignin. It is desired to see a step where lignin as pure as possible so that lignin is a valued added product in a nonfeed area as lignin is not considered to be a feed additive. The cellulose and hemicellulose are processed in bioreactors that are assumed to be continuous with enzyme additions and proper controls for temperature and pH. This is assumed to be a high temperature system employing thermophilic enzymes. A high temperature will promote mass transfer and allow better mixing so that impediments to reaction rate are minimized.

These substances are blobs of cellulose and hemicellulose polymers. Enzymes are suspended in fluids and perhaps associated with a thermal protectant that will protect them from denaturing. The fluid containing enzymes diffuses into the cellulosic and hemicellulosic masses where the enzyme removes cellobiose (beta dimer of glucose) and glucose (using cellulobiohydrolase and beta-glucosidase) and also sections of beta-linked polymers to be subsequently degraded further in this slurry solution (using beta-glucanase). The beta-glucosidase catalyzes the hydrolysis of the glycosidic bonds to terminal nonreducing end of residues in beta-D-glucosides and oligosaccharides, with the release of glucose. The hemicellulase component breaks up the relevant polymer into xylose and related carbohydrates such as arabinose, which are expected also to be converted to ethanol in the fermentor bioreactor. This combined enzyme system is described in principle in the literature and also noted in the form of a commercial enzyme product (Kamp, P. 2012).

The high temperature and optimized low-cost mixing systems ensure that the reaction proceeds optimally fast. This starts off as an unsteady-state diffusion process with reaction that approaches a steady state as the enzyme system diffuses into the mass of mixed cellulose and hemicellulose. It can be assumed for simplicity that parallel reactions to form glucose and xylose and related molecules occur at fixed maximal rate constants because it is expected that with high temperature and optimal mixing the reactive interferences will be reduced. For simplicity, this is assumed to be a continuous process though it can easily be discontinuous involving staggered batch reactor units if necessary. Preferably, parallel units are in place so that a system can be taken down for repair and cleaning when necessary.

6.4.1 Model for Ethanol Process Development and Cost Analysis

A model for the fermentation is built to be used to understand the system and present a suggested approach for cost analysis and cost minimization. The next step in the process (after liquefaction and saccharification) as shown in Figure 6.4 is to present this carbohydrate mixture to the ethanol bioreactor system. Here, a yeast system has the capability to both form ethanol from glucose and some glucose-bearing chains that are small in number and also form ethanol from xylose and related entities. If separate yeast systems are required to manage glucose and xylose supply, they may be in series. Preferably, the yeast can be grown and ethanol produced at higher temperature and enhanced mixing that alleviate much of the product resistance to the reaction.

The bioreactor is assumed to be a continuously stirred tank reactor that allows the recycle of the biomass (Purwadi, P. et al. 2007; Sarks, C. et al. 2014). This allows the building of the biomass to a maximum extent and lower residence time so that the ethanol formed can be taken from the bioreactor at the separator to minimize both the repression on biomass growth and ethanol generation by the yeast biomass. Parallel reactors are used so that one reactor can be taken down and cleaned, without significantly disturbing the production continuity. It is assumed that there is an

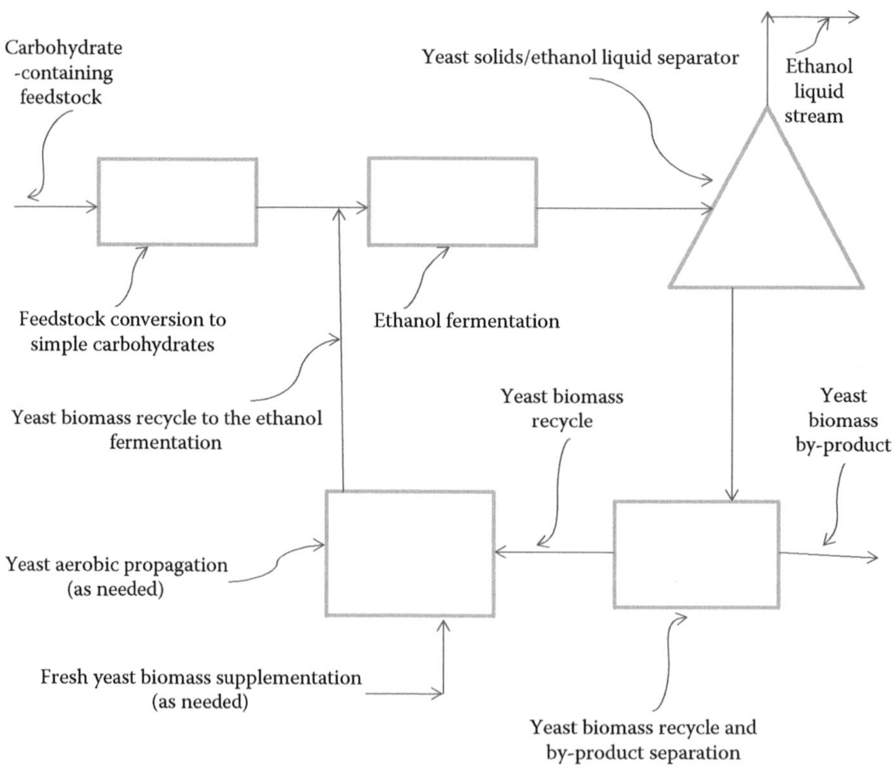

FIGURE 6.4 Ethanol process flow chart.

optimum environment formed that allows the yeast to propagate sufficiently to make up for the biomass losses as well as produce ethanol. If this does not occur, yeast biomass makes up for losses and ethanol formation by the yeast will occur in separate reactors as indicated in Figure 6.4. The process is represented mathematically.

The glucose- and xylose-bearing streams enter yeast biomass flocs that may have an associated yeast-stabilizing entity if that makes sense from an economic standpoint (Goldstein, W.E. 2014). This phenomenon may also be represented by particle forms in other analogous systems such as those used to produce citric acid (Papagianni, M. and Mattey, M. 2006). Yeast is replenished in the bioreactor system through new biomass growth to match the yeast biomass removed and the yeast biomass that expires or becomes insufficient as a biocatalyst. The yeast when needed (for startup and to replenish the system anew) comes from a seed tank system.

The model presented in this chapter assumes that yeast flocs represent diffusion barriers for substrate influx and ethanol exit flux. The yeast flocs are expected to be stable biocatalyst entities to benefit the ethanol production process. They have been studied in the literature and can have positive fermentation and separation qualities (Domingues, L. et al. 2000; Liu, C-G. et al. 2012). The flocs may be protective to the yeast due to association within the floc. The entry of carbohydrate feedstock to the floc is by diffusion. The reaction occurs within the floc to deplete the carbohydrate and form ethanol. The ethanol in turn diffuses from the small flocs. The reaction is assumed to be maximal and not inhibited except for diffusion influences and the inhibitory effect of ethanol on the yeast. The higher concentration of ethanol in the interior of the floc due to diffusion will inhibit the processes of yeast maintenance, growth, and ethanol production. The model can then be used to derive means to reduce ethanol inhibition and improve yeast productivity.

Therefore, the reactant depleted in the biomass produces biomass and supplies the substrate that forms ethanol, and the ethanol exits the yeast biomass flocs. Biomass grows to the extent that biomass is removed from the reaction system. The system is run continuously and biomass is separated from the ethanol fluid in a centrifuge/efficient settler. Some of the biomass is recycled and some exits the system.

The recycled biomass facilitates biomass regrowth in a manner that conserves glucose and xylose to optimally form ethanol with the minimal formation of biomass. Sufficient biomass is withdrawn and replenished to keep the biomass from becoming insufficiently active. Fresh yeast is presented to the system when necessary to shut one of the production lines for cleaning and to start the parallel production line.

Parallel systems are used so that these continuously running systems can be shut down for repair and cleanup. Note the continuous operation can be set up as batch reactors running in a staggered fashion. The biomass steady state and analysis of the ethanol steady system analogously follow that developed in the author's previous work involving the analysis of formation and removal steps in waste treatment for a different objective (Goldstein, W.E. 2014) and is also covered in the literature (Kunduru, M.R. 1994). This development is based on steady-state diffusion and reaction within yeast floc particles and macroscopic mass balances in the fermentation bioreactor system as a whole.

This is suggested to be an optimal design and of course can be adapted as a module for any feedstock anywhere. The principles can be applied to batch reactor systems of various designs. Modular systems can be constructed that can supply components of any size for any locale in the world. This will facilitate ethanol production and permit the introduction of the lowest cost process anywhere for any feedstock.

The mathematical model for the ethanol portion of the process as shown further is taken from the model for formation and elimination of a contaminant in a waste treatment process (see Goldstein, W.E. ed. 2014). This model fits this situation in a sufficiently precise manner.

As shown in Figure 6.5, yeast is assumed to cluster as a spheroid-like floc consisting of many yeast cells. The floc formation is dictated by solution conditions and degree of mixing similar to fungal spheroid structures that can be induced to be formed in other applications as yeast is a form of mold. The yeast cells die and disintegrate and also form spheroid structures that are assumed to achieve a steady size.

Glucose, maltose, and so on, diffuse into the structure to help form the biomass, maintain the biomass, and also form ethanol. As indicated in Figure 6.5, there is a glucose diffusion gradient from the periphery into the particle that depletes as the glucose is used up. As indicated in Figure 6.5, the ethanol formed in the interior of the biomass floc builds up in concentration to diffuse out at the periphery of the biomass floc. There is a buildup of ethanol concentrations in the interior of the yeast biomass floc. After sometime, the increased ethanol concentrations in the interior cause cells to stop growing, to die, and to disintegrate perhaps by native enzymes of the yeast. This can be visualized as a spheroid ring shape of the floc at steady state.

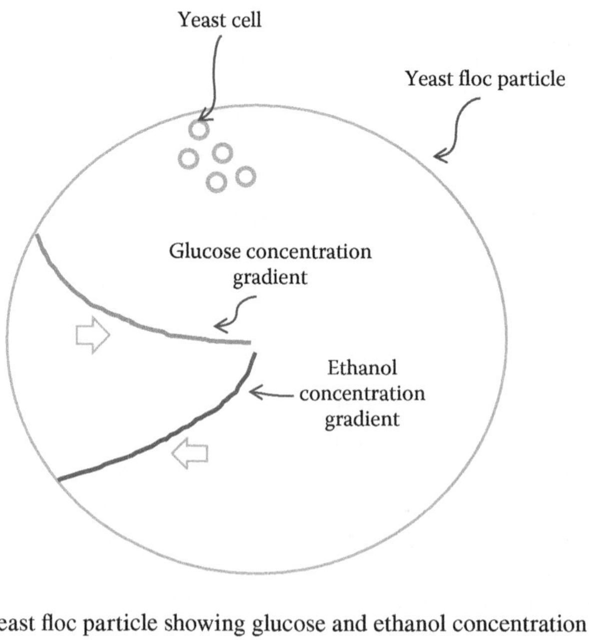

FIGURE 6.5 Yeast floc particle showing glucose and ethanol concentration gradients.

6.4.2 Process Model for Feedstock Supply, Ethanol Formation, and Biomass and Ethanol Separation

The volumetric flow is steady in the system of supply of substrate to sustain the yeast biomass and ethanol production (Figure 6.4). The biomass is inoculated and built up to reach a steady-state level and the feed adjusted so that the yeast biomass will produce ethanol and also replenish itself to compensate for cell death and biomass discharge. The yeast biomass inoculum has an initial cell content that is uniform. This culture may be modified and augmented in ways to make an active biomass better equipped to produce ethanol and also maintain itself. Certainly, the yeast has to adapt to the feedstock to which it is supplied. If the feedstock changes, the adaptation must continue so that the yeast will be modified to adapt and follow the change.

The volumetric stream is made up of constituents in low concentrations compared to the dominant components, which are largely the glucose feedstock. However, this feedstock may be composed of converted starch and converted lignocellulose. The system should also be able to handle a variety of feedstocks from different sources, such as corn, rice, or cassava. There are other possibilities that are already noted previously in Chapter 5. Presumably, the yeast system will be able to accommodate and adjust to such variations.

The size of the bioreactor depends on the expected influent, past, present, and future. It is assumed that the bioreactor system can be changed in a modular way if it is worthwhile to do so. The biomass is developed to have enhanced reaction properties in the following manner.

The biomass propagates in the bioreactor. Some is discharged from a postreactor separator and the liquid containing the product ethanol is discharged so that the biomass and liquid build up to a steady-state level.

The biomass is recycled at a particular ratio to sustain the biomass that has built up. During this process, the process is at a steady state. The material entering is consistently altered or removed in a steady fashion so that the output is the same unless the feedstock changes. The biomass is assumed to be capable of uptake of nutrients and perhaps concentrating them to an extent so that conditions for transport and reaction within the biomass particles are favorable. The biomass is sized to optimize self-bathing with external fluid, with sufficient mixing externally.

It is assumed that the feedstock diffuses into the floc of formed and maintained yeast cells to nurture and sustain the yeast cells and to cause new cells to develop to replace those that no longer function. The ethanol that the yeast cell overproduces (perhaps for protective purposes) forms in the interior and diffuses out just as feedstock diffuses into the yeast floc.

Ethanol damages the yeast cell membranes when it is at the highest concentration in the interior of the floc and lowest at the external extremity of the floc as it is swirled away constituting the product concentration in the ethanol solution. If the concentration of ethanol becomes too high, the cells will cease to function and die excessively by virtue of membrane damage and a drying effect that may cause the yeast cells to wither, become less functional, propagate to a lesser extent, and not produce much ethanol (Dombek, K.M. and Ingram, L.O. 1987).

If the cell can tolerate more ethanol, it may be a more productive cell and part of a more productive yeast cell floc. The floc may be very desirable if it is for brewing beer and is suggested as a concept for the industrial production of ethanol (Prince, I.G. and Barford, J.P. 1982; Carlsen, H.N. et al. 1991; Xu, T.J. et al. 2005). It may be less so if the objective is a mutant strain that overproduces ethanol (which of course is desired). This can set a pattern for the internal model for the yeast floc that helps define ethanol inhibition and toxicity and suggests ways to deal with and relieve that toxicity.

The equation depicting feedstock diffusion into the yeast biomass floc is given as Equation 6.1, and the equation for ethanol flux exiting the floc is given as Equation 6.2.

$$\frac{D_F \partial^2 F}{\partial x^2} = \frac{\partial F}{\partial t} + R_F(x) \tag{6.1}$$

$$\frac{D_E \partial^2 E}{\partial x^2} = \frac{\partial E}{\partial t} + R_E(x) \tag{6.2}$$

where D_F and D_E are the diffusion coefficients for feedstock (F) entering and ethanol (E) exiting the yeast biomass floc, respectively. The ∂ defines a partial derivative, x is position, t is time, $R_F(x)$ is the reaction rate for the disappearance of feedstock (e.g., glucose) in the floc, and $R_E(x)$ is the reaction rate for the appearance of ethanol within the floc that will diffuse out.

The terms are shown as partial derivatives since there is time variation in the system as well as position variation. However, once the system is at steady state, and given that the reactions in the floc are fast relative to the buildup of yeast in the system, then Equations 6.1 and 6.2 become equations at steady state and nonvarying with time and only reflect the change with position x. Conditions within the yeast floc are assumed to adapt and change rapidly compared to changes in the entire bioreactor system, which is external to the flocs.

Therefore, at steady state, Equations 6.1 and 6.2 can be written as Equations 6.3 and 6.4 as follows:

$$\frac{D_F d^2 F}{dx^2} = R_F(x) \tag{6.3}$$

$$\frac{D_E d^2 E}{dx^2} = R_E(x) \tag{6.4}$$

Here d defines an ordinary derivative. Again, note that the feedstock F (e.g., glucose) related concentration proceeds from the exterior to the interior of the floc and is decreasing while the ethanol product (E) proceeds from the central interior of the floc where it is typically at higher concentrations to the extremity of the floc where it exits to form the bioreactor ethanol solution.

As covered in Appendix 6A, assumptions are made and we can solve Equations 6.3 and 6.4, so we see the pattern and dependence of F and E with x in the biomass floc and understand relationships to overall concentrations in the recirculation main bioreactor system. We define the reactions as follows and then apply modifications using assumptions as described. Starting with the below equations

$$\frac{D_F d^2F}{dx^2} = R_F(x) \tag{6.3}$$

$$\frac{D_E d^2E}{dx^2} = R_E(x) \tag{6.4}$$

$R_F(x)$ is defined as

$$R_F(x) = \gamma_{FB} B\, \mu_{FO}\left(1 - \frac{E}{E_{MAX}}\right) + \gamma_{EB} B\, \mu_{EO}\left(1 - \frac{E}{E_{MAX}}\right) \tag{6.5}$$

$$R_E(x) = -\gamma_{EB} B\, \mu_{EO}\left(1 - \frac{E}{E_{MAX}}\right) \tag{6.6}$$

where μ_{FO} is the specific rate of reaction in reciprocal time units to maintain the yeast biomass. The use of μ_{FO} follows from the expression $\mu_{FO}\ F/(K_F + F)$ with F assumed >> than the feedstock inhibition factor (K_F) so that K_F is eliminated and the resulting F in the denominator cancels the F in the numerator. This is assuming that feedstock inhibition is basically nonexistent for all relevant purposes within the yeast biomass floc (Blanch, H.W. and Clark, D.S. 1997; Goldstein, W.E. 2014).

In Equations 6.7 through 6.9, we define the parameters

$$\frac{\gamma_{EB} B \mu_{EO}}{E_{MAX}} = \alpha \tag{6.7}$$

and

$$\gamma_{EB} B \mu_{EO} = \beta \tag{6.8}$$

$$\text{such that } \frac{\beta}{\alpha} = E_{MAX} \tag{6.9}$$

Equations 6.5 through 6.9 show that the reaction rate of the feedstock applies to maintain the biomass and also provide for the conversion of feedstock to ethanol. Equation 6.6 specifically is the rate of formation of ethanol by the yeast biomass. In these equations, μ_{EO} is the specific rate of conversion of the feedstock to ethanol by the biomass. Again, μ_{FO} is the specific rate of reaction in reciprocal time units to maintain and propagate the yeast biomass. Both the yeast biomass maintenance/growth and the yeast biomass ability to produce ethanol are inhibited by the presence of ethanol so this must be considered to be a modification of the specific reaction rates (μ_{FO} and μ_{EO}) applied to the biomass performance capability.

The ethanol inhibition of the conversion of feedstock by the biomass is basically a poisoning of the yeast due to ethanol so that the yeast functions less effectively to repair itself and also to produce ethanol as ethanol concentration increases in the yeast floc.

One of the functions of the yeast is to replace the biomass lost due to death to maintain a steady-state biomass concentration in the fermentation broth as a whole and the biomass that remains. The second function of the yeast that is inhibited as ethanol increases is to continue to produce ethanol effectively. Therefore, in Equations 6.5 and 6.6, the factor $(1 - E/E_{MAX})$ represents this inhibition of the biomass maintenance and that of the capability of the biomass to produce ethanol effectively. The factor E_{MAX} is the ethanol concentration where this inhibition causes the cessation of the yeast biomass' ability to maintain itself and also causes it to cease ethanol production. This ethanol inhibition may be due to effects on yeast transport membranes as part of the yeast's biological structure and biochemical function (Bauer, E.F. and Pretorius, L.S. 2000).

In Equation 6.5, γ_{FB} is the factor for the stoichiometric conversion of feedstock to biomass in units of g/g for maintenance, and B is the concentration of biomass in the system in g/cc. Similarly, in Equation 6.6, γ_{EB} is the factor for stoichiometric conversion of feedstock to ethanol by the biomass.

Equations 6.3 and 6.4 (using Equations 6.5 and 6.6) are solved under the boundary conditions such that at the outside of any floc particle consisting of the yeast-associated aggregate, at position $x = 0$, F is equal to F_O and E is equal to E_O. These are the conditions in the solution of the entire bioreactor (ignoring, for the present, mass transfer restrictions from the broth liquid to the surface of the yeast particle due to inadequate mixing). In the center of the representative yeast floc, it is assumed that the gradients of F and E with distance (dF/dx and dE/dx) are zero due to symmetry of the floc.

Equation 6.4 is solved first since this can be integrated to give an expression for E as a function of x, E(x). This can then be substituted into Equation 6.3 to obtain the expression for F as a function of x, F(x). The details of this are provided in Appendix 6A. As covered in Appendix 6A, Equation 6.10 is obtained.

$$\frac{1 - E/E_{MAX}}{1 - E_O/E_{MAX}} = \exp(x\sqrt{(\alpha/D_E)}) + \frac{\exp(2L\sqrt{(\alpha/D_E)})\exp(-x\sqrt{(\alpha/D_E)})}{1 + \exp(2L\sqrt{(\alpha/D_E)})} \tag{6.10}$$

For convenience, we define a parameter, ε, as collection of constants in Equation 6.3, expressed as Equation 6.11,

$$\varepsilon = \frac{\gamma_{FB} B \mu_{FO} + \gamma_{EB} B \mu_{EO}}{D_F} \tag{6.11}$$

Continuing, the differential equation defining the feedstock F as a function of x is given by substituting into Equation 6.3 to obtain Equation 6.12.

$$\frac{d^2F}{dx^2} = \varepsilon\left(1 - \frac{E}{E_{MAX}}\right) \tag{6.12}$$

Substituting for $(1 - E/E_{MAX})$ into Equation 6.12 using Equation 6.11 results in a new expression for Equations 6.6 and 6.12, since we now have an expression for E in terms of position x. Therefore, Equation 6.12 is transformed to Equation 6.13.

$$\frac{d^2F}{dx^2} = \varepsilon\left(1 - \frac{E_O}{E_{MAX}}\right) * \exp(x\sqrt{(\alpha/D_E)}) + \frac{\exp(2L\sqrt{(\alpha/D_E)})\exp(-x\sqrt{(\alpha/D_E)})}{1+\exp(2L\sqrt{(\alpha/D_E)})} \tag{6.13}$$

where * indicates the multiplication sign. For convenience again, let the constant parameters Z_1 and Z_2 be defined as follows:

$$Z_1 = \frac{\varepsilon(1-(E_O/E_{MAX}))}{1+\exp(2L\sqrt{(\alpha/D_E)})} \tag{6.14}$$

$$Z_2 = \frac{\varepsilon(1-(E_O/E_{MAX}))[\exp(2L\sqrt{(\alpha/D_E)})]}{1+\exp(2L\sqrt{(\alpha/D_E)})} \tag{6.15}$$

Substituting Equations 6.14 and 6.15 into Equation 6.12 results in

$$\frac{d^2F}{dx^2} = Z_1\exp\left(x\sqrt{\frac{\alpha}{D_E}}\right) + Z_2\exp\left(-x\sqrt{\frac{\alpha}{D_E}}\right) \tag{6.16}$$

Integrating Equation 6.16 and solving using the boundary conditions of values at x− = 0 and the no flux condition of dF/dx = 0 at x = L deemed to be the center of the yeast bloc, as covered in Appendix 6A, we obtain an expression for the feedstock F as a function of position x in the floc

$$F - F_O = \frac{Z_1}{(\alpha/D_E)}[\exp(x\sqrt{(\alpha/D_E)}-1)] + \frac{Z_2}{(\alpha/D_E)}[\exp(-x\sqrt{(\alpha/D_E)}-1)] + x\left[\frac{-Z_1}{\sqrt{(\alpha/D_E)}}[\exp(L\sqrt{(\alpha/D_E)})]\right] + \frac{Z_2}{\sqrt{(\alpha/D_E)}}[\exp(-L\sqrt{(\alpha/D_E)})] \tag{6.17}$$

Looking at the change in feedstock value from x = 0 to x = L, we obtain

$$F_L - F_O = \frac{Z_1}{(\alpha/D_E)}[\exp(L\sqrt{(\alpha/D_E)}-1)] + \frac{Z_2}{(\alpha/D_E)}[\exp(-L\sqrt{(\alpha/D_E)}-1)] + L\left[\frac{-Z_1}{\sqrt{(\alpha/D_E)}}[\exp(L\sqrt{(\alpha/D_E)})]\right] + \frac{Z_2}{\sqrt{(\alpha/D_E)}}[\exp(-L\sqrt{(\alpha/D_E)})] \tag{6.18}$$

Similarly, for the gradient in ethanol from x = 0 to x = L, from before,

$$\frac{1-E/E_{MAX}}{1-E_O/E_{MAX}}=\exp(x\sqrt{(\alpha/D_E)})+\frac{\exp(2L\sqrt{(\alpha/D_E)})\exp(-x\sqrt{(\alpha/D_E)})}{1+\exp(2L\sqrt{(\alpha/D_E)})} \tag{6.10}$$

Expressing this as the gradient in ethanol from any point x to the surface of the yeast biomass floc at x = 0,

$$E-E_O=(E_{MAX}-E_O)*\left[1-\left\{\frac{\exp(x\sqrt{(\alpha/D_E)})+\exp(2L\sqrt{(\alpha/D_E)})\exp(-x\sqrt{(\alpha/D_E)})}{1+\exp(2L\sqrt{(\alpha/D_E)})}\right\}\right] \tag{6.19}$$

Then at x = L

$$E_L-E_O=(E_{MAX}-E_O)*\left[1-\left\{\frac{\exp(L\sqrt{(\alpha/D_E)})+\exp(2L\sqrt{(\alpha/D_E)})\exp(-L\sqrt{(\alpha/D_E)})}{1+\exp(2L\sqrt{(\alpha/D_E)})}\right\}\right] \tag{6.20}$$

This then gives the change in ethanol from the center of the yeast floc to the surface.

The fluxes at the outer surface of the yeast floc particle (ethanol out and feedstock in) are relevant for coupling to the main bioreactor equations. Following Appendix 6A flux calculations, these are given as follows:

For the ethanol flux at the outer surface of each floc particle that is entering the main solution, the equation is as follows:

$$D_E\frac{dE}{dx_{@x=0}}=\frac{D_E((E_{MAX}-E_o)\sqrt{(\alpha/D_E)})[-1+\exp(2L\sqrt{(\alpha/D_E)})]}{1+\exp(2L\sqrt{(\alpha/D_E)})} \tag{6.21}$$

For the feedstock flux entering the floc particle at the outer surface, the equation is as follows:

$$D_F\frac{dF}{dx_{@x=0}}=D_F\left(\frac{Z_1}{\sqrt{(\alpha/D_E)}}\right)\left(1-\exp\left(L\sqrt{\frac{\alpha}{D_E}}\right)\right)+D_F\left(\frac{Z_2}{\sqrt{(\alpha/D_E)}}\right)\left(-1+\exp\left(-L\sqrt{\frac{\alpha}{D_E}}\right)\right) \tag{6.22}$$

With α, Z_1, and Z_2 defined as previously noted in Equations 6.7, 6.14, and 6.15, respectively.

6.4.3 Calculation Example for Transport in a Yeast Biomass Floc

Starting with Equation 6.7, and ending with Equation 6.22, the following provides a useful example to examine factors important to the productivity of the yeast floc particle. This is input to the steady-state balance to be developed for the bioreactor based on macroscopic flows.

From the literature, the ethanol concentration that can be inhibitory can start being significant perhaps at 50 g/L or 0.05 g/cc, so this will be set equal to E_{MAX} in the equations. This is, of course, strain dependent. However, this is a reasonable figure for this example. Feedstock concentration might be between 100 and 200 g/L, or 0.1–0.2 g/cc, so this is set equal to F_0.

Once the ethanol conversion is up and running at steady state, and the biomass is stabilized at a concentration in the floc, this may be as an example at 20 g/L or 0.02 g/cc, so this value is set for B_0. The stoichiometric conversion of feedstock to ethanol by the biomass is about 0.5 g/g ethanol of feedstock so $\gamma_{EB} = 0.5$ is assumed. Also, the yield of biomass from feedstock is also assumed to be 0.5, so $\gamma_{FB} = 0.5$.

The mobility of ethanol and feedstock within the biomass floc will vary according to the tightness of the floc, pore size, and so on. However, assuming rather quick motion within the pores, the effective diffusion coefficients within the pores of a yeast floc particle might be about 10 times slower than if diffusing unhindered in water so D_E and D_F may be equal to 10^{-6} cm²/s for this example. If facilitated to an extent, then 10^{-5} is more appropriate as an estimate.

Yeast may double about once every 0.5 hours, a doubling time once every 1800 s. In the simple biomass growth equation, biomass growth is exponential or $B/B_0 = \exp k\theta_{2x}$ where θ_{2x} is the doubling time. Since B/B_0 is 2 as the biomass is doubled in the time period θ_{2x} (1800 s) then k is computed to be 3.85×10^{-4} s^{-1} ($k = \log_e B/B_0/\theta_{2x}$), so for this calculation, it is assumed that the rate constant for ethanol formation μ_{0E} is a little higher, 4.5×10^{-4} s^{-1}. It is assumed that the reaction rate for maintenance of the biomass and replacement of degraded biomass is 5.5×10^{-4} s^{-1}. The distance from the periphery of the biomass floc particle to its center is assumed to be 0.1 cm or 1 mm, which will of course contain a very large number of yeast cells. The rate figures may be higher if reaction rates are faster. The floc size can also vary according to what is observed for the yeast strain in use.

In this calculation, we are interested in the ethanol concentration that results in the interior of the floc as well as the feedstock concentration in the interior of the floc particle, since ethanol is inhibiting both biomass maintenance and ethanol formation.

Substituting into Equation 6.20, for this example, we determine that $(E_L - E_0)$, the ethanol concentration in the interior of the particle less the ethanol concentration at the surface of the floc particle is

$$E_L - E_O = (E_{MAX} - E_O) * \left[1 - \left\{\frac{\exp(L\sqrt{(\alpha/D_E)}) + \exp(2L\sqrt{(\alpha/D_E)})\exp(-L\sqrt{(\alpha/D_E)})}{1 + \exp(2L\sqrt{(\alpha/D_E)})}\right\}\right]$$

or 0.013 to 0.04 g/cc. The higher number represents a greater assumed available area for the floc particle. In this example, E_0 is 0.01 g/cc so E_{MAX} is 0.05 g/cc as given before.

Continuing, the figures for change in feedstock in the floc particle are computed using Equation 6.18. Therefore,

$$F_L - F_O = \frac{Z_1}{(\alpha/D_E)}[\exp(L\sqrt{(\alpha/D_E)}-1)] + \frac{Z_2}{(\alpha/D_E)}[(\exp(-L\sqrt{(\alpha/D_E)}-1)]$$
$$+L\left[\frac{-Z_1}{\sqrt{(\alpha/D_E)}}[\exp(L\sqrt{(\alpha/D_E)})]\right] + \frac{Z_2}{\sqrt{(\alpha/D_E)}}[\exp(-L\sqrt{(\alpha/D_E)})] \quad (6.18)$$

or $F_L - F_O$ is computed to be −0.029 to −0.0889 g/cc, again depending on the assumption in available catalytic area. The gradient across the floc particle is of course negative since F_L is less than F_O.

The fluxes for ethanol and feedstock at the outer boundary of the floc in this example are computed to be 2.81×10^{-7} to 1.2×10^{-4} g/cm²-s (latter at elevated accessible area near the surface of the floc particle) for ethanol, and -6.23×10^{-7} to -2.67×10^{-4} g/cm²-s (latter number for higher catalytic area and other assumptions in regard to performance) for feedstock entering the floc particle. These are examples that will of course change depending on the main bioreactor system balances. The choice of the specific rates is set in order that the feedstock flux equals the sum of the ethanol flux, the carbon dioxide flux (which is about equal to the ethanol flux by idealized stoichiometry) and about 10% allotted of the diffusive feedstock flux set to maintain the biomass with some regeneration for degradation and loss. This is an estimate that will require experimental confirmation but is certainly reasonable for the purposes of this example and in general.

6.4.4 Process Modeling in the Main Bioreactor System

Consider the unsteady-state equation where yeast biomass (X, wt/vol.) is building up and developing in circulation in the bioreactor system depicted in Figure 6.4. The bioreactor portion system consists of a multitude of yeast biomass floc particles. All of these particles are assumed to be identical for the purposes of simplicity in the analysis.

Showing the mass balance for biomass formation in unsteady state and defining terms, we express the derivative of yeast biomass concentration with time, depicting its buildup with time in terms of parameters relevant to Figure 6.4. This begins with a mass balance around the ethanol bioreactor where yeast biomass growth is occurring. In this case, we assume that flows are started and reach a steady state.

The bioreactor is inoculated with a starting organism culture concentration (X_0) derived from the seed tank shown in Figure 6.4. This starting concentration will propagate (expand) in the system with time. We would like to know, for example, how long this will take, since we want to reach a condition where yeast biomass is steady within the bioreactor so that what grows with time is equal to that which exits with time plus that which is destroyed or is no longer functional (following the diagram, Figure 6.4, to build biomass in the system, starting with an inoculum from a seed tank).

Therefore, we write the following mass balance around the ethanol bioreactor system (not including the liquid solid separation) in order to build yeast biomass:

$$\frac{VdX}{dt} = rF_DX_R + \mu XV - k_dXV - F_D(1+r)X \qquad (6.23)$$

This equation follows the format given in the literature (Blanch, H.W. and Clark, D.S. 1997; Goldstein, W.E. 2014). In this equation, the rate of change of biomass in the bioreactor (VdX/dt) equals that which is recycled to the bioreactor from the biomass separator (rF_DX_R), plus that which is formed in the bioreactor by propagative reaction (μXV), less the rate of biomass that dies in the bioreactor ($-k_d$ XV), and less the rate of biomass that exits the bioreactor ($-F_D(1 + r)X$).

In this equation, V is the volume of the bioreactor where biomass growth and reactions take place, for example, where consumption of nutrients, such as carbohydrates, occurs; X is the biomass content in weight/unit volume (allowed to be different from the biomass concentration in the yeast floc, B); t is time; and d represents the derivative operator from calculus so that V dX/dt expresses the rate of biomass (e.g., kilograms) forming with time, where r represents the fraction of the liquid stream containing yeast solids that is recycled to the entry point to bioreactor and not discharged from the system, F_D represents the liquid volumetric flow into the system, X_R represents the biomass concentration in the stream that is recycled to the bioreactor entry, μ represents the specific growth rate of the biomass in reciprocal time units, and k_d represents the specific death rate (or rate of nonfunctioning) of the yeast biomass in reciprocal time units.

Biomass growth is autocatalytic, meaning its formation is dependent on the biomass already present. Therefore, the recycling of a portion (r) of the incoming stream back increases the rate of formation of the biomass. This is useful since faster rates of reaction reduce the volume of the bioreactor to lessen costs that result with smaller bioreactor size.

It is worth noting that solutions to such equations result using the principles of calculus, discovered by Sir Isaac Newton and Gottfried Wilhelm von Leibniz in the seventeenth century (von Leibniz, G.W. 1684; Newton, I. 1687; Zegarelli, M. 2008). This area of mathematics significantly changed civilization as we know it. In order to do this, some assumptions and substitutions are in order.

A mass balance around the solids settler/centrifuge can be expressed as follows:

$$F_D(1+r)X = (1-w)F_DX_E + (r+w)\,F_DX_R \qquad (6.24)$$

where F_D (1 + r) is the volumetric flow of liquid exiting that bioreactor that contains solids and feeds the settler/centrifuge. Multiplying this term by X (the solids content in the liquid stream in weight per unit volume) provides the rate of the weight of solids entering the settler.

The term w is the fraction of inlet volumetric flow F_D (volume/time) that is discharged from the system in the yeast biomass exit stream after biomass settling and liquid ethanol/yeast biomass separation. Therefore, 1 – w is the liquid fraction of inlet volumetric flow F that is discharged from the yeast biomass settling and separation in the overhead ethanol product discharge stream.

The term $(1 - w)\ F_D X_E$ is the fraction of suspended yeast biomass solids exiting the settler in the liquid exit stream. If the settler is perfect, then there are no suspended solids in this stream so that X_E is zero.

The term wF_D is the volume of ethanol-containing liquid exiting in solids discharge from the settler.

The term $wF_D X_R$ is the rate of yeast biomass solids exiting the system from the settler.

Since X_E is taken to be zero, then the mass balance can be solved for X_R such that

$$X_R = \frac{(1+r)X}{(r+w)} \tag{6.25}$$

Then, substituting into Equation 6.23, replacing X_R, and dividing by V, letting the k_d term be incorporated in μ, we arrive at

$$\frac{dX}{dt} = -\left(\frac{w}{(r+w)}\right)\left(\frac{F_D}{V}\right)(1+r)\ X + \mu X \tag{6.26}$$

Furthermore, we express the specific growth rate (μ, units of reciprocal time) of the biomass using the Monod expression strictly (at this point) in terms of the biomass controlling substrate S (Henze, M. et al. 2000). For present purposes, we let

$$\mu = \frac{\mu_{max} S}{K_s + S} \tag{6.27}$$

where μ_{max} is the maximum specific growth rate of the biomass, S is the critically controlling substrate (e.g., glucose concentration) for the biomass, and K_s is the Monod constant for biomass growth (Blanch, H.W. and Clark, D.S. 1997, p. 193).

In reality, the specific biomass growth rate is a function of potentially many inhibitors and activators, which can all be expressed in terms of Monod type functions (Henze, M. et al. 2000). This includes inhibition by ethanol, which is presumed to reduce the yeast growth rate and ability to produce ethanol depending on the concentration of ethanol in the biomass floc particle interstices in contact with the yeast cells. However, if biomass growth proceeds aerobically (as it would starting from a seed tank when the bioreactor yeast is replaced following cleaning) with little or no formation of ethanol, then ethanol will not be present to inhibit the biomass formation.

This simplified representation is adequate for present purposes though we still have to account for the effect of the ethanol as a form of product inhibition as noted previously when ethanol is being formed by yeast biomass floc particles at steady state. We will look at the carbohydrate-related substrate first and ignore the ethanol inhibition for the time being. Note from before that assuming the biomass floc model proposed herein, the concentration of ethanol in the interstices of the floc model should be higher and, therefore, more inhibitory to the stability and maintenance of the yeast and its ability to produce the ethanol product.

Therefore, continuing with examining a way to estimating the time to grow the yeast, we replace μ by the expression in Equation 6.27, and then obtain

$$\frac{dX}{dt} = -\left(\frac{w}{(r+w)}\right)\left(\frac{F}{V}\right)(1+r)\,X + \left(\frac{\mu_{max}S}{(K_s+S)}\right)X \tag{6.28}$$

Inspection indicates that solving Equation 6.28 to obtain an exact and not implicit analytical solution for X is not possible since S is a function of time and affects X in a nonlinear manner. However, it can be assumed in this case that the Monod constant (K_s) is likely small relative to S, except for very small values of S (Blanch, H.W. and Clark, D.S. 1997, p. 193). Assuming that is the case, the expression can be rewritten since S cancels out.

$$\frac{dX}{dt} = -\left(\frac{w}{(r+w)}\right)\left(\frac{F_D}{V}\right)(1+r)\,X + \mu_{max}X \tag{6.29}$$

The equation is solved through means of calculus to examine its approach to steady state and the steady-state value, predicated on reaching a limiting lower and final substrate (S) value.

$$\frac{X}{X_0} = \exp\left[-\left(\frac{w}{(r+w)}\right)\left(\frac{F_D}{V}\right)(1+r) + \mu_{max}\right]t \tag{6.30}$$

In this expression, X represents an estimate of the yeast biomass concentration term in an unsteady-state condition as it increases from an initial inoculum from the yeast seed tank. From this equation, as long as μ_{max} (the specific growth rate of the biomass) exceeds the net loss of biomass from the system, the biomass will grow (from its initial inoculated value of X_0).

This equation is only valid of course as long as S (the substrate remaining) sufficiently exceeds K_s and there is no inhibition by ethanol or anything else during the initial yeast propagation. This assumption may be valid if the solution for yeast growth does not potentiate the growth of ethanol and that the substrate solution and conditions (e.g., aerobic rather than anaerobic) prevent the formation of ethanol until the system is ready to produce ethanol. Note that the growth of yeast during ethanol formation (under mostly anaerobic conditions) must occur since yeast must grow at the rate equal to the rate of yeast discharge and loss from the system.

When the yeast is ready to start ethanol production, the conditions change to a perfect environment for ethanol formation and maintain the yeast biomass. There will likely be an optimum oxygen environment.

The switchover from growth to production of ethanol must be done in a way so that the organism acclimates to the change in the least stressful manner. This adjustment period might be considered to be an extension of the yeast growth to near steady state using inoculum. Otherwise, the biomass will be mathematically predicted to grow indefinitely which is, of course, impossible. Actually, what happens in this case, is that S will become negative if Equation 6.22 is allowed to proceed too far which is, of course, also impossible. What is done here, as a practical expedient, is to set a lower limit on S in order to allow us to estimate the time to reach a steady-state condition.

There is also an equation for substrate depletion that must be considered since that will allow the estimation of the lowest substrate level, with the assumption made. Expressing the differential equation for depletion of substrate as a function of time, that is, the rate of depletion of substrate in the bioreactor equals the net of substrate fed to the system $(F_D/V)\ S_F$ less that leaving the system $(F_D/V)S$ plus that depleted in the system by use by the growing biomass $(1/Y)\ \mu$, and that depleted through formation of ethanol, or

$$\frac{dS}{dt} = \left(\frac{F_D}{V}\right)(S_F - S) - \left(\frac{1}{Y_{SX}}\right)\mu_{SX}X - \left(\frac{1}{Y_{SE}}\right)\mu_{SE}E \qquad (6.31)$$

Here S_F is the substrate concentration in the feed stream (which, by the way, does not contain any active biomass by the assumptions of this model).

In this equation, Y_{SX} is the stoichiometric coefficient expressed as the weight yield of biomass per unit of substrate, so $1/Y_{SX}$ is the weight of substrate used up per unit of biomass formed. Similarly, Y_{SE} is the stoichiometric coefficient expressed as weight yield of ethanol per unit of substrate, so $1/Y_{SE}$ is the weight of substrate used up per unit of ethanol formed in the bioreactor.

From before, for the case where S is sufficiently greater than K_S, and for conditions where ethanol is not yet allowed to be produced,

$$\frac{X}{X_0} = \exp\left[-\left(\frac{w}{(r+w)}\right)\left(\frac{F}{V}\right)(1+r) + \mu_{MAX}\right]t \qquad (6.30)$$

Substituting this equation into the previous equation for dS/dt, replacing X, and for the situation before ethanol is being formed,

$$\frac{dS}{dt} = \left(\frac{F}{V}\right)(S_F - S) + \left(\frac{1}{Y}\right)\mu_{MAX}\, X_0 \exp\left[-\left(\frac{w}{(r+w)}\right)\left(\frac{F}{V}\right)(1+r) + \mu_{MAX}\right]t \qquad (6.32)$$

So, now we have an equation expressing dS/dt as a function of S and t, with the limitation that K_s must be sufficiently less than S so that it can be ignored. In this case, this equation is for the period prior to ethanol production as noted.

With the term for ethanol production also drawing on the use of the feedstock, covering the period after when the biomass has been built up in the system prior to ethanol production, we again revert to Equation 6.31 as the steady solution for biomass, feedstock, and ethanol.

$$\frac{dS}{dt} = \left(\frac{F}{V}\right)(S_F - S) + \left(\frac{1}{Y}\right)\mu_{SX}X - \left(\frac{1}{Y_{SE}}\right)\mu_{SE}E \qquad (6.31)$$

Since the biomass has already been built up on the system before ethanol production is started, then it has to adjust to ethanol production as a next step. In this case, the biomass concentration is governed by a system mass balance with biomass present and feedstock coming in producing ethanol formed in the biomass flocs. For this condition, the equation for unsteady-state biomass growth in the system to replenish that lost is from before

$$\frac{dX}{dt} = -\left(\frac{w}{(r+w)}\right)\left(\frac{F}{V}\right)(1+r)X + \mu_{MAX}X \tag{6.29}$$

Note that biomass flocs that produce ethanol are subject to conditions where ethanol produced in the interior of the biomass flocs is implicitly higher than that in the external solution. Given this, yeast will be inhibited and perhaps destroyed in part. As yeast is being destroyed in the floc, it is also being maintained as discussed previously.

In reality, the breakdown products will also exit the flocs from the pores in the floc surrounding the yeast particles. The breakdown products represent yeast death and/or inactivation. The yeast maintenance rate will compensate for this loss.

If the flux of this inactive or dead yeast could be measured, it might be possible to find ways to optimize yeast activity and survival to maximize culture activity. This does mean that yeast activity and productive capability in the floc will likely reach an optimal or maximal value at some position in the floc; so this tells us that this floc size should be optimized. The net yeast maintenance in the floc must vary with position within the floc particle given this is correct. This point may be more important when research is conducted to study this subject.

At this point, a balance on ethanol in the system is required. Ethanol is presumed not to enter the overall system in the feedstock. It does recirculate in the system as part of the liquid and biomass in recycle. It does exit as part of the liquid stream at the separator as the ethanol product stream. Some ethanol also exits in the biomass this is discharged from the system as part of the process. Some of the ethanol is recycled with the biomass unless efforts are made to reduce or eliminate this.

Note we are allowing for the ethanol concentration in the biomass to potentially be different from ethanol in solution which is E. The ethanol concentration at the exterior of the biomass may be E_0, potentially different from E due to change in concentration going from liquid phase externally to the interior of the biomass floc as noted in equations solved for the floc particle containing yeast cells. Furthermore, the concentration of feedstock (e.g., glucose) in the biomass is F while that in the exterior solution is S since, perhaps the feedstock could be more soluble (reach a higher concentration) in the biomass phase than in the exterior solution. Similarly, the biomass may be considered to be more concentrated in the floc at concentration B and more dilute per unit of exterior solution, which is X. This concentration effect may bear on biocatalyst properties and perhaps may be accentuated to increase the catalytic "power" of the biomass. Perhaps the association of yeast cells in the floc creates a synergistic effect to improve product output. This happens in brewing so perhaps it will occur in production geared to ethanol production. This could occur by isolation of yeast species with more attractive properties (Bauer, E.F. and Pretorius, L.S. 2000).

Solving Equation 6.32 (without the ethanol term) to express S as a function of t,

$$S = S_F + \left\{ \frac{[(1/Y)\,\mu_{max} X_0]}{[-w/(r+w)]} \left[\frac{F_D}{V}(1+r) + \mu_{max} + \frac{F_D}{V} \right] \right\}$$
$$* \left\{ -\exp\left[\frac{-w}{(r+w)} \frac{F_D}{V}(1+r) + \mu_{max} \right] t + \exp\left[-\left(\frac{F_D}{V} \right) t \right] \right\} \tag{6.33}$$

(* = multiplication)

The substrate (glucose) in the feed is taken to be 200 g/L. The substrate leaving the bioreactor is 1 g/L (Vieira, E.D. et al. 2013). For now, we assume K_s simply can be taken to be much smaller than the substrate concentration S so that it can be ignored for solving the unsteady-state equation estimating the time to steady state.

This is assuming the K_s for oxygen given in the reference is proportional to the K_s for substrate as if it were glucose. The ratio of S to K_s is approximately 40:1.

The inoculum concentration (X_0) is arbitrarily taken to be 1 mg/L or 10^{-6} g/cc.

The volumetric flow rate (F_D) is 3.6 mgd (million gallons per day) or $13.63{*}10^6$ L/day or $1.58{*}10^5$ cc/s.

The volume of the reactors (V) is 7,200,000 gallons.

The ratio of F_D to V (F_D/V) is 0.5 days^{-1}.

Letting Y = 0.5, so 1/Y is 2, then 2 g of substrate are used up per gram of biomass formed.

Let w = 0.015, meaning about 1.5% of the liquid feeding the settler exits in the biomass exit. Let the recycle ratio be 0.2 as an example only.

Therefore, w/(r + w) is 0.07, meaning that the fluid exiting with the yeast biomass exit stream is 7% of the total of fluid exiting with the biomass stream plus that recycled.

The literature indicates that the biomass may double in 0.2 days or every 4.8 hours (Prince, I.G. and Barford, J.P. 1982; Vieira, E.D. et al. 2013). Therefore, let $\mu_{max} = 5\ \text{day}^{-1}$

Inserting numbers, it can be shown that the substrate will go well below 1 g/L in 6–7 days, so the closest approach to steady state is somewhat beyond 6–7 days.

Thus, at dS/dt = 0, in Equation 6.30,

$$0 = \left(\frac{F_D}{V} \right)(S_F - S) - \left(\frac{1}{Y} \right) \mu X \tag{6.31}$$

Solving for X, the biomass concentration at steady state,

$$X = \frac{F}{V} \frac{(S_F - S)}{(1/Y)\mu} \tag{6.32}$$

Putting the numbers in the equation and assuming μ is approximately μ_{max}, meaning K_s is zero, and S is very small, X is calculated to be approximately 40 g/L, rising from 1 mg/L arbitrarily set at 1 mg/L initially. This is in range of expected biomass concentration but probably a little high. Putting K_s back in the equation as nonzero, we can calculate the steady-state concentration of substrate, S, from the steady-state equation where the change in biomass is zero with time.

Therefore, using Equation 6.25 with ethanol not produced as yet

$$0 = -\left(\frac{w}{(r+w)}\right)\left(\frac{F}{V}\right)(1+r)X + \left(\frac{\mu_{max}S}{(K_s+S)}\right)X \tag{6.34}$$

The biomass cancels out of this equation at steady state. Therefore, solving for the substrate concentration at steady state, we obtain,

$$S_{ss} = -\left(\frac{w}{(r+w)}\right)\left(\frac{F}{V}\right)(1+r)\ \frac{K_s}{[\mu_{max} - (w/(r+w))(F/V)(1+r)]} \tag{6.35}$$

Inserting the numbers for this example,

$$S_{ss} = 0.034\,K_s \tag{6.36}$$

If K_s is zero, then the final substrate concentration is zero. In practice, this is not attained. If K_s is approximately 20, then the final substrate concentration is approximately 0.7 in this example.

6.4.5 Ethanol Formation Balance

As noted, Figure 6.4 presents a block diagram for the processing of feedstock to ethanol in the fermentation facility. The ethanol will exit the settler as indicated. Some remaining ethanol may also recirculate as part of the liquid stream unless it preferentially separated by a membrane or other process such as pressing it out of the biomass. Some ethanol and feedstock will also exit with the biomass solids that are discharged. Assume the ethanol is removed in the main product stream plus a small residual in the biomass and is not part of the biomass recycle returning to the fermentor.

Given these assumptions, a mass balance for the ethanol across the fermentation system is given by the following:

$$\frac{VdE}{dt} = F_D E_F + rF_D E_R - F_D(1-w)E - wF_D E_W + A_{floc}(\text{Ethanol Flux})\ n_p V \tag{6.37}$$

In Equation 6.37, the volume V of the ethanol production fermentation unit, multiplied by the change in the ethanol concentration with time in the fermentor operating as a continuous stirred tank reactor, dE/dt, is equal to: the volumetric flow entering the system per unit time times the concentration of the ethanol (zero unless there a several fermentation stages) entering ($F_D E_F$), plus the ethanol in recycle rF_D E_R, which is also assumed to be zero assuming that the ethanol exits in the product stream and a little in the biomass discharged and the ethanol is extracted or washed out to avoid it being recycled. It is inhibitory, so eliminating may be desirable.

The ethanol in the solid yeast discharge stream entrapped in the solid yeast biomass exits the system (E_W), which will be the same as in the concentration in the biomass recycled X_R unless this is eliminated. To this, we add the rate of formation of the ethanol per unit time through the reaction in the interior of the yeast biomass floc particles equal to A_{floc} (Ethanol Flux) n_p V. The rate of ethanol generation is the diffusive flux exiting each yeast floc particle times the effective transport area of each yeast floc particle times the number of yeast floc particles per unit volume of the fermentor times the volume of the fermentor. The diffusive flux of ethanol from the floc particle is given by Equation 6.21 as already noted.

At steady state, with ethanol not entering the system, nor recirculating, the below equation becomes

$$0 = -F_D(1-w)E - wF_DE_w + A_{floc}(\text{Ethanol Flux})\, n_pV \tag{6.37}$$

From Equation 6.25, the concentration of biomass in the recycle stream, X_R, equals

$$X_R = \frac{(1+r)X}{(r+w)} \tag{6.25}$$

A mass balance in biomass regenerated in the system per Figure 6.4 is

$$F_DX_D + rF_DX_R + \mu XV = F_D(1+r)X \tag{6.38}$$

Substituting for X_R, noting that X_D is zero as there is no biomass in the feed at steady state, and solving for μ results in Equation 6.39 or

$$\mu = \left(\frac{F_D}{V}\right)(1+r)\left(\frac{w}{(r+w)}\right) \tag{6.39}$$

This equation points out that the system regenerates biomass to replace that discharged so that biomass remains constant (ignoring aging) and is assumed to be inherently maintained as noted in the biomass floc particles model presented previously.

Given this, and returning to Equation 6.37, rearranging this equation and solving for E results in

$$E = \frac{(-wE_W)}{[1-w]} + \frac{A_{floc}(\text{Ethanol Flux})n_pV}{[F_D(1-w)]} \tag{6.40}$$

Equation 6.40 points out that the ethanol produced by the system is a function of the ethanol generated in the yeast biomass flocs (A_{floc} (Ethanol Flux) n_p V) less the ethanol removed from the fermentation system as product or in discarded biomass.

Biomass removed from the fermentation system must be replaced by biomass being generated. From before, Equation 6.23 shows the differential equation for formation of biomass in this situation.

$$\frac{dX}{dt} = -\left(\frac{w}{(r+w)}\right)\left(\frac{F}{V}\right)(1+r)X + \left(\frac{\mu_{max}S}{(K_s+S)}\right)X \tag{6.23}$$

At steady state, $dX/dt = 0$. The generation of biomass (represented by $(\mu_{max} S/(K_s + S))X$ must equal the biomass exiting the system $(-(w/(r + w))\ (F/V)\ (1 + r)\ X)$. By inspection, the biomass term X cancels out, so that we obtain

$$\left(\frac{w}{(r+w)}\right)\left(\frac{F_D}{V}\right)(1+r) = \left(\frac{\mu_{max}S}{(K_s+S)}\right) \tag{6.41}$$

Note also that the reaction for formation of biomass intrinsically is a function of many variables besides substrate or feedstock represented by S. These include pH, ion concentrations, and so on that must be determined experimentally. This is represented by the yeast biomass growth rate μ. Therefore, we rewrite Equation 6.41 as Equation 6.42 or

$$\mu = \left(\frac{w}{(r+w)}\right)\left(\frac{F_D}{V}\right)(1+r) \tag{6.42}$$

From Equation 6.42, the yeast biomass growth rate at steady state is governed by the flow of substrate F_D into volume V (F_D/V, reciprocal time units, inversely residence time), the recycle ratio for the biomass (r), and the biomass removal ratio, $w/(r + w)$. This points out that operating continuously requires good process control in regard to flow and separation and maintaining good control of biomass in the system. The absence of such control will result in washout of the biomass and instability in operation. However, achieving such control is well worth it in regard to being able to tune the system to maximum productivity and yield in actual operation.

Continuous operation in this manner will allow the yeast biomass to form to reach steady state. As noted and assumed, the yeast will form into biocatalytic flocs where the yeast will be maintained and produce ethanol. There will be a continual dynamic of such biomass forming and becoming flocs consisting of yeast cells that are aggregated for optimal performance and assumed stability. There is much evidence of such flocs in the brewing industry as noted and also papers on this in the industrial ethanol area. Also, the sister area of fungal growth and dense pellet formation is already known in the area of biomass formation and stabilization to produce citric acid over many decades—a well-developed historical process (Schweiger, L.B. and Snell, R.L. 1949).

6.4.6 Aspects of Yeast Floc Formation and the Floc Particles

The aggregation of the floc should be rapid and follow a form of Ficks' law of unsteady-state diffusion applied to aggregation of yeast cells into flow particles of the form

$$\frac{D\partial^2 Y_{cell}}{\partial x^2} = \frac{\partial Y_{cell}}{\partial t} + kY_{cell} \tag{6.43}$$

where D is the diffusion of yeast cells toward a forming aggregate expressed as the floc, Y_{cell} is the concentration of free yeast cells, and k Y_{cell} is the first-order representation of capture of the yeast cells into the floc aggregate. The yeast cells may aggregate as a preferred mode to carry out their task of survival and production of ethanol which is a toxic entity to them. Equation 6.43 is a representation of the formation of a yeast floc particle from yeast cells as a function of time and distance in the floc particle.

For the yeast floc particle, the total biomass in a particular volume (e.g., grams per liter dry basis) can be assumed to be equal to the average number of biomass particles in a liter of broth times the average biomass weight per particle. The average weight and the variance of this particle distribution will be determined experimentally. The biomass particle weight is equal to the biomass particle density times its volume. If the biomass occurs in clusters, as expected, then this volume will vary. However using the average is convenient and acceptable for our purposes.

The total biomass concentration per liter or per cc relates to the number of particles per unit volume times the average weight of a biomass particle (Haddad, S.A. and Lindegren, C.C. 1953).

$$B = n_p b_p \tag{6.44}$$

where B = biomass concentration in the ethanol bioreactor, grams/liter or per cc; n_p = number of yeast floc particles per liter (average as noted); and b_p is the weight of a yeast floc particle in grams

$$\text{Further, } b_p = \rho_p v_p \tag{6.45}$$

where ρ_p and v_p are the yeast biomass particle density (grams/cc) and volume (cc), respectively. Let d_p = yeast floc particle size, equivalent diameter, in cm. Therefore, for the floc particle volume

$$v_p = \frac{\pi\,(d_p)^3}{6} \tag{6.46}$$

and

$$b_p = \frac{\rho_p \pi\,(d_p)^3}{6} \tag{6.47}$$

Therefore, the biomass concentration per liter of fermentation volume can be expressed as

$$B = \frac{n_p \rho_p \pi (d_p)^3}{6} \tag{6.48}$$

Equation 6.48 allows us to express the yeast floc particles per unit volume of the fermentor, the biomass particle density, and the biomass particle size

Putting in some numbers we can see cases that reflect ethanol formation.

Let B = 0.02 g/cc, let d_p = 0.2 cm, and ρ_p = 1.5 g/cc

With these we can calculate the average number of biomass particles in the 1000 L of broth volume. Using Equation 6.48, solving for n_p, we obtain 3.183 particles per cc or 3183 particles per liter. If d_p = 0.1 cm, 25.465 particles per cc result or 25,465 particles (assumed to spherical) per liter. This shows the sensitivity of n_p to particle size.

The geometric surface area of each particle is given by

$$a_p = \frac{\pi(d_p)^2}{4} \tag{6.49}$$

If the inner areas of the floc near the surface are accessed well due to porosity, the area term is enhanced. This may well be the case since restricting this surface which is important to mass transfer to the geometric outer surface seems restrictive and perhaps too conservative. We can call this area enhancement factor A_{ENH}.

Therefore in our example, a_p = 0.00785 cm^2. Thus, the total surface area per liter for transport from the broth into the particles and out of them is 25,465 particles per liter times 0.00785 cm^2 per particle or 200 cm^2/L of broth in one case without area enhancement.

In a second case, the biomass floc may be intrinsically more concentrated, and biomass particle slightly larger which will affect biocatalysis. If we let B = 0.1 g/cc for the above, so that 15.9 particles per cc result, with the surface area intrinsically enhanced since some internal area at the surface comes into play as if it were more like activated carbon in a sense, then perhaps the effective area for transport is 100 times greater so that a_p = 0,031416 cm^2 is multiplied by an area enhancing factor (A_{ENH}) of 100, then we then obtain a_p times A_{ENH} = 0.031416 cm^2 times 100 or 3.1416 cm^2 per particle. In this case, the total surface area transport is from the broth into and out of the particles is 15,920 particles per liter times 3.1416 cm^2 per particle or 50,014 cm^2/L of broth. This is a considerable magnification effective in more convective and less diffusive situations. This has to be evaluated experimentally.

6.4.7 Mass Transfer Resistance Effects and Responses

Transfer to the solution and from the particle may involve mass transfer resistance, depending on the level of mixing which may be simply based on the rising carbon dioxide generation and not mechanical agitation for reasons of cost. Therefore, the mass transfer coefficient from the external solution to the surface of the particle can be a factor to productivity. We count on adequate mixing to reduce this barrier (disrupting a liquid film that surrounds the particle) to the extent that intraparticle reaction and diffusion is controlling. The mass transfer coefficient depends on many properties, such as surface tension and viscosity.

Correlations in the literature help us predict the magnitude of this barrier (McCabe, W.L. et al. 2005). If this barrier is water like and the solution is relatively dilute (e.g., biomass particles do limit convective mass transfer), then the mass transfer barrier from this aspect may be relatively small. This means small relative to other barriers to transport to reaction sites inside the particle, such as diffusion, which usually presents a much larger barrier. This point has to be checked in specific cases since mixing in the ethanol fermentor may not be particularly vigorous, and may only be based on carbon dioxide bubbling which is only enough to move things about and stir them, and no more. Perhaps that is all that is needed if the yeast biomass is not particularly high in concentration and if floc particle sizes are not too small. Mixing a high quantity of relatively small yeast biomass flocs can raise effective viscosity considerably and thus inhibit mass transfer to a notable degree.

In general, introducing mass transfer resistance external to the biomass particles means that the boundary condition at the outer surface of the particle has to be examined relative to the assumption that we can accept the concentration at the particle surface to be the concentration in the broth and not something much lower due to mass transfer resistance. Examining this, we wish to relate K_{mt} $(C_0 - C_{s0})$, the mass transfer coefficient in cm/s times the effective concentration difference externally at the particle surface (philosophically only since there are many interpretations of this gradient) to $-D_E(dE/dx)_0$, the diffusive flux of ethanol at the surface of the floc particle (with a similar comment for the external mass transfer and diffusive flux of feedstock into the particle).

Another possible barrier is a slime coating (if such a thing exists) that adheres to the biomass particle and blocks it. Mixing alone may not do much to such a barrier. In this case, the organism may produce enzymes to attack this barrier, since the barrier is a form of nutrition for the organism (and resistance to accessing external nutrition beyond the slime barrier). Such enzymes that are secreted will be produced according to activators and stimulants and retarded by inhibitors. A true model may then have an expression for reduction of this slime barrier due to enzyme attack.

The existence of such a slime barrier in-effect imposes a substantial barrier to access to the external concentration of say solution ethanol or feedstock. Therefore, broadly thinking, inducing the organism to provide the enzymatic means to remove this barrier is important if we have to keep the surface of the yeast floc clean. Consider numerical values examining the relationship,

$$K_m(E_{SOLN} - E_0) = D_E \frac{dE}{dX_{@x=0}} \tag{6.50}$$

This is the alternative flux boundary condition at the surface of the particle. This is only an approximation, since the internal reaction equation will vary if the boundary condition is a flux rather than a concentration.

From the literature let K_m equal 0.01 cm/s (McCabe, W.L. et al. 2005). Let E_0 be 0.02 g/cc and let E_{SOLN} be 0.01 g/cc, let the flux be 1.2×10^{-4}g/cm^2-s per floc particle. The left-hand side is 1×10^{-4} g/cm^2-s. This is proximate to the flux out of the particle so the external mass transfer resistance is comparable in this case. Each case must be checked since it is entirely possible then that enhanced solution mixing can improve the ethanol productivity of the system. Research in this area is warranted looking at productivity gains for the cost involved;

For one case, continuing, proceed with sample calculation to estimate ethanol concentration achieved. The flux of ethanol exiting a floc particle is 1.2×10^{-4} g/cm^2-s. The flux of feedstock (glucose) into the flux particle for the example is $-2{,}67 \times 10^{-4}$ g//cm^2-s. For the enhanced area case, A_{ENH} $a_p = 3.14$ cm^2.

The biomass concentration assumed to be concentrated in the floc is 0.1 g/cc. The fluid flow to the fermentor is 3.6×10^5 gal/day or 1.58×10^{-6} cc/s. The fermentor volume is 7.2×10^5 gal or 2.726×10^9 cc. The residence time is two days or 172,800 s.

The fraction of ethanol carried out with the biomass is 0.017 of the provided to the separator and no ethanol is recycled as that would be counterproductive.

From Equation 6.40,

$$E = \frac{-wE_W}{(1-w)} + \frac{A_{ENH}A_{floc}(\text{Ethanol Flux})n_pV}{[F_D(1-w)]}$$

Substituting values, $E = 0.012$ g/cc. This case is developed for a very porous yeast floc, the enhanced area assumption, and accentuated diffusion due to the porosity. This assumes that there is cooperation among yeast cells in the floc and good access to the solution so that transport effects are maximized. External mass transfer resistance may be relevant and worth some action and investment fi justified.

This model is interesting. If the physical concept is changed, where the floc is so small as to be adrift as a minute entity in solution, the parameters can be altered to the extent as to be less concerned about diffusion and thinking more of the entity without that. This is actually very useful to be used as a guide for experiments to improve the ethanol process in terms of rate, yield, and lower costs.

Back to the floc model, if the mass transfer rate is possibly significant in cases, then the mathematical model should be modified. The boundary condition at the outer surface of the floc should be changed to a flux boundary for both ethanol and feedstock. It is useful to look at that in closing to discuss how this conceptually may be used to guide experiments. Starting with Equations 6.3 through 6.6 as repeated below,

$$\frac{D_F d^2F}{dx^2} = R_F(x) \tag{6.3}$$

$$\frac{D_E d^2E}{dx^2} = R_E(x) \tag{6.4}$$

$R_F(x)$ is defined as

$$R_F(x) = \gamma_{FB}B\,\mu_{FO}\left(1-\frac{E}{E_{MAX}}\right) + \gamma_{EB}B\,\mu_{EO}\left(1-\frac{E}{E_{MAX}}\right) \tag{6.5}$$

$$R_E(x) = -\gamma_{EB}B\mu_{EO}\left(1-\frac{E}{E_{MAX}}\right) \tag{6.6}$$

The boundary conditions previously were $E = E_0$ and $F = F_0$ at $x = 0$ (the outer surface of the floc) and $dE/dx = 0$ and $dF/dx = 0$ at $x = 1$ (the center of the floc).

With the mass transfer from the fermentor solution to the surface of the particle, the first boundary condition changes more precisely so that at x = 0,

$$K_{mE}(E_{@x=0} - E_0) = D_E \frac{dE}{dX_{@x=0}} \tag{6.51}$$

and

$$K_{mE}(F_0 - F_{@x=0}) = D_E \frac{dE}{dX_{@x=0}} \tag{6.52}$$

The change in boundary conditions means that Equations 6.3 through 6.6 would be rederived and an alternative model used. What we would want to do experimentally is to improve mixing so that Equations 6.3 through 6.6 apply with the boundary conditions as used in the derivation of this chapter. Imposing mechanical agitation may be costly. A better way would be to assess the agitation provided by carbon dioxide bubbles and flow to see if clever ways can be found to use that energy to mix the fermentor contents just enough without excessive cost. That is good engineering and a worthy endeavor to improve process economics

APPENDIX 6A: DERIVATION OF FEEDSTOCK AND ETHANOL FLUX-RELATED EQUATIONS FOR THE YEAST FLOC

Starting with the differential equations for diffusion and reaction of feedstock and ethanol in a yeast floc—these equations are derived from mass balances across a differential shell in the floc (Goldstein, W.E. and Schreuder, W.A. 2011).

$$\frac{D_F d^2F}{dx^2} = R_F(x) \tag{6.3}$$

$$\frac{D_E d^2E}{dx^2} = R_E(x) \tag{6.4}$$

The feedstock F-related concentration (e.g., glucose) proceeds from the exterior to the interior of the floc and is decreasing while the ethanol product (E) proceeds from the central interior of the floc where is typically at higher concentrations to the extremity of the floc where it exits to form the bioreactor ethanol solution.

We define

$$R_F(x) = \gamma_{FB} B\mu_{FO}\left(1 - \frac{E}{E_{MAX}}\right) + \gamma_{EB} B\mu_{EO}\left(1 - \frac{E}{E_{MAX}}\right) \tag{6.5}$$

$$R_E(x) = -\gamma_{EB} B\mu_{EO}\left(1 - \frac{E}{E_{MAX}}\right) \tag{6.6}$$

The parameters in these equations are defined as

$$\frac{\gamma_{EB} B \mu_{EO}}{E_{MAX}} = \alpha \tag{6.7}$$

and

$$\gamma_{EB} B \mu_{EO} = \beta \tag{6.8}$$

$$\text{such that } \frac{\beta}{\alpha} = E_{MAX} \tag{6.9}$$

Equations 6.5 through 6.9 show that the reaction rate of the feedstock applies to maintain the biomass to account for losses and biomass removal and also provide for conversion of feedstock to ethanol. Equation 6.6 specifically is the rate of formation of ethanol by the yeast biomass. In these equations, μ_{EO} is the specific rate of conversion of the feedstock to ethanol by the biomass. Again, μ_{FO} is the specific rate of reaction in reciprocal time units to maintain the yeast biomass. Both the yeast biomass maintenance and growth and the yeast biomass ability to produce ethanol are inhibited by the presence of ethanol itself so this must be considered to be a modification of the specific reaction rates (μ_{FO} and μ_{EO}) applied to the biomass performance capability.

The ethanol inhibition of the conversion of feedstock by the biomass is basically a poisoning of the yeast due to ethanol so that the yeast functions less effectively to repair itself for growth and maintenance, and also to produce ethanol as ethanol concentration increases in the yeast floc.

One of the functions of the yeast is to replace biomass lost due to death and discharge to maintain a steady-state biomass concentration in the fermentation broth as a whole and maintain the biomass that remains. The second function of the yeast that is inhibited as ethanol increases is to continue to produce ethanol effectively. Therefore, in Equations 5.5 and 5.6, the factor $(1 - E/E_{MAX})$ represents this inhibition of the biomass maintenance and growth and that of the capability of the biomass to produce ethanol effectively. The factor E_{MAX} is the ethanol concentration where this inhibition is such as to cause cessation of the yeast biomass' ability to maintain itself and also cause it to cease to produce ethanol.

In Equation 6.5, γ_{FB} is the factor for stoichiometric conversion of feedstock to biomass in units of gram per gram for maintenance, and B is the concentration of Biomass in the system in g/cc. Similarly, in Equation 6.6, γ_{EB} is the factor for stoichiometric conversion of feedstock to ethanol by the biomass.

Equations 6.3 and 6.4 (using Equations 6.5 and 6.6) are solved under the boundary conditions such that at the outside of any floc particle consisting of the yeast-associated aggregate, at position $x = 0$, F is equal to F_O and E is equal to E_O. These are the conditions in the solution of the entire bioreactor (ignoring, for the present, mass transfer restrictions from the broth liquid to the surface of the yeast particle due to inadequate mixing). In the center of the representative yeast floc, it is assumed that the gradients of F and E with distance (dF/dx and dE/dx) are zero due to symmetry of the floc.

Equation 6.4 is solved first since this can be integrated to give an expression for E as a function of x, E(x). This can then be substituted into Equation 6.3 to obtain the expression for F as a function of x, F(x). The details of this are as follows:

Rewriting Equation 6.4, substituting Equation 6.6, and using parameters in Equations 6.7 through 6.9

$$D_E \frac{d^2E}{dx^2} - \alpha E = -\beta \tag{6A.1}$$

The solution to this equation is

$$E = C_1 \exp\left(x\sqrt{\frac{\alpha}{D_E}} \right) + C_2 \exp\left(-x\sqrt{\frac{\alpha}{D_E}} \right) + \frac{\beta}{\alpha} \tag{6A.2}$$

(see Rainville, E.D. 1958).

Since there is a no flux condition at the center of the floc (L), the change in ethanol concentration at that point must be zero. Therefore,

$$\frac{D_E dE}{dx} = 0 \quad \text{at } x = L \tag{6A.3}$$

Differentiating Equation 6A.2 with respect to x, setting x = L results in being able to express the constant C_2 in terms of C_1 as follows:

$$C_2 = C_1 \exp\left(2L\sqrt{\frac{\alpha}{D_E}} \right) \tag{6A.4}$$

$$\text{At } x = 0, \quad E = E_0 \tag{6A.5}$$

Substituting Equation 6A.5 into Equation 6A.5, and substituting for C_2 using Equation 6A.4 results in the equation for the constant C_1.

$$C_1 = \frac{E_0 - (\beta/\alpha)}{1 + \exp(2L\sqrt{(\alpha/D_E)})} \tag{6A.6}$$

Substituting back into Equation 6A.2 using Equations 6A.6 and 6A.4 and rearranging, results in the equation for the ethanol concentration in the floc as a function of distance and the other parameters previously given as

$$\frac{1 - E/E_{MAX}}{1 - E_O/E_{MAX}} = \exp(x\sqrt{(\alpha/D_E)}) + \frac{\exp(2L\sqrt{(\alpha/D_E)})\exp(-x\sqrt{(\alpha/D_E)})}{1 + \exp(2L\sqrt{(\alpha/D_E)})} \tag{6.10}$$

Next, Equation 6.10 is used to derive the equation for the gradient in feedstock as a function of position in the flock.

For convenience, we define a parameter, ε, as collection of constants in Equation 6.3, expressed as

$$\varepsilon = \frac{\gamma_{FB} B \mu_{FO} + \gamma_{EB} B \mu_{EO}}{D_F} \tag{6.11}$$

Continuing, the differential equation defining the feedstock F as a function of x is given by substituting into Equation 6.3 to obtain

$$\frac{d^2F}{dx^2} = \varepsilon\left(1 - \frac{E}{E_{MAX}}\right) \tag{6.12}$$

Substituting for $(1 - E/E_{MAX})$ into Equation 6.12 using Equation 6.11 results in a new expression for Equations 6.6 and 6.12, since we now have an expression for E in terms of position x. Therefore, Equation 6.12 is transformed to

$$\frac{d^2F}{dx^2} = \varepsilon\left(1 - \frac{E_O}{E_{MAX}}\right) \\ {}^*\exp(x\sqrt{(\alpha/D_E)}) + \frac{\exp(2L\sqrt{(\alpha/D_E)})\exp(-x\sqrt{(\alpha/D_E)})}{1+\exp(2L\sqrt{(\alpha/D_E)})} \tag{6.13}$$

where * indicates the multiplication sign. For convenience again, let the constant parameters Z_1 and Z_2 be defined as follows:

$$Z_1 = \frac{\varepsilon(1 - E_o/E_{MAX})}{1+\exp(2L\sqrt{(\alpha/D_E)})} \tag{6.14}$$

$$Z_2 = \frac{\varepsilon(1 - E_o/E_{MAX})[\exp(2L\sqrt{(\alpha/D_E)})]}{1+\exp(2L\sqrt{(\alpha/D_E)})} \tag{6.15}$$

Substituting Equations 6.14 and 6.15 into Equation 6.12 results in

$$\frac{d^2F}{dx^2} = Z_1 \exp\left(x\sqrt{\frac{\alpha}{D_E}}\right) + Z_2 \exp\left(-x\sqrt{\frac{\alpha}{D_E}}\right) \tag{6.16}$$

Integrating Equation 6.16 once results in

$$\frac{dF}{dx} = \left(\frac{Z_1}{\sqrt{(\alpha/D_E)}}\right)\exp\left(x\sqrt{\frac{\alpha}{D_E}}\right) - \left(\frac{Z_2}{\sqrt{(\alpha/D_E)}}\right)\exp\left(-x\sqrt{\frac{\alpha}{D_E}}\right) + C_3 \tag{6A.7}$$

Integrating Equation 6A.7 again results in

$$F = \left(\frac{Z_1}{\alpha/D_E}\right)\exp\left(x\sqrt{\frac{\alpha}{D_E}}\right) + \left(\frac{Z_2}{\alpha/D_E}\right)\exp\left(-x\sqrt{\frac{\alpha}{D_E}}\right) + C_3x + C_4 \quad (6A.8)$$

At $x = 0$,

$$F_0 = \left(\frac{Z_1}{\alpha/D_E}\right) + \left(\frac{Z_2}{\alpha/D_E}\right) + C_4 \quad (6A.9)$$

Therefore,

$$C_4 = F_0 - \left(\frac{Z_1}{\alpha/D_E}\right) - \left(\frac{Z_2}{\alpha/D_E}\right) \quad (6A.10)$$

At $x = L$,

$$\frac{dF}{dx} = 0 \quad (6A.11)$$

So that Equation 6A.7 becomes at $x = L$

$$0 = \left(\frac{Z_1}{\sqrt{(\alpha/D_E)}}\right)\exp\left(L\sqrt{\frac{\alpha}{D_E}}\right) - \left(\frac{Z_2}{\sqrt{(\alpha/D_E)}}\right)\exp\left(-L\sqrt{\frac{\alpha}{D_E}}\right) + C_3$$

and then solving for the constant C_3,

$$C_3 = -\left(\frac{Z_1}{\sqrt{(\alpha/D_E)}}\right)\exp\left(L\sqrt{\frac{\alpha}{D_E}}\right) + \left(\frac{Z_2}{\sqrt{(\alpha/D_E)}}\right)\exp\left(-L\sqrt{\frac{\alpha}{D_E}}\right) \quad (6A.12)$$

Substituting Equations 6A.10 and 6A.12 into Equation 6A.8 and rearranging, results in Equation 6.17, the equation for the feedstock F as a function of position x in the floc.

$$F - F_O = \frac{Z_1}{(\alpha/D_E)}\exp(x(\sqrt{(\alpha/D_E)} - 1) + \frac{Z_2}{(\alpha/D_E)}\exp(-x\sqrt{(\alpha/D_E)} - 1)$$
$$+ x\left[\frac{-Z_1}{\sqrt{(\alpha/D_E)}}\exp(L\sqrt{(\alpha/D_E)})\right] + \frac{Z_2}{\sqrt{(\alpha/D_E)}}\exp(-L\sqrt{(\alpha/D_E)}) \quad (6.17)$$

Looking at the change in feedstock value from $x = 0$ to $x = L$, we obtain

$$F_L - F_O = \frac{Z_1}{(\alpha/D_E)}\exp(L\sqrt{(\alpha/D_E)} - 1) + \frac{Z_2}{(\alpha/D_E)}\exp(-L\sqrt{(\alpha/D_E)} - 1)$$
$$+ L\left[\frac{-Z_1}{\sqrt{(\alpha/D_E)}}\exp(L\sqrt{(\alpha/D_E)})\right] + \frac{Z_2}{\sqrt{(\alpha/D_E)}}\exp(-L\sqrt{(\alpha/D_E)}) \quad (6.18)$$

REFERENCES

Bauer, E.F. and Pretorius, L.S. 2000. Yeast stress response and fermentation efficiency: How to survive the making of wine—A review. *South African Journal of Enology and Viticulture*, 21, Special Issue, p. 27–51.

Bisson, L. Copyright 2001. Section 3—The Alcoholic Fermentation, Lesson 8, Introduction, Yeast Biology. University of California at Davis, University Extension.

Blanch, H.W. and Clark, D.S. 1997. *Biochemical Engineering*, Marcel Dekker, New York, 1997, 193.

Carlsen, H.N., Degn, H., and Lloyd, D. 1991. Effects of alcohols on the respiration and fermentation of aerated suspensions of baker's yeast. *Journal of General Microbiology*, 137, 2879–2883.

Corn Refiners Association. 2013. *Corn Starch*, 12th Edition. http://www.corn.org.

Curry, J. April 19, 2009. *Yeast and Fermentation*. Dukes of Ale BJCP Preparation Course.

Dombek, K.M. and Ingram, L.O. 1987. Ethanol production during batch fermentation with *Saccharomyces cerevisiae*: Changes in glycolytic enzymes and internal pH. *Applied and Environmental Microbiology*, June, 1286–1291. Ethanol damages the cell not just its presence.

Domingues, L., Vicente, A.A., Lima, N., and Teixeira, J.A. 2000. Applications of yeast flocculation in biotechnological processes. *Biotechnology. Bioprocess Engineering*, 5, 288–230.

DuPont. 2013. Accellerase® Trio TM. http://www.accellerase.dupont.com.

Frölander, A. and Rødsrud, G. 2011. Conversion of cellulose, hemicellulose, and lignin into platform molecules: Biotechnological approach. *Euro Bio Ref Summit School*, Lecce, Italy.

Galletti, A.M.R. and Antonetti, C. Sept. 19, 2011. Biomass pretreatment: Separation of cellulose, hemicellulose, and lignin. Exiting Technologies and Perspectives. UROBIOREF.

Genencor. 2011. Accellerase® TrioTM. A Division of Danisco.

Goldstein, W.E. ed. 2014. *Pharmaceutical Accumulation in the Environment: Prevention, Control, Health Effects, and Economic Impact*, Taylor & Francis Group. CRC Press, Boca Raton, FL. Chapter 9.

Goldstein, W.E. and Schreuder, W.A. 2011. Mathematical model of mold propagation and product formation in building materials, inherent transport phenomena and applications, in *Sick Building Syndrome and Related Illness: Prevention and Remediation of Mold Contamination*, W.E. Goldstein, Ed., Taylor & Francis Group, CRC Press, Boca Raton, FL, 58.

Haddad, S.A. and Lindegren, C.C. 1953. A method for determining the weight of an individual yeast cell. *Applied Microbiology*, February 23, 153–156.

Harmsen, P.F.H., Huijgen, W.J.J., Bermúdez López, L.M., and Bakker, R.R.C. September, 2010. Literature Review of Physical and Chemical Processes for Lignocellulose Biomass. Report ECN-E-10-013. Wageningen University & Research Centre. The Netherlands.

Henze, M., Gujer, W., Mino, T. et al. 2000, *Activated Sludge Models, ASM1, ASM2, ASM2d, and ASM3, Task Force Group on Mathematical Modeling for Design and Operation of Biological Waste Water Treatment*, IWA Publishing, Scientific and Technical Report Series, London, England.

Kamp, P. 2012. *Inbicon Biomass Refining. Cellulosic Ethanol Technology Platforms. Growth and Sustainability through Biomass Refining*. Technology Review. Inbicon.

Kunduru, M.R. 1994. Development of a biofilm bioreactor for enhanced ethanol production. Retrospective theses and dissertations. Paper 11280. http://lib.dr.iastate.edu/rtd.

Lake, M.A. 2010. Potential commercial uses for lignin. Southeastern Bioenergy Conference, Tifton, Georgia.

Liu, C.-G., Wang, N., Lin, Y.-H., and Bai, F.-W. 2012. Very high gravity ethanol fermentation by flocculating yeast under redox potential-controlled conditions. *Biotechnology for Biofuels*, 5, 61. http://www.biotechnologyforbiofuels.com/content/5/1/61.

Martínez-Moreno, R., Morales, P., Gonzalez, R., Mas, A., and Beltran, G. 2012. Biomass production and alcoholic fermentation performance of *Saccharomyces cerevisiae* as a function of nitrogen source. *FEMS Yeast Research*, 12 (2012), 477–485.

McCabe, W.L., Smith, J.C., and Harriott, P. 2005. *Principles of Diffusion and Mass Transfer between Phases, Unit Operations of Chemical Engineering*, 7th Edition, McGraw-Hill, New York, Chapter 17, 557.

Newton, I. 1687. *Principia*, Cambridge University, London, England.

Noller, C.R. 1958. *Textbook of Organic Chemistry*, 2nd Edition. W.B. Saunders Company, Philadelphia, PA, 59.

Novozymes. 2013. *Enzymes at Work*. Novozymes A/S. Denmark. http://www.novozymes.com.

Papagianni, M. and Mattey, M. 2006. Morphological development of *Aspergillus niger* in submerged citric acid fermentation as a function of the spore inoculum level. Application of neural network and cluster analysis for characterization of mycelial morphology. *Microbial Cell Factories*, 5, 3. doi:10.1186/1475-2859-5-3.

Prince, I.G. and Barford, J.P. 1982. Continuous tower fermentation for the power ethanol production. *Biotechnology Letters*, 4(4), 263–268.

Purwadi, P., Brandberg, T., and Taherzadeh, M.J. 2007. A possible industrial solution to ferment lignocellulosic hydrolyzate to ethanol: Continuous cultivation with flocculating yeast. *International Journal of Molecular Sciences*, 8, 920–932.

Rainville, E.D. 1958. *A Short Course in Differential Equations*, 2nd Edition. The MacMillan Company, New York.

Rubio-Arroyo, M.F., Vivanco-Loyo, P., Juárez, M., Poisot, M., and Guillermo, R.-G. 2011. Bio-ethanol obtained by fermentation process with continuous feeding of yeast. *Journal of Mexican Chemical Society*, 55(4), 242–245.

Sarks, C., Mingjie, J., Satos, T.K., Balani, V., and Dale, B.E. 2014. Studying the rapid bioconversion of lignocellulosic sugars into ethanol using high cell density fermentations with cell recycle. *Biotechnology for Biofuels*, 7, 73. http://www.biotechnologyforbiofuels.com/content/7/1/73.

Scharf, M. and Sethi, A. 2013. Novel lignases and aldo keto reductases for conversion of lignin-containing materials to fermentable products. US 20150184211 Patent Application Publication. Purdue Research Foundation, West Lafayette, Indiana.

Schill, S.R. 2015. Yearning for new yeasts. *Ethanol Producer Magazine*.

Schweiger, L.B. and Snell, R.L. 1949. Fermentation process. U.S. Patent 2,476,159.

Shapley, P. 2012. Cellulose and lignocellulose. http://buane.chem.uiuc.edu/pshapley/genchem2/bi013.html.

Snoep, J.L., Mrwebi, M., Schuurmans, J.M., and Rohwer, J.M. 2009. Control of specific growth rate in *Saccharomyces cerevisiae*. *Microbiology*, 155, 1699–1707.

Vieira, E.D., Andrietta, M.G.M.G/S, and Andrietta, S.R. 2013. Yeast biomass production: A new approach in glucose-limited feeding strategy. *Brazilian Journal of Microbiology*, 44(2) 551–558.

von Leibniz, G.W. 1684. *Acta Eruditorum*.

Xu, T.J., Zhao, X.Q., Zhou, Y.C., and Bai, F.W. Jan. 2005. Continuous ethanol fermentation using self-flocculating yeast strain and bioreactor system composed of multi-stage tanks in series. *Sheng Wu Gong Cheng Xue Bao*, 21(1), 113–117.

Zegarelli, M. 2008. *Calculus II for Dummies*, Wiley Publishing, Inc., Hoboken, NJ.

Zhang, W. and Geng, A. 2012. Improved ethanol production by a xylose fermenting recombinant yeast strain constructed through a modified genome shuffling method. *Biotechnology for Biofuels*, 5, 46, http://www.biotechnologyforbiofuels.com/content/5/1/46.

7 Cost Analysis for Optimal Ethanol Manufacture

7.1 COST ANALYSIS OBJECTIVES TO IMPROVE ETHANOL PROSPECTS

The main areas in cost analysis are the capital cost, the manufacturing cost, and the overhead costs. The overhead costs can sometimes be included as part of the manufacturing costs (Goldstein, W.E. et al. 1980; Goldstein, W.E., 1999).

Capital cost is that required to construct the ethanol facility and other facilities such as those to produce utilities or prepare raw materials required to manufacture the product. The cost to get a plant started involving special capital, and so on, is called the working capital.

Manufacturing cost consists of raw material costs, utilities costs, labor, and capital charges required to produce, in this case, ethanol and any by-products such as carbon dioxide that is captured or biomass product sold as distillers dry (or wet) grains. It can include overhead costs supportive of manufacturing including quality control, supervision of manufacturing, purchasing, human resources, and other costs depending on the preferences in a particular organization.

Capital charges include the depreciation on capital costs. Deprecation is the deterioration of the equipment that is allowed by the Internal Revenue Service (IRS) as a deduction since the equipment may have to be replaced. Capital charges can be set up in an interest-bearing fund to support such replacement. Capital charges also include the maintenance of capital (keeping the equipment and buildings operating and usable), taxes on the facility, and insurance costs. Insurance costs can cover protection in case any of the equipment goes bad and ceases to function during its usable life. The insurance costs can also cover liability as well as damage due to events such as earthquakes or fire.

The overhead costs (beyond those strictly associated with manufacturing) are those that are required to market and sell the product, the cost of inventory, company supervision outside of manufacturing, quality control, quality assurance outside of manufacturing, general purchasing, human resources, and so on. Research and development (R&D) is typically a separate entity and not charged to manufacturing, though certain functions such as process development may be chargeable to manufacturing. Sometimes, R&D is considered an investment and paid off by profits from manufacturing. Considering R&D as an investment may only be valid if tax credits for the investment in R&D are available.

The cost analysis for a particular organization (and how it handled) is basically governed by IRS requirements and the desire to maximize deductions and minimize tax payments. It is also governed by organization strategy and planning. As one can see, cost analysis is a complex and difficult area with many unknowns where data are often not available or difficult to estimate with reasonable certainty.

This uncertainty applies to the ethanol area. Much of the cost data are proprietary within the industrial sector and considered inside information not well shared with outsiders. A particular tact taken in this book is to show the nature of costs and ways in which they can be improved and controlled from a scientific and engineering standpoint. To do this, existing cost data available from public sources are analyzed in order to interpret cost patterns for the ethanol industry.

There are certain precepts that are important in regard to the production and sale of ethanol. First, in the raw materials area, feedstock cost is important. For ethanol, this involves any feedstock one can use anywhere that will produce an acceptable carbohydrate to make ethanol (Felix, E. et al. 2007; Jessen, H. 2014). In most of the present cases, this will involve farmers providing crops like corn, crop components that are normally waste such as corn stover, or wheat chaff, as examples. In the United States, it is not a secret that the farmers are doing well supplying corn and maybe in the near and distant future, corn stover, to make ethanol (DuPont. 2015).

Doing well means they are making a good profit doing so. Whatever is done in this area, the development must ensure that farmers continue to profit well and at the same time provide what is needed to supply food, such as corn. These do not have to be mutually exclusive objectives. The situation simply has to be well managed. This involves growing the crop and providing it in a way that is cooperative and sustainable. This involves taking contracts on pricing that are smart and also getting help to negotiate prices. This is not a small matter. This is a major endeavor that the writer thinks needs to be well looked at for suppliers anywhere on the globe for whatever crops are available.

Raw materials of importance to produce ethanol include enzymes to convert the feedstock to a form suitable for yeasts to economically make ethanol, and yeasts to produce ethanol from the feedstock. Enzyme and yeast manufacturers invest significantly into research and development to produce their product. Just like the farmers, they deserve an adequate price for their product to cover their costs and profit needs. Driving the price of the enzyme down should be based on technology advance and linked to improving the economic availability of ethanol. The enzyme and yeast suppliers should be encouraged to drive technology and product improvements that solve problems in the industry. This should allow enzyme and yeast manufacturers to expand in capacity and increase their sales and profits.

The writer remembers a situation in which he was involved where technology advance improved enzyme yield thereby allowing the cost to go down. The product formulators using enzymes unfortunately applied pressure to reduce prices for enzymes sold to them, and the enzyme suppliers complied much too complacently. The reduction in price should have resulted in increased sales at a price that allowed investment in facilities and research and development for the enzyme suppliers plus allowed their business to profit. Unfortunately, this was so poorly managed at the time that it simply forced certain manufacturers out of the business and others to sustain unacceptable losses. This is not suggesting collusion, but simply smart business practices that protect all participants. This can be managed so that all industries thrive if this is properly handled.

Utilities represent a major cost to produce ethanol. Arrangements need to be made so that natural gas (or oil or coal) and electricity are provided in a cost-effective manner by negotiation with the utilities. This can be linked to protecting the environment by having investment included to minimize pollution.

If alternative sources of electrical power are possible, they should be looked at seriously by financing parties and designers. For example, a hard look should be taken at geothermal sources and also solar power if these are conveniently and reliably available. In the extreme, if a plant could receive its electricity solely from a captive solar installation, then it is economically worthwhile and also attractive from a pollution control standpoint.

It may be possible, for example, that reliable and inexpensive wood pellet–fired boilers will work for certain locales and be competitive for steam generation with natural gas (Portz, T. 2016). Or, perhaps the facility can be designed to allow a multitude of raw material sources and options to generate energy needed depending on which resource is most favorable from an economic and environmentally friendly basis. The ethanol plant should allow for such multioptimization possibilities if this is practical from an economic standpoint (Kwiatkowski, J.R. et al. 2006). This needs to be assessed in design and planning phases considering future prospects (Duff, B. 2013; Vogelbusch. 2015; Wirt, A. 2014).

Ethanol is produced through the use of corn in the United States at present. However, this is changing to see if lignocellulose sources such as the corn crop waste (or corn stover) can be used (DuPont. 2013; Kazi, F.K. et al. 2010). Elsewhere, Brazil focuses on the use of sugar derived primarily form sugarcane. Worldwide, many crop sources can suffice as discussed in earlier chapters.

7.2 COST ANALYSIS AND SOURCES OF ETHANOL

The literature contains some information on cost analyses to produce ethanol from different crop sources (Shapouri, H. and Gallagher, P. 2005; Shapouri, H. and Salassi, M. 2006; Whims, J. 2002). Most of the experience with ethanol in the United States involves the use of corn. This is changing to include lignocellulose sources to produce ethanol. One such lignocellulose source is (logically) a by-product of corn growth and processing, corn stover. Lignocellulose as a category is covered in this cost analysis.

In the United States, the primary route to produce ethanol from corn kernels is through use of dry milling processing. Since wet milling of corn has also been considered, that is included.

A most logical choice for some might be to expand/modify an existing corn wet milling plant so that it could produce ethanol as well as the normal array of corn products. This could (in theory) save a tremendous amount of capital and perhaps lower manufacturing costs due to the cooperation that should result. This should save manufacturing costs since labor might be shared, utility capital costs may be lower, and some overhead functions like quality assurance and control could be in part shared. Reuse of equipment that has been "mothballed" or put in storage, and more leverage resulting for purchasing may result. If "bare bones, green field" construction is required, then one needs a new facility and the attendant costs. However, if one is getting into the business, using an existing facility with existing personnel and resources makes a lot of sense.

Actual data from the corn wet milling and dry milling industries on capital and manufacturing costs are basically nonexistent since this information is kept private since it is proprietary. However, one can use past data from different sources to a least approximate expectant costs as a guideline. This tactic is used in this book to help assess costs. It can serve as a valuable method to help assess technology needs and costs to prepare ones' own captive facility.

Costs of corn wet milling, dry milling, and lignocellulose processes are compared using charts prepared based on industry data. The data sources are covered in Table 7A.1 in Appendix 7A. References used to compile these tables are noted in Appendix 7A and covered in the references at the end of this chapter.

Figures 7.1 through 7.17 in Chapter 7 show graphs developed through the analysis of data provided by specialists in industry, government, and academia. Corn wet milling, corn dry milling, and lignocellulose feedstock–related analyses are covered in succession.

7.3 COST OF ETHANOL PRODUCED USING FEEDSTOCK FROM CORN WET MILLING FACILITIES

The data sources for the cost analyses in the literature are based on different assumptions and published in different years. It is assumed that these can be treated as a composite where best mean values can be assessed within this variability. The costs presented in the different analyses are extrapolated to 2014 in order to attempt to compensate for this variable in the analysis. The variability introduced by these experts based on their assumptions is inherent to this complex area and applies to all facilities based on the writer's experience. Assessments within this variability are still quite useful.

For capital costs, the extrapolation to 2014 is based on industry-provided trend charts (Ulrich, G.D. and Vasudevan, P.T. 2002; Chemical Engineering Plant Cost Index. 2011). For extrapolation of manufacturing costs (where possible) to 2014, a variety of references were used (Landress, L. and Loo, F. 2014; U.S. Bureau of Labor Statistics. 2015a,b; U.S. Energy Information Administration. Annual Energy Review. 2013a,b).

Figure 7.1 presents a representative curve for the capital cost of ethanol facilities based on feedstock supplied from corn wet milling. These capital cost data only apply to the ethanol facility and do not include the corn wet milling facility that supplies the feedstock to produce ethanol. The best fit is shown as an exponential function for the available data. Since individual capital costs for equipment are expected to follow a power law rule (e.g., capital cost proportional to a capacity factor raised to a power such as 0.6) based on the literature, this was examined as a first attempt to fit this limited data since an assembly of capital costs for a facility may in total, logically follow a power law rule (Aries, R.S. and Newton, R.D. 1955; Goldstein, W.E. 2014).

The power law did not fit all the data properly as desired, so the exponential fit was chosen since all data are assumed to have equal weight in the analysis. What is apparent is that the capital cost decreases nonlinearly as the productive capacity increases. As indicated, the mean value of the capital cost per gallon of ethanol

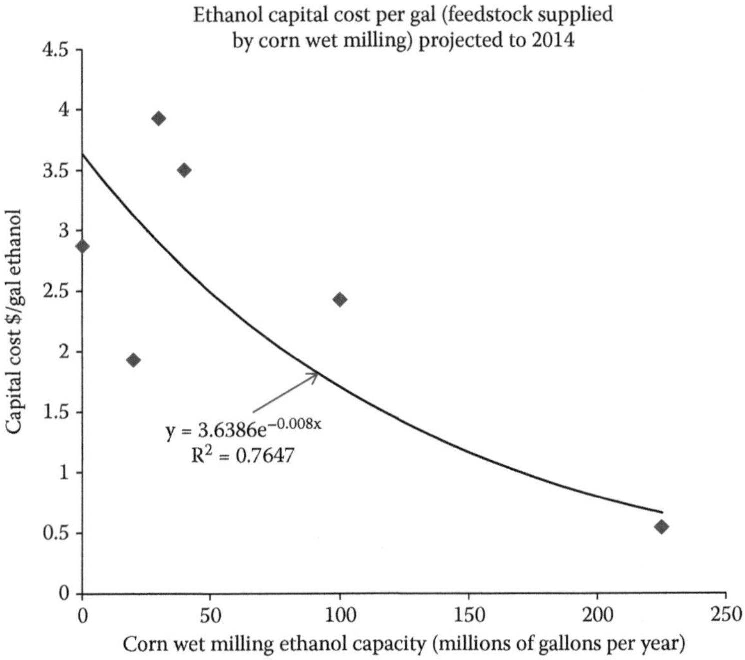

FIGURE 7.1 Ethanol capital cost (feedstock from corn wet milling) as a function of capacity.

seems to approach \$3.6/gal well below 10 million gallons per year and approach \$0.5/gal just in excess of 200 million gallons per year.

Thus, the value of economies of scale is evident here. Larger facilities generally have a lower capital cost per unit of product (Aries, R.S. and Newton, R.D. 1955; Goldstein, W.E. 2014). The aspect requiring thorough analysis is that one has to expend in excess of \$100 million to build such a facility (200 million gallons or nearly 800 million liters or 0.8 million cubic meters of ethanol) with the expectation it will be profitable.

Figure 7.1 also indicates that a small facility, for example, one producing 100,000 gallons or 400,000 liters (400 cubic meters) per year will do so spending \$350,000 in capital with the expectation that this will also be profitable when other costs are taken into account. This exercise must go on at each site globally and each specific situation evaluated. A methodical procedure is involved to develop a capital cost for a facility starting with the creation of a process flow chart and follow-up with many considerations germane to a specific case in a specific world locale. (Goldstein, W.E. et al. 1980; Hunt, V.D. et al. 1982; Durante, D.A. and Sneller, T.C. 2006).

Figure 7.2 shows utility cost data for ethanol manufacturing using corn wet milling feedstock. Utilities costs were projected to 2014 as indicated. The composite costs at each level of ethanol capacity are used in computing manufacturing costs. Increases in prices are taken into account.

Figure 7.3 shows costs for chemicals and biochemicals. The chemical costs show a downward trend with increasing capacity as might be expected. Since technology

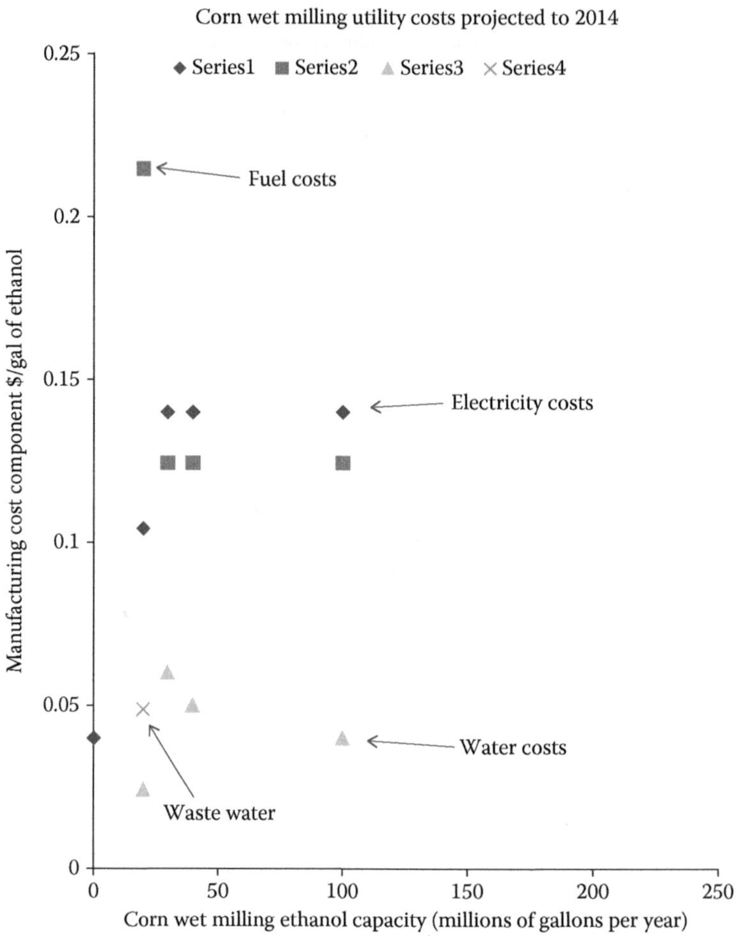

FIGURE 7.2 Utility costs for ethanol manufacturing from corn wet milling as a function of capacity.

advances in yeast and enzymes are expected, a cost increase factor of 1.5 is projected into the analysis to cover that aspect in estimating prices in 2014 for the case of corn wet milling feedstock applied to ethanol manufacture.

Figure 7.4 covers projected costs for labor and overhead. Since personnel are more effectively utilized from a cost standpoint as ethanol capacity is increased, these costs are expected to decrease in $/gal of ethanol as capacity increases (Goldstein, W.E. et al. 1980; Goldstein, W.E. 2014).

Figures 7.5 and 7.6 are composite charts for ethanol manufacture (using corn wet milling feedstock) that illustrate the impact of feedstock costs, nonfeedstock costs, and capital charges (due to the effect on manufacturing cost capital investment charges). The sum of feedstock costs and nonstock costs equals the total manufacturing cost.

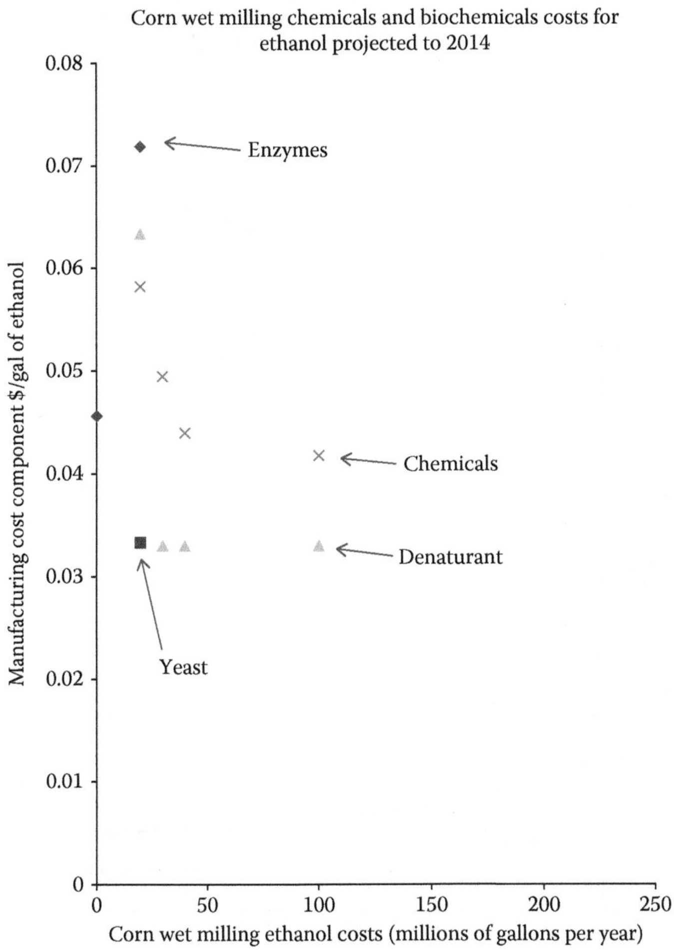

FIGURE 7.3 Nonfeedstock material costs for ethanol manufacturing from corn wet milling as a function of capacity.

In Figure 7.5, the total manufacturing cost is computed including capital charges. It ranges from an extrapolated value of about $2.5/gal at very small ethanol capacity near 150,000 gallons per year to $1/gal at 225 million gallons per year. Feedstock shows a slight change with capacity but is basically constant at $0.90/gal of ethanol. This is expected since its availability and price are dictated by agricultural and bartering factors. Nonfeedstock costs (nonfeedstock raw materials like chemicals, enzymes, and yeast; utilities; labor and overhead; and capital charges in the case of Figure 7.5) range from about $1.90/gal at very low ethanol capacities (150,000 gallons per year) to near $0.1/gal at about 225 million gallons per year.

In Figure 7.6, the total manufacturing cost (comprising feedstock and nonfeedstock costs) is computed with capital charges not included in manufacturing costs. As indicated, the manufacturing cost is lower compared to the case where capital

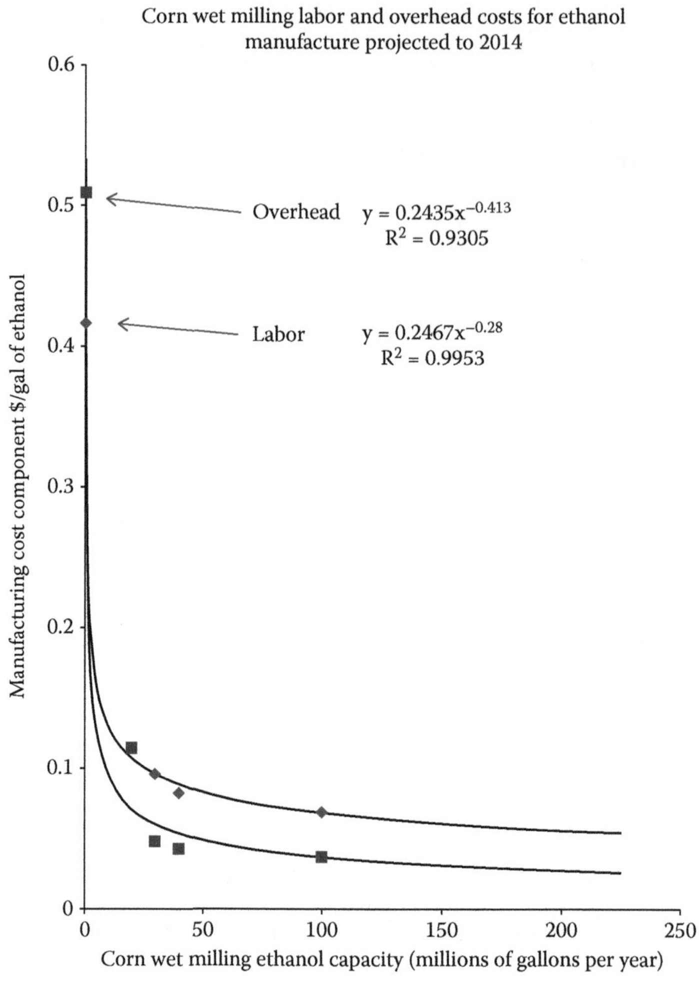

FIGURE 7.4 Labor and overhead costs—corn wet milling to ethanol facility—as a function of capacity.

charges are included (these costs might have to be shown and justified to obtain tax deductions are pointed out earlier). The chart only projects to 100 mm gal/year ethanol due to lack of data at 225 mm gal/year.

7.4 COST OF ETHANOL PRODUCED USING FEEDSTOCK FROM CORN DRY MILLING FACILITIES

The costs related to dry corn milling are considered in Figures 7.7 through 7.13. The capital costs from the literature for an ethanol facility supplied by feedstock provided by corn dry milling are shown in Figure 7.7. As indicated, the costs derived by various workers are found to be lower than those shown in Figure 7.1 for ethanol

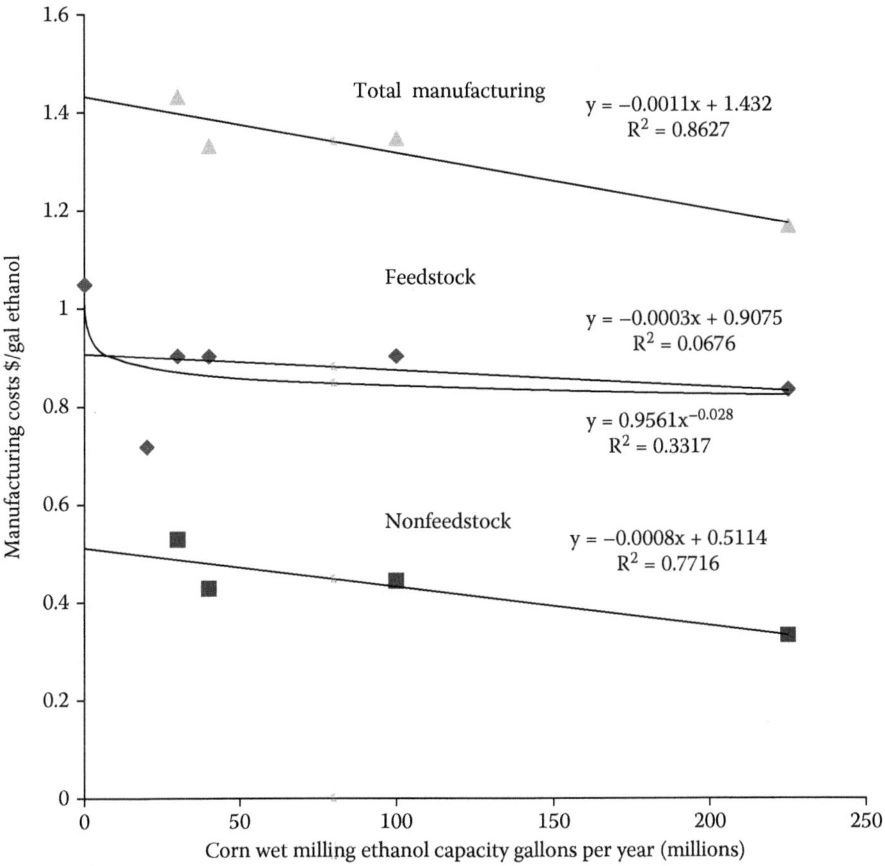

FIGURE 7.5 Feedstock and nonfeedstock costs—corn wet milling—as a function of ethanol capacity.

produced using corn wet milling feedstock. Basic corn dry milling involves the use of whole corn ground and milled and treated with enzymes (the same ones used for corn wet milling) and is a relatively simpler process. Note that reviewing options of by-product recovery may result in consideration of a facility that is a hybrid of corn wet and dry milling as discussed earlier. Such an option is well worth evaluating as another way to make this area more profitable.

An ethanol facility using corn dry milling feedstock that the writer and his team help develop to successful and presumably profitable operational status (New Energy of Indiana) in the mid-1980s required an investment of $1.8/gal for a 100 million gallons per year facility (Yang, R.D. et al. 1981; Voegele, E. 2014a,b). This cost is at a higher capacity than shown in Figure 7.7 and would be above the curve to about $2.5/gal if projected to 2014. The New Energy of Indiana facility was designed and constructed when this industry was more in the inception phase so costs would be higher since not as much was known and this was relatively new territory as an enterprise.

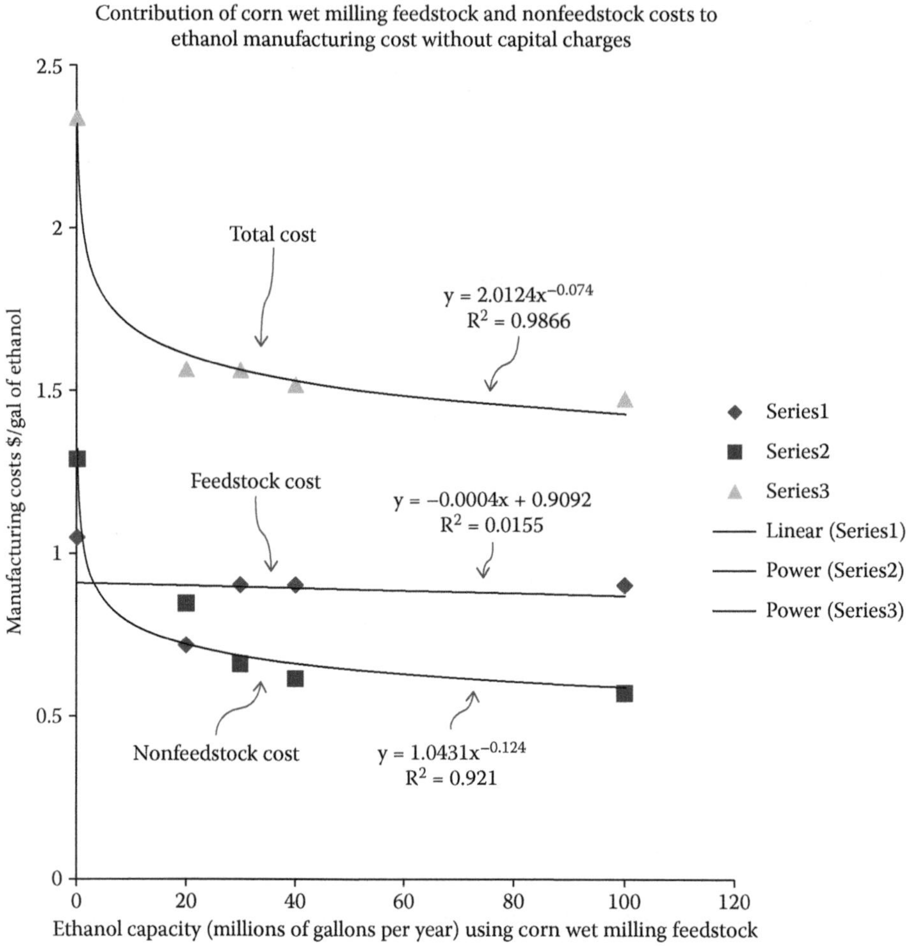

FIGURE 7.6 Manufacturing cost of ethanol based on using corn wet milling feedstock.

Figure 7.8 shows the utility costs for ethanol production where the feedstock is provided by corn dry milling. Considering the scatter among data from several workers, the utility costs from corn wet (Figure 7.2) and drying milling are not necessarily distinct from each other statistically. The utility costs ($/gal of ethanol) for wet and dry milling feedstock cases are in the same order of relative magnitude.

Figure 7.9 shows the evident dependence of overhead and labor costs ($/gal of ethanol) for corn dry milling feedstock sources. These costs are very comparable to those for corn wet milling sources of ethanol in Figure 7.4.

Costs ($/gal ethanol) for enzymes, yeast, and chemicals shown in Figure 7.10 are again comparable to those found in the case for corn wet milling (Figure 7.3).

Figures 7.11 through 7.13 show the dependence of manufacturing cost on variables where feedstock is supplied by corn dry milling for situations where capital charges are included or excluded. Manufacturing costs with capital charges (Figure 7.11) are

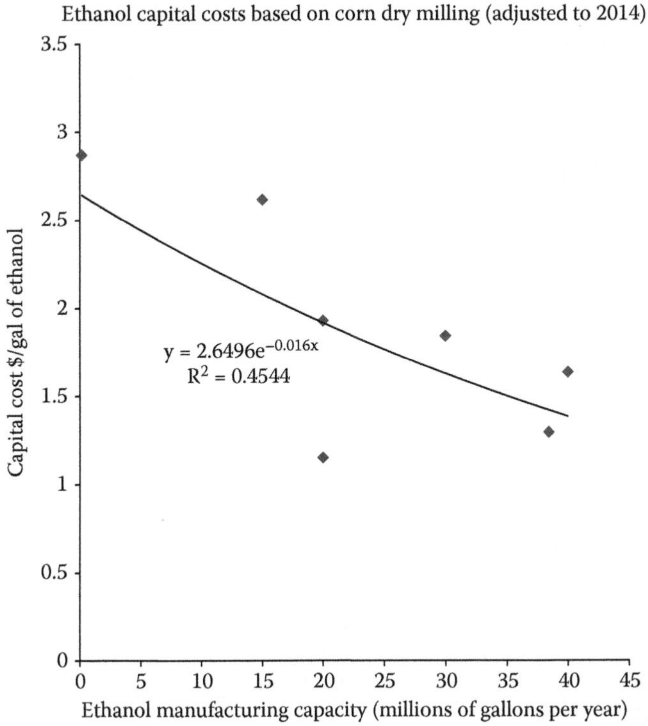

FIGURE 7.7 Capital costs for ethanol manufacture based on corn dry milling feedstock.

about \$0.4/gal above those without capital charges, representing a considerable cash flow opportunity.

As shown in Figure 7.12, exclusion of capital charges and inclusion of feedstock indicates the significant influence of feedstock on costs compared to nonfeedstock costs. Figure 7.13 shows the total cost (\$/gal of ethanol) with inclusion of feedstock (supplied by corn dry milling) and capital charges. As evident comparing this figure to the same case for Figure 7.5 (feedstock from corn wet milling), the manufacturing cost of ethanol produced using feedstock from corn dry milling is less expensive than the ethanol manufacturing cost using feedstock from corn wet milling. This is why corn dry milling is preferentially used to provide a feedstock for ethanol manufacture. However, there are profitable opportunities as noted in using a hybrid process that provides additional by-products including those with differences that may allow some new opportunities for the industry.

7.5 COST OF ETHANOL PRODUCED USING FEEDSTOCK FROM LIGNOCELLULOSE SOURCES

Figure 7.14 provides the graph of ethanol capital cost as a function of ethanol capacity. The curve has a sharp upward trend indicative perhaps of an emerging technology with cost reduction due to technology advance expected to occur in the future.

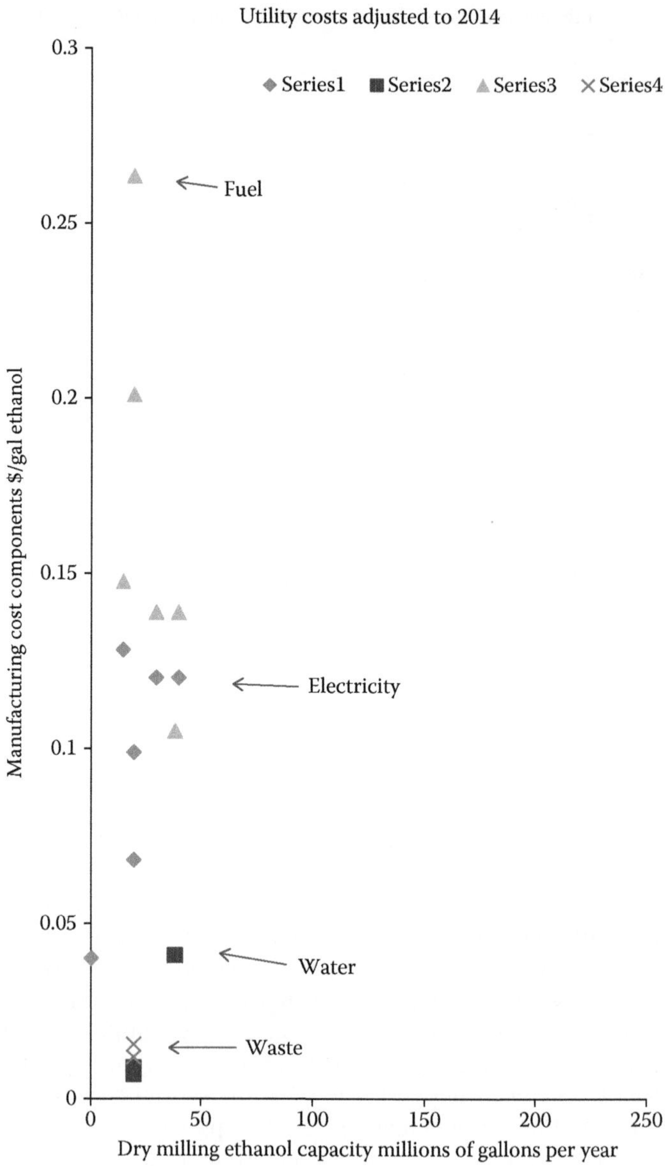

FIGURE 7.8 Utility costs for ethanol manufacturing based on corn dry milling.

Low production rates and the need for larger fermentation vessels, other equipment, and addition runs to meet capacity needs raise costs. The points are quite clustered and additional data to be gathered as this technology develops should show lower costs and reduction in the variance in the data (Whims, J. 2002; Kwiatkowksi, J.R. et al. 2006; Shapouri, H. and Gallager, P. 2005; Duff, B. 2013; Shapouri, H. and Salassi, M. 2006).

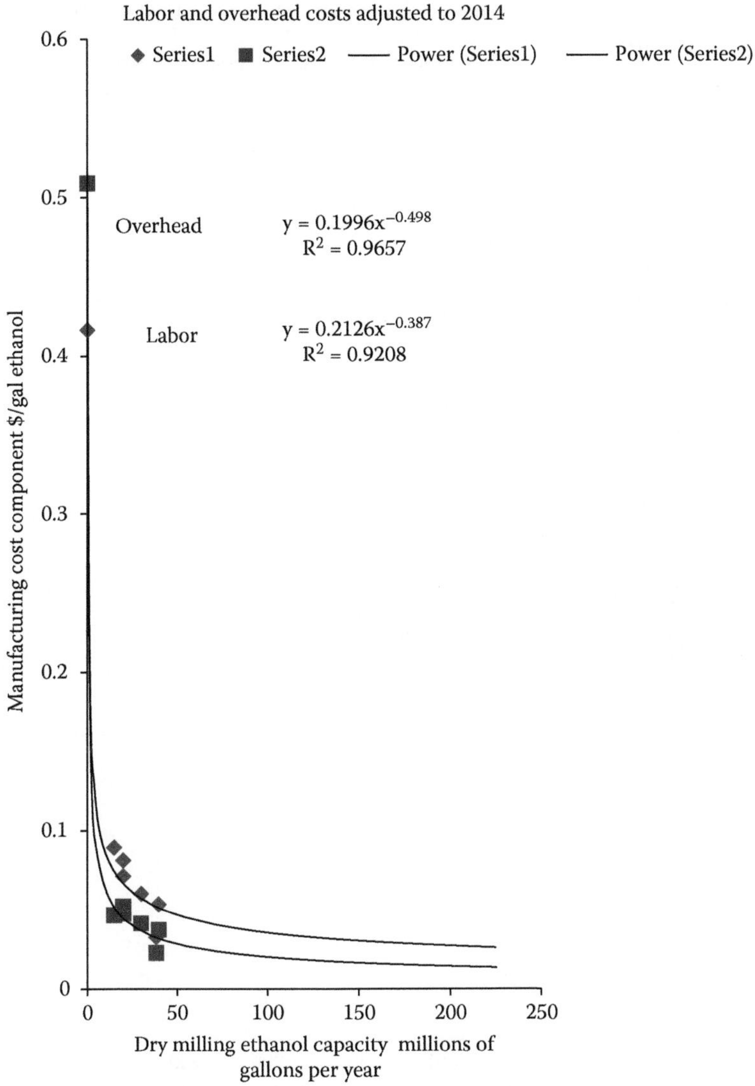

FIGURE 7.9 Labor and overhead costs for ethanol manufacture based on corn dry milling.

Figure 7.15 shows the graphs of total manufacturing cost as a function of capacity. Even though more feedstock is used since productivity is lower the feedstock cost is natively less than for corn so higher feedstock costs due to using more raw material may be balanced somewhat by a lower basic raw material cost per unit of feedstock. The basis cost of lignocellulose is not insignificant since once it is discovered to be of use, more is charged for it (Aden, A. et al. 2002; Kumar, D. and Murthy, G.S. 2011; Wooley, R. et al. 1999; Wright, M.M. and Brown, R.C. 2007). The elevation of these cost numbers in estimates represents work-in-progress since

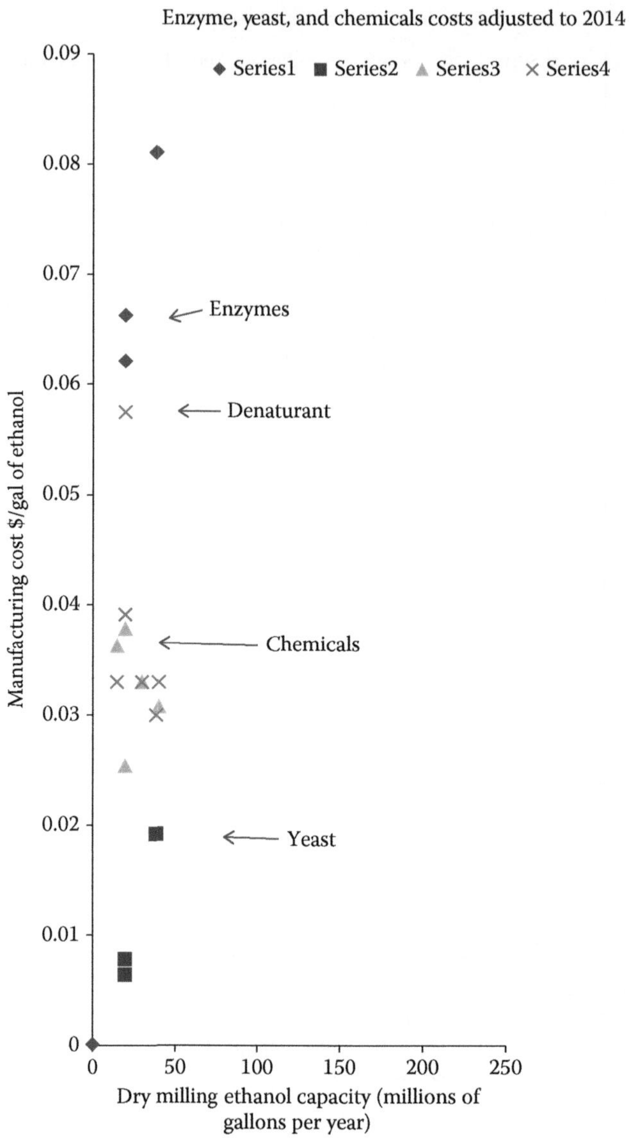

FIGURE 7.10 Enzyme, yeast, and chemicals cost in ethanol manufacture using corn dry milling feedstock.

these costs are expected to drop as productivities improve with manufacturing experience and progress in research and development that is expected to occur. The variances in data and correlation coefficients are expected to improve as the experience with lignocellulose sources occurs.

Figure 7.16 shows the significant effect of capital charges on manufacturing costs relative to feedstock costs. Residual costs (representing nonfeedstock raw materials,

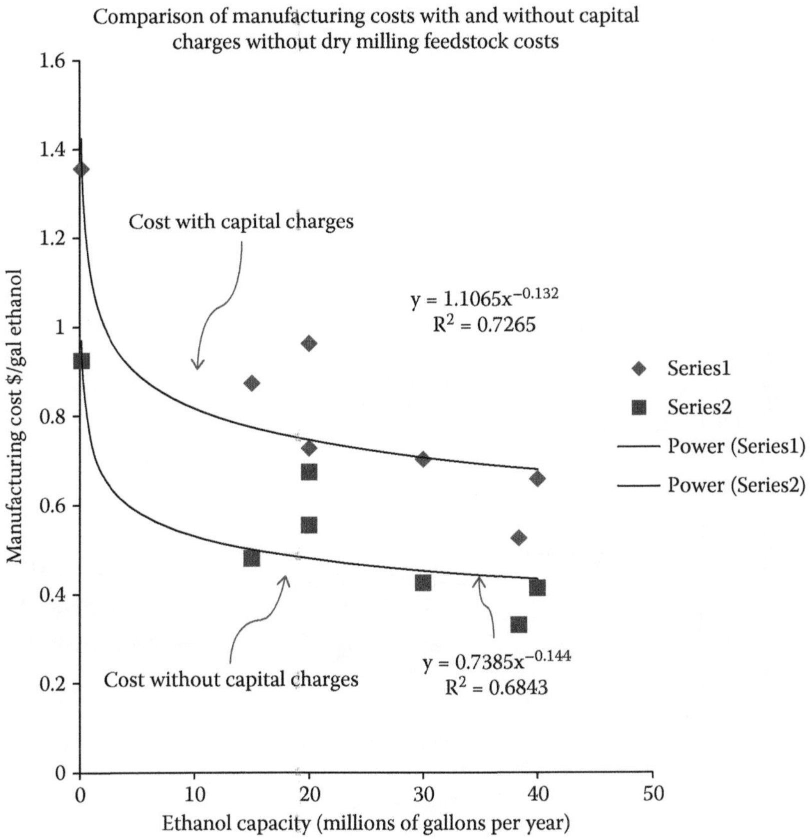

FIGURE 7.11 Comparison of manufacturing costs with and without capital charges (not including feedstock).

utilities, labor, and overhead), as shown in Figure 7.17, are in the proximate ranges of such costs in corn wet and dry milling after subtraction of feedstock costs and capital charges. These costs will follow suit and drop as productivity (better use of raw materials) due to more output per unit time and per unit volume improves. This will result in lignocellulose conversion processes getting closer to wet and dry mill feedstock sources in terms of productive use of raw materials.

7.6 CONTRAST IN COSTS TO PRODUCE ETHANOL FROM FEEDSTOCKS

Table 7.1 provides estimated costs to contrast three ways to select and process feedstock to make ethanol (by corn dry milling, corn wet milling, and lignocellulose processes). The thinking applies to considering any feedstock.

Based on charts of cost in this section as a function of capacity, the writer has examined costs as provided from treatment of data in the literature plus judges (from

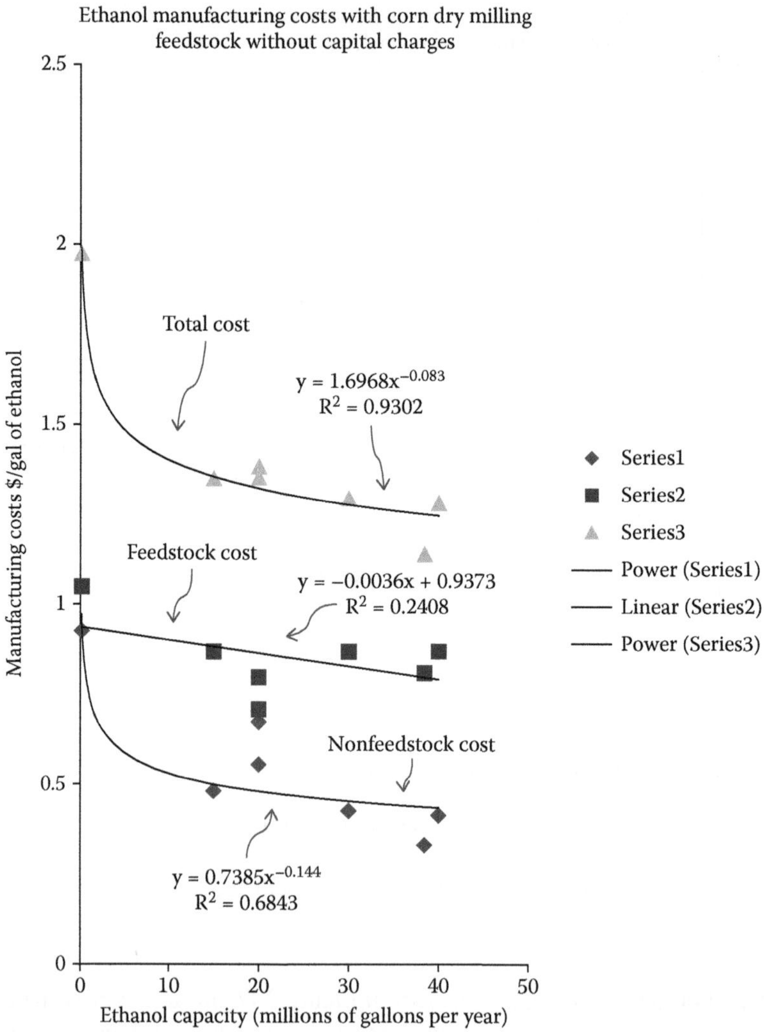

FIGURE 7.12 Ethanol manufacturing costs with corn dry milling feedstock without capital charges.

his experience) and anticipates improvements in technology and cost that will result from attaining a next level of development. A manufacturing capacity of 50 million gallons per year is selected since that seems to be a level (from examining Figures 7.1 through 7.17) where the advantage of cost reduction as a function of ethanol capacity is more evident.

Corn dry milling is the most widely accepted method in the United States to produce ethanol at present. Based on Figure 7.7 and associated predictive thinking, capital costs in the range of \$1.3–\$1.5/gal ethanol are expected. Enzyme technology will show some modest improvement in this area plus bioreactor design and

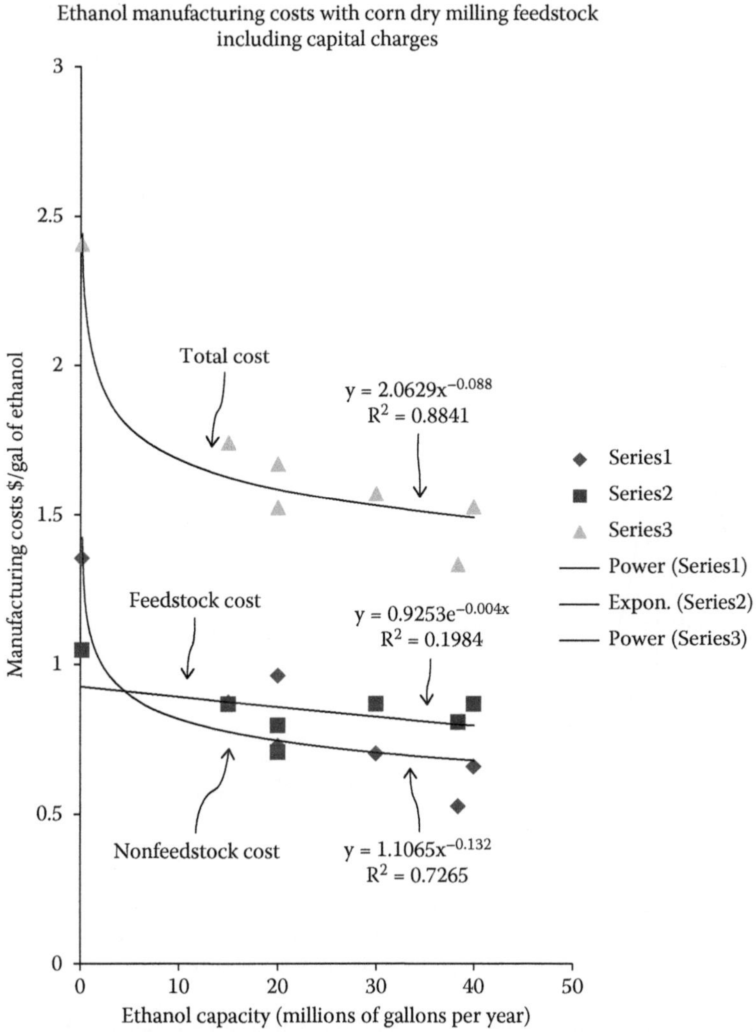

FIGURE 7.13 Ethanol manufacturing costs using corn dry milling feedstock as a function of ethanol capacity.

operation will improve. From these steps and positive impact on other cost areas, \$1.6/gal ethanol manufacturing cost before any by-product sales credit is anticipated. Currently, from Table 7.1 it is clear that it is the least expensive route.

The second route, corn wet milling, is the next costliest (estimated \$2.25–\$2.6/gal capital cost and \$2/gal ethanol manufacturing cost). However, as noted previously, corn wet milling is designed and operated to produce an array of products from the corn kernel, for one or another type of human food and animal feed. It is designed and has evolved to be cost optimal for that. Therefore, it can show reduction against cost for sale of by-products. The same can be said for corn dry milling except corn

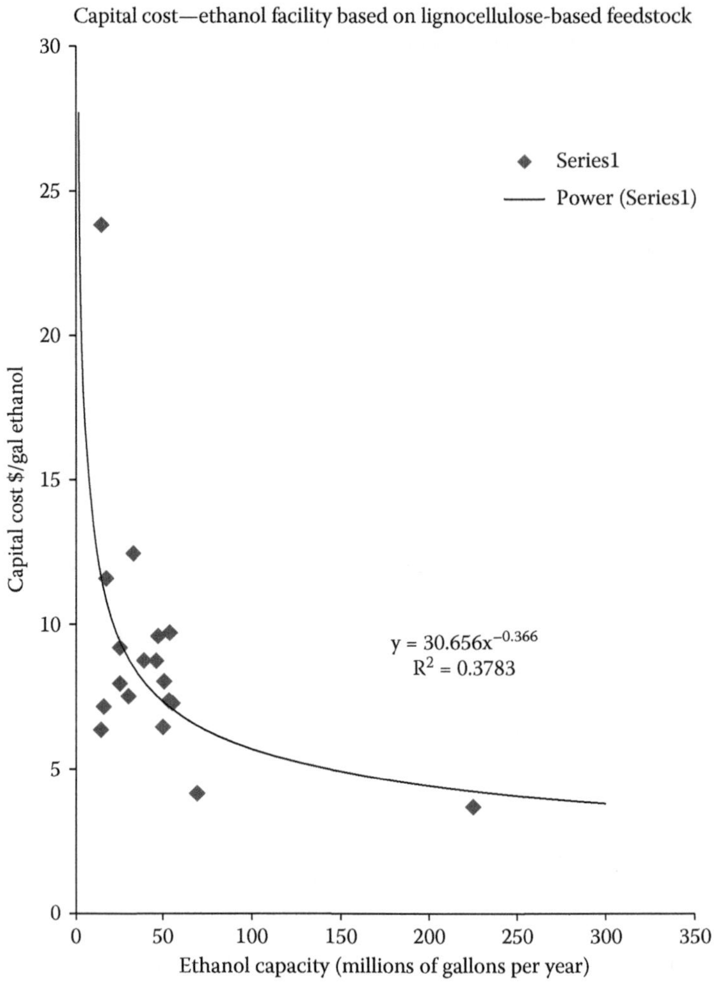

FIGURE 7.14 Capital cost of ethanol facility (lignocellulose-based feedstock) as a function of capacity.

dry milling is relatively a simpler process than corn wet milling, so it is expected to be relatively less expensive for producing ethanol. Ethanol is only an advantageous product for corn wet milling if there is extra capacity for it and where it can be an attractive product for this type of processing as well.

The third type of feedstock, just coming into being as a commercial industry, is the use of lignocellulose sources to produce ethanol. Its predicted capital costs are relatively higher ($4–6/gal ethanol capital cost). However, this cost is a slight drop below the data set in Figure 7.14 since it is expected that there will be technology advances to improve lignocellulose processes to a next level, in terms of feedstock processing and digestion linked to enzymology, resulting in improved bioreactors impacting to help lower capital costs. Further, these same improvements will impact

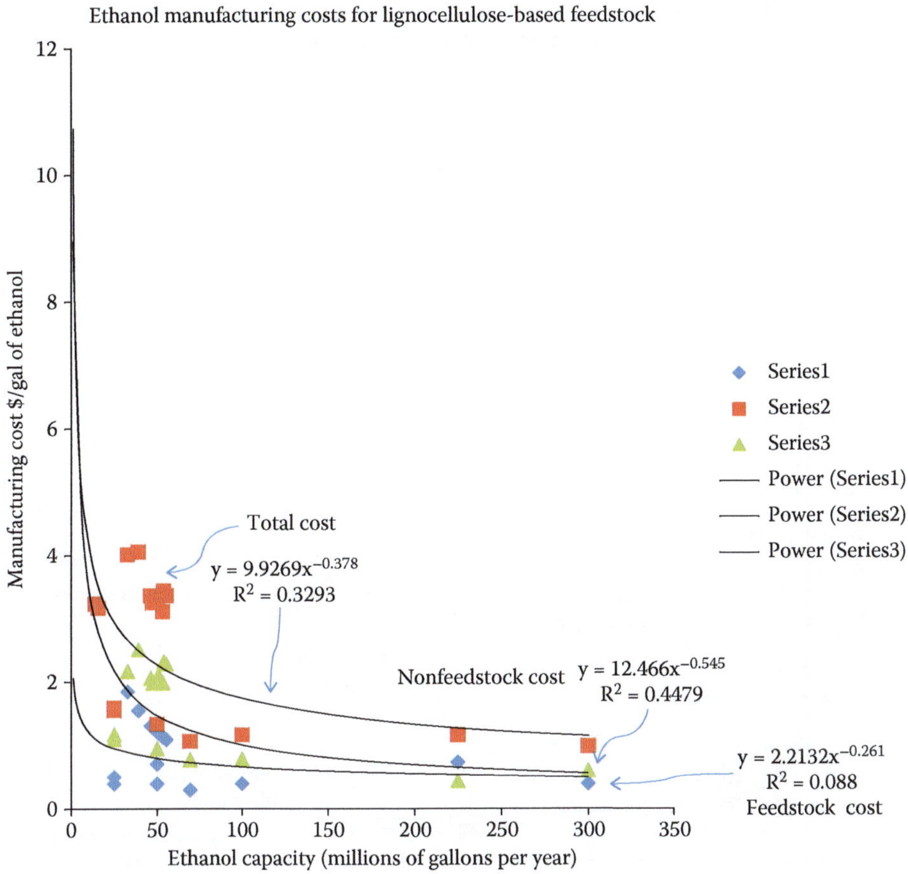

FIGURE 7.15 Ethanol manufacturing costs (lignocellulose-based feedstock) as a function of ethanol capacity.

on manufacturing costs resulting in a manufacturing cost of $2.25/gal (on the low end of the curve in Figure 7.15). The industry is new and is evolving commercially and from a technological basis. A best way to look at this in order to develop lignocellulose processes is to set a goal to try to get as close as possible to corn dry milling costs.

One needs to construct a facility where the capital investment is predicted to be paid off in an acceptable period of time and with an acceptable return on the investment. It is clear that the investment amount cannot be too high so as to heighten the risk of the investment. So what may happen is that the facility is designed to start at a lower level of capacity that is increased in time to take advantage of technology advances in the process that reduce capital costs. The facility size is therefore a judgment call to meet risk plus best anticipate satisfying target markets.

The costs shown in Table 7.1 are based on predictions in this analysis as to what they are expected to be in the 2014–2018 period. Between 30 and 100 mm gal/year,

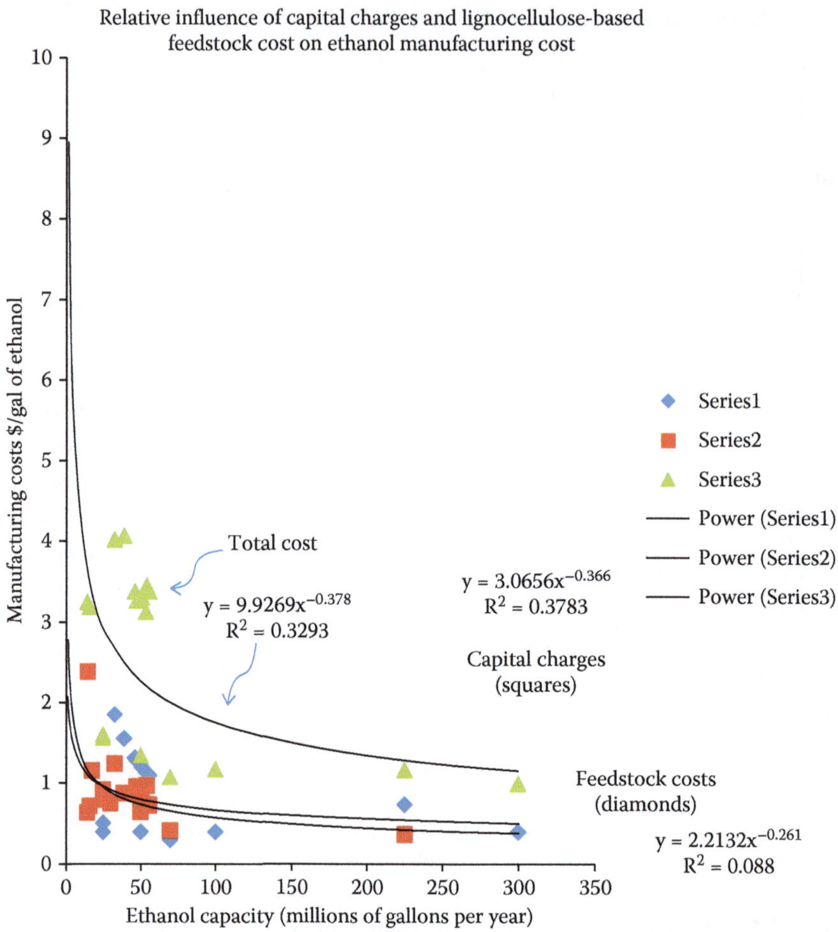

FIGURE 7.16 Relative influence of capital charges and lignocellulose-based feedstock costs on ethanol manufacturing costs.

from the historical literature analyzed, the capital cost per gallon of ethanol produced, shows a tendency to decrease as capacity increases, so 50 mm gal per year is a nice in-between point, as too small a plant results in too high a predicted capital cost per gallon. From Table 7.1, dry milling at $1.4/gal is the least expensive resulting in a $70 mm facility at 50 mm gal per year. Corn wet milling is next at $2/gal on the low end, resulting in a $100 mm facility at 50 mm gal per year. Note, that by definition, both corn dry milling and wet milling involve significant multiproduct facilities with many ways to pay off the investment besides producing ethanol, some still to be identified and produced.

As indicated in Table 7.1, based on historical analyses, projected to the present, lignocellulose-related processes are predicted to be the most expensive in terms of capital invested (between $4–6/gal of ethanol, capital costs between $200 and

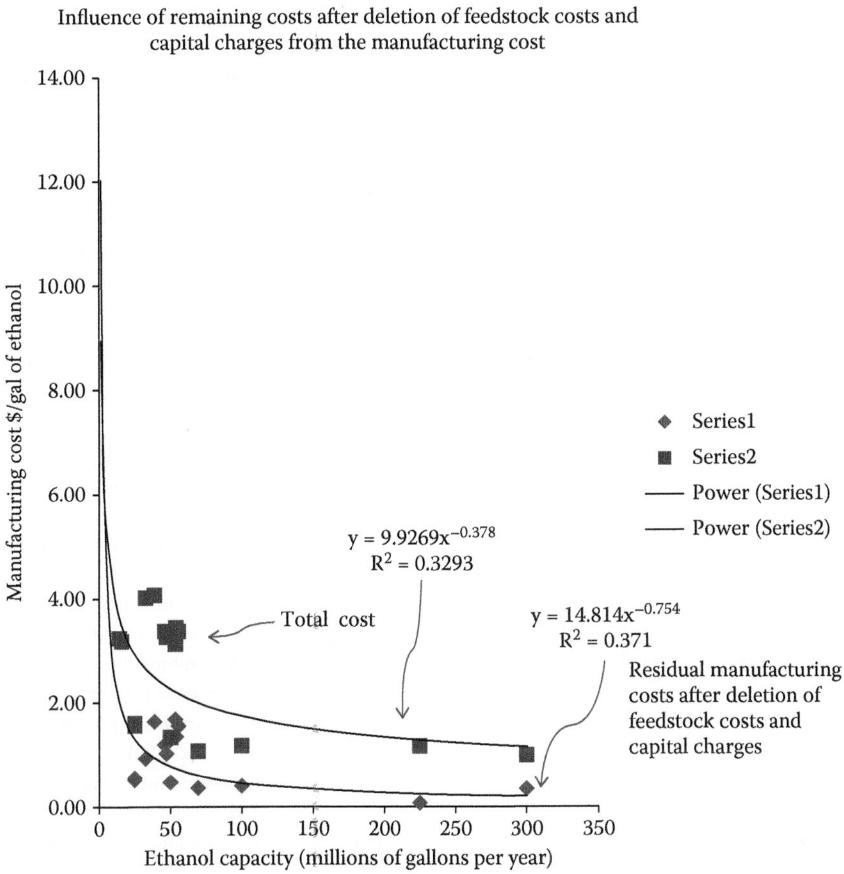

FIGURE 7.17 Influence of remaining costs after deletion of feedstock costs and capital charges from the ethanol manufacturing cost using lignocellulose feedstock.

$300 million at 50 million gallons per year capacity). Based on past analysis, sharp rises in capital costs for lignocellulose-supplied ethanol plants are predicted in regions below 100 million gal/yr as shown in Figure 7.14. Basically, this makes prediction of such costs at early stages subject to considerable variance and difficult to apply. However, advances in technology are expected to bring about cost reductions as noted. One way they are reflected is the increase in enzyme costs that are expected to be compensated for by reduced capital and manufacturing costs that arise from improvements in productivity. This is the "win–win" scenario discussed in this book that benefits ethanol producers and their suppliers.

This variance can be reduced by including many factors in design, which have to be proven but can impact on making such predictive costs decidedly much lower. So, since the process is new anyway, it is best to factor in such improvements using research and development and process development to prove and implement such positive inputs to manufacturing.

TABLE 7.1
Comparison of Ethanol Costs (50 mm gal/year) as a Function of Feedstock

Costs $/gal Ethanol	Corn Dry Milling	Corn Wet Milling	Lignocellulose Source	Comments as regards to Improvements
Capital invested	1.3–1.5	2.25–2.3	4.0–6.0	Improved productivity lowers cost extensive enzyme reaction development lignocellulose (ligno) area Bioreactor design to lessen capital costs Special lignomilling linked to continuous ethanol fermentation Special processing and enzyme digestion and milling, extend reaction time Extend/retrofit existing facilities or build anew Implementing research and pilot plant development
Capital charges	0.21	0.435	0.6	New plant more insurance, less experience more maintenance Clever continuous process design for lignocellulose Lower capital costs lowers capital charges
Feedstock	0.78	0.9	0.65	Lignocellulose feedstock has less starch content, harder to convert, natively less expensive
Denaturant	0.04	0.04	0.04	Should be same
Chemicals	0.035	0.043	0.03	Wet milling a little more required
Enzymes	0.075	0.06	0.15	More sophisticated enzymes should double cost in ligno area
Yeast	0.02	0.033	0.02	
Electricity	0.12	0.14	0.12	
Fuel	0.2	0.14	0.2	
Water	0.04	0.05	0.04	
Wastewater	0.02	0.05	0.02	
Labor	0.04	0.08	0.03	More continuous for ligno and more controls lessens cost
Overhead	0.035	0.05	0.03	Less cost due to automation
Operating manufacturing cost estimate	1.6	2	2.25	
DDG by-product credit	**0.45**	***0.45***	***0.45***	

(*Continued*)

TABLE 7.1 (*Continued*)
Comparison of Ethanol Costs (50 mm gal/year) as a Function of Feedstock

Costs $/gal Ethanol	Corn Dry Milling	Corn Wet Milling	Lignocellulose Source	Comments re. Improvements
Adjusted manufacturing cost for by-product credit	1.15	***1.55***	***1.80***	
Projected nominal profit ($/gal)	0.35	***–0.05***	***–0.30***	
		DDG Analysis and Impact on Gross Profit		
By-Product DDG Credit				
DDG prices January 2016 ($/ton)	150	***150***	***150***	Necessary for profitable operation
$/kg	0.165	***0.17***	***0.17***	Preserve through sanitation and nutritive additives
mm bushels of corn at 2.8 gal/bushel	17.86	***17.9***	***17.9***	New era of nutritive research on animal food and health
mm kg of corn needed	455	***455***	***455***	
mm kg of DDG at 0.3 per unit of corn	136	***136***	***136***	Ligno has to prove its feed quality, lignin is nonnutritive
$mm provided from DDG sales	22.5	***22.5***	***22.5***	New markets for lignin new forms like nano
DDG by-product product for corn ($/gal)	0.45	***0.45***	***0.45***	Materials testing and development needed
Ethanol prices December 2015 $/gal	1.5	1.5	1.5	*Ethanol Producer Magazine* (Bryan, T. 2015)

Note: ***Bold and italics*** emphasize and "what if" scenario.

The capital charges are composed of several factors. One is depreciation, a cash flow that allows capital costs to be deducted from IRS reported profits. However, depreciation is a real cost as well since equipment and processes deteriorate requiring replacement. Depreciation is an incentive to put profits away in a sinking fund to be recovered later when needed. Capital charges also include insurance, taxes, and maintenance. One may seek savings in these areas. However, higher investments involve higher charges in these areas. So, as expected, capital charges are least for corn dry milling, more for corn wet milling, and highest for lignocellulose processes. As capital costs decrease due to technologically related improvements, capital charges will decrease as well. A major factor to consider is utilities such as electricity, fuels, water, and waste treatment (Walton, B. 2014). Attention must be made to optimize these and consider captive utilities to optimize costs. There are many ways to accomplish this as mentioned in this book.

In this example, feedstock costs involve corn in one case and a variety of lignocellulose sources in the other. Feedstock costs based on historical analyses are lower for corn dry milling (Table 7.1). They are the highest for corn wet milling. They are predicted to be lower for lignocellulose sources such as corn stover, the waste from the corn plant after the corn kernel is harvested for use. As corn stover is used more, it will rise in value, still remaining well below basic corn costs.

Parties involved in the use of corn stover will manage pricing to protect this fledging lignocellulose industry and cooperate but still push it to keep their share of revenue. Other arrangements are possible such as to share in ethanol revenues.

The cost is a function of the price of corn, which will vary up and down as a consequence of market forces. A stabilizing factor is the increase in corn yield per acre that has continuously occurred that can help keep prices down. Since the carbohydrate content in corn stover is less than that in the corn kernel, this will be a factor to make the corn stover source relatively more costly. Demand for ethanol is a push to raise corn prices. In general, there is no particular trend in feedstock cost as a function of ethanol capacity (Figures 7.5, 7.6, 7.12, 7.13, 7.15, and 7.16 in Chapter 7 show feedstock Cost). However, persons in the business will acquire contracts for corn futures to be able to purchase corn at favorable prices in this complex area.

Table 7.1 also shows the sale of by-products to help reduce cost. A proven saleable by-product is Distillers Dry Grains (DDG). The sample calculation made in Table 7.1 shows that in the case of corn dry milling, with ethanol sales at $1.50 per gallon, it results in a DDG by-product credit of $0.45/gal of ethanol. A reduced manufacturing cost for ethanol based on using feedstock from corn milling equal to $1.15/gal results by applying the by-product credit. Since ethanol is selling in this example for $1.5/gallon, it results in a gross profit of $0.35/gal.

If one assumes that ethanol from corn wet milling-produced feedstock also can benefit equivalently and analogously from the DDG by-product sales, then the manufacturing cost for the ethanol via this route is reduced to $1.55/gal point or a near breakeven with ethanol sales revenue with ethanol priced at $1.50 gallon.

If one assumes that the ethanol from lignocellulose sourced feedstock can also produce an equivalent DDG by-product, then the $0.45/gal by-product credit applies to this case resulting in reduction of the manufacturing cost to $1.80/gal. This represents a $0.30/gal loss, which is getting close to breakeven. Given that some of the improvements noted in this book are implemented in all ways possible, then it is anticipated that the lignocellulose-based process can be made cash flow positive earning a positive gross profit in the next stage perhaps after the 2014–2018 period.

7.7 PROCESS OPTIMIZATION AND COST REDUCTION TO REACH COMMERCIAL OBJECTIVES

From the manufacturing cost standpoint, the cost of feedstock is dependent how efficiently it is used in processes to make ethanol. Corn is simply wasted and not converted properly at times, which is a cost managed by ever improving process quality management. So, one would work to raise the yield of ethanol per unit of feedstock through process improvement. One would make the processes as continuously operating as possible since this can tend to promote better use of feedstock and lessen losses. Furthermore, if continuous processes in making ethanol can be employed, optimizing the amount of ethanol per unit of biomass as discussed earlier in Chapter 6, less biomass is created and used. There is a lower limit on this since one wants to produce enough biomass so that eventually a stable by-product Distillers Dry Grains Solubles (DDGS) is produced (important for profitability).

In the case of lignocellulose-based processes, this is a grand opportunity to push continuous processing with optimal biomass use since this is believed to result in a very competitive lignocellulose-based process that can rival corn processes in many respects, if (and this is a big if) the technology development proceeds to its maximum. Lignocellulose has not been proven to be a proper source of animal DDGS feed. Certainly, part of the lignocellulose, the lignin, may be nonsuitable, antinutritive, and even toxic. So the lignin should be separated perhaps early in the lignocellulose processing.

Continuous enzyme reactors should be made to act on "lignin-less" biomass earlier with economic milling using the mix of enzymes and pH/temperature conditions optimal for the feedstock as discussed earlier. Once sufficiently solubilized, the liquefied mix can enter into the continuous conversion (continous stirred tank reactors [CSTR] to plug flow reactors) where biomass is being created in a controlled way prior to the primarily anaerobic ethanol formation. This may allow one to practice EVOP (evolutionary operating practice) and reduce residence times and achieve sufficient ethanol yields concordant with reaction characteristics discussed earlier for ethanol. Given this, the effective capital cost per gallon of ethanol may be pushed to the \$4/gal lower limit in the author's opinion or even further depending on success in process research and improvement tasks. If the government decides to provide loan financing for the facility at a low interest rate without penalties for default, lessening capital charges (if the company does not take title to the capital equipment for a period), then the capital charges may be reduced to almost half of what is seen in Table 7.1 for lignocellulose (or to \$0.3/gal ethanol). Process innovations across the board could reduce the capital to as much as \$3.5/gal ethanol. It depends on how much the government will do to help such facilities get "kick-started." Loan forgiveness has been allowed for solar firms, popular or not so popular with various parties.

In regard to manufacturing cost as projected and possible improvements, costs for dry milling, wet milling, and lignocellulose are in the range of each other. The enzyme cost for lignocellulose has to be higher as indicated since more sophisticated enzyme processes are involved. That is factored in Table 7.1 based on the estimate based on the author's experience.

Utility costs such as electricity are a function of providing what is needed for agitation, for example, in the fermentors to possibly achieve conditions desired for yeast flocs as this writer postulated. If these practices are well done, good chemical engineering is practiced. The writer believes that ethanol will be profitable from corn and lignocellulose sources if the market holds and make ethanol advantageous for use. It is the writer's contention that such profits will extend to parties worldwide who successfully translate such practices to their own feedstock and country.

APPENDIX 7A

TABLE 7A.1
Ethanol Plant Capacities and Costs

Year	Type of Process	Ethanol Capacity mmgal/year	Feedstock	Ethanol 2.8 gal ethanol/bush	gal ethanol/kg feedstock	Feedstock (mmbush/year)	lbs/bushel 56 $/bushel	lbs/kg 2.2 Feedstock (mm kg/year)	Days/year 365 Feedstock ($/kg)	L/gal 3.786 Feedstock ($mm/year)	kg/mg 1000 Feedstock ($/gal ethanol)
2002	Dry mill	15	Corn			5.66	2.3	144.0727273	0.090357143	13.018	0.867866667
2002	Dry mill	30	Corn			11.32	2.3	288.1454545	0.090357143	26.036	0.867866667
2002	Dry mill	40	Corn			15.1	2.3	384.3636364	0.090357143	34.73	0.86825
2002	Wet mill	30	Corn			11.77	2.3	299.6	0.090357143	27.071	0.902366667
2002	Wet mill	40	Corn			15.69	2.3	399.3818182	0.090357143	36.087	0.902175
2002	Wet mill	100	Corn			39.22	2.3	998.3272727	0.090357143	90.206	0.90206
2006	Dry mill	38.4	Corn			14.10181818	2.2	358.9553719	0.086428571	31.024	0.807916667
2007	Cassava	14	Cassava								
2007	Cassava	26.7	Cassava								
1999	Grain	25	Corn							17	0.68
2007	Grain	225	Corn				2.12		0.083285714	187.5	0.833333333
2005	Ligno	50	Cellulose source							35.4	0.708
2007	Ligno	225	Cellulose source						0.05	164.925	0.733
2010	Ligno	53.4	Corn stover					730	0.083	60.59	1.134644195
2010	Ligno	50.7	Corn stover					730	0.083	60.59	1.195069034
2010	Ligno	32.8	Corn stover					730	0.083	60.59	1.847256098

(Continued)

TABLE 7A.1 (*Continued*)
Ethanol Plant Capacities and Costs

Year	Type of Process	Ethanol Capacity mmgal/ year	Feedstock	Ethanol 2.8 gal ethanol/ bush	gal ethanol/ kg feedstock	Feedstock (mmbush/ year)	lbs/ bushel 56 $/bushel	lbs/kg 2.2 Feedstock (mm kg/ year)	Days/year 365 Feedstock ($/kg)	L/gal 3.786 Feedstock ($mm/year)	kg/mg 1000 Feedstock ($/gal ethanol)
2010	Ligno	39.1	Corn stover					730	0.083	60.59	1.549616368
2010	Ligno	46.2	Corn stover					730	0.083	60.59	1.311471861
2010	Ligno	53.9	Corn stover					730	0.083	60.59	1.124118738
2010	Ligno	55.5	Corn stover					730	0.083	60.59	1.091711712
2010	Ligno	47.3	Corn stover					730	0.083	60.59	1.280972516
2002	Ligno	69.3	Corn stover					730	0.028	20.44	0.294949495
1999	Ligno	839.5	Corn stov/switch					730	0.0143	10.439	0.012434783
2002	Dry mill (sm)	20	Corn								0.7965
2002	Dry mill (lg)	90	Corn								0.8095
2011	Ligno least ethanol	14.14	Tall fescue								
2011	Ligno most ethanol	15.76	Tall fescue								
2013	BBI Intern dry mill	50	Corn		0.023191095	17.5		2156			
2013	BBI Intern dry mill	40	Corn								
2013	BBI Intern dry mill	30	Corn								
2013	BBI Intern dry mill	20	Corn								
2013	Ligno DuPont	17.25	Corn stover								

(*Continued*)

TABLE 7A.1 (*Continued*)
Ethanol Plant Capacities and Costs

Year	Type of Process	Ethanol Capacity mmgal/year	Feedstock	Ethanol 2.8 gal ethanol/bush	gal ethanol/kg feedstock	Feedstock (mmbush/year)	lbs/bushel 56 $/bushel	lbs/kg 2.2 Feedstock (mm kg/year)	Days/year 365 Feedstock ($/kg)	L/gal 3.786 Feedstock ($mm/year)	kg/mg 1000 Feedstock ($/gal ethanol)
2015	Ligno DuPont	30	Corn stover					750			
2013	Ligno Abengoa	14.56	Corn stover, grass								
2014	Ligno Abengoa	25	Corn stover, grass +					1474			
2014	Ligno Abengo Eur	1.32	Cellulose source								
2014	Wht bar Abengo Eu	52.84	Wheat, barley								
2013	Fiberight	2.3	Msw cellulose								
2013	Fiberight	6	Msw cellulose					1044			
2013	INEOS Bio	4.6	Gasification msw								
2013	Poet	14.56									
2014	Poet Royal DSM	20						1254			
2015	Vogelbusch	28.9	Corn					257.3			
2015	Vogelbusch	28.9	Wheat					295.65			
2015	Vogelbusch	28.9	Sugar beet					1095			

(*Continued*)

TABLE 7A.1 (*Continued*)
Ethanol Plant Capacities and Costs

Year	Type of Process	Ethanol Capacity mmgal/year	Feedstock	Ethanol 2.8 gal ethanol/bush	gal ethanol/kg feedstock	Feedstock (mmbush/year)	lbs/bushel 56 $/bushel	lbs/kg 2.2 Feedstock (mm kg/year)	Days/year 365 Feedstock ($/kg)	L/gal 3.786 Feedstock ($mm/year)	kg/mg 1000 Feedstock ($/gal ethanol)
2015	Vogelbusch	28.9	Sugar cane					1204.5			
2005	Dry mill	20	Corn	2.75	0.108	7.273	2.16	185.1851852	0.085	15.74074074	0.7074
	dm other table	20		2.75	0.108			185.1851852	0.085	15.74074074	0.7855
2005	Wet mill	20	Corn	2.65	0.104	7.55	2.16	192.3076923	0.085	16.34615385	0.7172
		20		2.65	0.104			192.3076923	0.085	16.34615385	0.866
2005	Cane sugar (US)	20	Sugar from cane		0.02145			932.4009324	0.03179	29.64102564	1.482051282
2005	Raw sugar (US)	20	Raw sugar		0.14894			134.2822613	0.4642	62.3338257	3.116691285
2005	Refined sugar (US)	20	Refined sugar		0.1551			128.9490651	0.5599	72.19858156	3.609929078
2005	Molasses (US)	20	Molasses		0.0763			262.1231979	0.0693	18.16513761	0.908256881
2005	Beet sugar (US)	20	Sugar from beet		0.0273			732.6007326	0.0431	31.57509158	1.578754579
2005	Cane sugar (Brazil)	20	Sugar from cane bz								
2005	Beet sugar (Europe)	20	Sugar from beet eu								
1982	Farmer corn	0.15	Small farmer corn	2.5	0.0982	0.0682	2.3	1.74	0.09036	0.1572264	1.048176

(Continued)

TABLE 7A.1 (*Continued*)
Ethanol Plant Capacities and Costs

Year	Enzymes ($)	Enzymes ($/gal ethanol)	Yeast ($)	Yeast ($/gal)	Enzymes ($/gal ethanol)	Denaturant ($)	Denat ($/gal ethanol)	Chemicals ($/gal ethanol)	Feed Stock Cost ($/gal ethanol)	Electricity Cost ($/gal ethanol)	Electricity ($)	Electricity (kWh)	($/kWh)
2002					0.055		0.03	0.033		0.064			0.04
2002					0.05		0.03	0.03		0.06			0.04
2002					0.048		0.03	0.028		0.06			0.04
2002					0.065		0.03	0.045		0.07			0.03
2002					0.06		0.03	0.04		0.07			0.03
2002					0.054		0.03	0.038		0.07			0.03
2006	2016,000	0.0525	477,000		0.0525	1,038,000	0.02703			0.027682	1063,000		0.05
2007													
2007													
1999													
2007													
2005													
2007													
2010													
2010													
2010													
2010													
2010													

(*Continued*)

TABLE 7A.1 (*Continued*)
Ethanol Plant Capacities and Costs

Year	Enzymes ($)	Enzymes ($/gal Ethanol)	Yeast ($)	Yeast ($/gal)	Enzymes ($/gal Ethanol)	Denaturant ($)	Denat ($/gal Ethanol)	Chemicals ($/gal Ethanol)	Feed Stock Cost ($/gal Ethanol)	Electricity Cost ($/gal Ethanol)	Electricity ($)	Electricity (kWh)	($/kWh)
2010													
2010													
2010													
2002													
1999													
2002		0.0377		0.0039	0.0377		0.0356	0.0231	0.7965	0.04			
2002		0.0365		0.0046	0.0365		0.0339	0.0228	0.8095	0.0349			
2011													
2011													
2013												3.50	
2013													
2013													
2013													
2013													
2015													
2013													
2014												7.67	
2014													

(*Continued*)

TABLE 7A.1 (*Continued*)
Ethanol Plant Capacities and Costs

Year	Enzymes ($)	Enzymes ($/gal Ethanol)	Yeast ($)	Yeast ($/gal)	Enzymes ($/gal Ethanol)	Denaturant ($)	Denat ($/gal Ethanol)	Chemicals ($/gal Ethanol)	Feed Stock Cost ($/gal Ethanol)	Electricity Cost ($/gal Ethanol)	Electricity ($)	Electricity (kWh)	($/kWh)
2014													
2013													
2013													
2013													
2013													
2014													
2015													
2015													
2015													
2015													
2005		0.0416		0.0049	0.0416		0.0541	0.0356	0.7074	0.0581			0.081
2005		0.0674		0.0312	0.0674		0.0594	0.0546	0.7172	0.0613			0.081
2005													
2005													
2005													
2005													
2005													
2005													
2005													
1982	6000	0.04							1.048176	0.02	3000		0.06

(Continued)

TABLE 7A.1 (*Continued*)
Ethanol Plant Capacities and Costs

Year	Fuel Cost ($)	Fuel Cost ($/gal ethanol)	Fuel	Gas Cost ($/mcf)	Btu/gal ethanol	Util Charge ($/mmBtu)	Steam ($)	Steam ($/gal	Steam ($/1000 lb)	Water (gal)	Water (gal)	Water ($)	Water ($/gal)	Water ($/1000 gal)	Waste .25 Wat	Sewage Cost ($/gal)
2002		0.118	Gas	3												
2002		0.111	Gas	3												
2002		0.111	Gas	3												
2002		0.084	Coal			2							0.03			
2002		0.084	Coal			2							0.025			
2002		0.084	Coal			2							0.02			
2006	3	0.0839	Gas			6	5	0.1316	7.76			922,000	0.02401	0.27		
2007																
2007																
1999																
2007																
2005																
2007																
2010																
2010																
2010																
2010																
2010																
2010																

(Continued)

TABLE 7A.1 (*Continued*)
Ethanol Plant Capacities and Costs

Year	Fuel Cost ($)	Fuel Cost ($/gal ethanol)	Fuel	Gas Cost ($/mcf)	Btu/gal ethanol	Util Charge ($/mmBtu)	Steam ($)	Steam ($/gal	Steam ($/1000 lb)	Water (gal)	Water (gal)	Water ($)	Water ($/gal)	Water ($/1000 gal)	Waste .25 Wat	Sewage Cost ($/gal)
2010																
2002																
1999																
2002		0.1607											0.0044			0.0077
2002		0.1099											0.0015			0.0041
2011																
2011																
2013					3.20					8	1.4				35,000,000	
2013																
2013																
2013																
2013																
2015																
2013																
2014																
2014																
2014																
2013																

(*Continued*)

TABLE 7A.1 (*Continued*)
Ethanol Plant Capacities and Costs

Year	Fuel Cost ($)	Fuel Cost ($/gal ethanol)	Fuel	Gas Cost ($/mcf)	Btu/gal ethanol	Util Charge ($/mmBtu)	Steam ($)	Steam ($/gal	Steam ($/1000 lb)	Water (gal)	Water (gal)	Water ($)	Water ($/gal)	Water ($/1000 gal)	Waste .25 Wat	Sewage Cost ($/gal)
2013																
2013																
2013																
2014																
2015																
2015																
2015																
2015																
2005		0.2107											0.0034			0.0067
2005		0.1449											0.0151			0.0305
2005																
2005																
2005																
2005																
2005																
2005																
2005																
1982																

(Continued)

TABLE 7A.1 (*Continued*)
Ethanol Plant Capacities and Costs

Year	Labor	Labor Cost ($)	Labor Cost $/gal	Capital Cost $mm	Capital Cost $/gal	Maint Cost $	Maint Cost $/gal	Deprec $mm	Deprec $/gal	Over-head $mm	Over-head $	Over-head ($/gal) ethanol	Other ($)	Other $/gal ethanol	Total ($/gal without feedstock)
2002			0.067	27	1.8		0.035	1.72	0.1147	0.524		0.03493			0.46
2002			0.045	38	1.26667		0.03	2.42	0.0807	0.93		0.031			0.41
2002			0.04	45	1.125		0.027	2.866	0.0717	1.12		0.028			0.39
2002			0.07	81.1	2.70333		0.06	5.165	0.1722			0.035		0.04	0.53
2002			0.06	96.4	2.41		0.055	6.14	0.1535			0.031		0.035	0.43
2002			0.05	167	1.67		0.05	10.64	0.1064			0.027		0.03	0.44
2006		1.04	0.02701			1,315,000	0.034245	4.664	0.1215		7	0.0188	496,000	0.01292	
2007															
2007															
1999				27.9	1.116										1.164
2007				111	0.49333										0.333
2005				294	5.88										1.518
2007				756	3.36										0.9
2010				376	7.0412										
2010				389	7.67258										
2010				391	11.9207										
2010				327	8.36317										
2010				386	8.35498										
2010				501	9.29499										
2010				386	6.95495										

(*Continued*)

TABLE 7A.1 (*Continued*)
Ethanol Plant Capacities and Costs

Year	Labor	Labor Cost ($)	Labor Cost $/gal	Capital Cost $mm	Capital Cost $/gal	Maint Cost $	Maint Cost $/gal	Deprec $mm	Deprec $/gal	Over-head $mm	Over-head $	Over-head ($/gal) ethanol	Other ($)	Other $/gal ethanol	Total ($/gal without feedstock)
2010				434	9.17548										
2002				197.4	2.84848										
1999				233.8	0.2785										
2002			0.0609	20	1		0.0319					0.0357		0.0035	
2002			0.0478	270	3		0.0474					0.0325		0.0043	
2011				91.36	6.4611										
2011				114.63	7.27348										
2013				55	1.1										
2013				48	1.2										
2013				40.5	1.35										
2013				30	1.5										
2013				197	11.4203										
2015	85 Full time			225	7.5										
2013				342	23.489										
2014	76 Fulltime	5.00	0.20	229.4	9.176										
2014															
2014															
2013				58.5	25.4348										
2013				60	10										
2013				126	27.3913										

(Continued)

TABLE 7A.1 (*Continued*)
Ethanol Plant Capacities and Costs

Year	Labor	Labor Cost ($)	Labor Cost $/gal	Capital Cost $mm	Capital Cost $/gal	Maint Cost $	Maint Cost $/gal	Deprec $mm	Deprec $/gal	Over-head $mm	Over-head $	Over-head ($/gal) ethanol	Other ($)	Other $/gal ethanol	Total ($/gal without feedstock)
2013				247	16.9643										
2014				225	11.25										
2015				48.9	1.69204										
2015				48.9	1.69204										
2015				48.9	1.69204										
2015				48.9	1.69204										
2005			0.0578	31.4	1.57		0.0616					0.0422		0.0044	
2005			0.0929	31.4	1.57		0.0882					0.0929		0	
2005															
2005															
2005															
2005															
2005															
2005															
2005															
1982		20,805	0.1387	0.232	1.54667			6730	0.0449	0.0255		0.16967	41747	0.2783	

(*Continued*)

TABLE 7A.1 (*Continued*)
Ethanol Plant Capacities and Costs

Year	Type	Feedstock	Manufacturing Cost $/gal ethanol	mmgal/year ethanol	Capital Cost ($/gal)	Reference
2002	Dry mill	Corn	1.33	15	1.8	Whims, J. (2002)
2002	Dry mill	Corn	1.28	30	1.2667	Whims, J. (2002)
2002	Dry mill	Corn	1.26	40	1.125	Whims, J. (2002)
2002	Wet mill	Corn	1.43	30	2.7033	Whims, J. (2002)
2002	Wet mill	Corn	1.33	40	2.41	Whims, J. (2002)
2002	Wet mill	Corn	1.35	100	1.67	Whims, J. (2002)
2006	Dry mill	Corn	1.38	38.4		Kwiatkowski, J.R. et al. (2006)
2007	Cassava	Cassava	1.78	14		Felix, E. et al. (2007)
2007	Cassava	Cassava	1.4	26.7		Felix, E. et al. (2007)
1999	Grain	Corn	1.16	25	1.116	Wright, M.M. and Brown, R.C. (2007)
2007	Grain	Corn	1.17	225	0.4933	Wright, M.M. and Brown, R.C. (2007)
2005	Ligno	Cellulose source	1.52	50	5.88	Wright, M.M. and Brown, R.C. (2007)
2007	Ligno	Cellulose source	1.17	225	3.36	Wright, M.M. and Brown, R.C. (2007)
2010	Ligno	Corn stover	3.12	53.4	7.0412	Kazi, F.K. et al. (2010)
2010	Ligno	Corn stover	3.3	50.7	7.6726	Kazi, F.K. et al. (2010)
2010	Ligno	Corn stover	4.02	32.8	11.921	Kazi, F.K. et al. (2010)
2010	Ligno	Corn stover	4.06	39.1	8.3632	Kazi, F.K. et al. (2010)
2010	Ligno	Corn stover	3.37	46.2	8.355	Kazi, F.K. et al. (2010)
2010	Ligno	Corn stover	3.44	53.9	9.295	Kazi, F.K. et al. (2010)
2010	Ligno	Corn stover	3.37	55.5	6.955	Kazi, F.K. et al. (2010)
2010	Ligno	Corn stover	3.26	47.3	9.1755	Kazi, F.K. et al. (2010)
2002	Ligno	Corn stover	1.07	69.3	2.8485	Aden, A. et al. (2002)

(*Continued*)

TABLE 7A.1 (*Continued*)
Ethanol Plant Capacities and Costs

Year	Type	Feedstock	Manufacturing Cost $/gal ethanol	mmgal/year ethanol	Capital Cost ($/gal)	Reference
1999	Ligno	Corn stov/switch	1.42	839.5	0.2785	Wooley, R. et al. (1999)
2002	Dry mill (sm)	Corn	0.9945	20	1	Shapouri, H and Gallagher, P. (2005)
2002	Dry mill (lg)	Corn	0.9207	90	3	Shapouri, H and Gallagher, P. (2005)
2011	Ligno least ethanol	Tall fescue	3.24	14.14	6.4611	Kumar, D. and Murthy, G.S. (2011)
2011	Ligno mostt ethanol	Tall fescue	3.18	15.76	7.2735	Kumar, D. and Murthy, G.S. (2011)
2013	BBI Intern dry mill	Corn		50	1.1	Duff, B. (2013)
2013	BBI Intern dry mill	Corn		40	1.2	Duff, B. (2013)
2013	BBI Intern dry mill	Corn		30	1.35	Duff, B. (2013)
2013	BBI Intern dry mill	Corn		20	1.5	Duff, B. (2013)
2013	Ligno DuPont	corn stover		17.25	11.42	DuPont. (2013)
2015	Ligno DuPont	Corn stover		30	7.5	DuPont. (2015)
2013	Ligno Abengoa	Corn stover, grass		14.56	23.489	Jessen, H. (2014)
2014	Ligno Abengoa	Corn stover, grass+		25	9.176	Jessen, H. (2014)
2014	Ligno Abengo Eur	Cellulose source		1.32		Voegele, E. (2014)
2014	Wht bar Abengo Eur	Wheat, barley		52.84		Voegele, E. (2014)
2013	Fiberight	Msw cellulose		2.3	25.435	
2013	Fiberight	Msw cellulose		6	10	
2013	INEOS Bio	Gasificationi msw		4.6	27.391	
2013	Poet			14.56	16.964	
2014	Poet Royal DSM			20	11.25	Wirt, A. (2014)
2015	Vogelbusch	Corn		28.9	1.692	Vogelbusch. (2015)
2015	Vogelbusch	Wheat		28.9	1.692	Vogelbusch. (2015)

(*Continued*)

TABLE 7A.1 (*Continued*)
Ethanol Plant Capacities and Costs

Year	Type	Feedstock	Manufacturing Cost $/gal ethanol	mmgal/year ethanol	Capital Cost ($/gal)	Reference
2015	Vogelbusch	Sugar beet		28.9	1.692	Vogelbusch. (2015)
2015	Vogelbusch	Sugar cane		28.9	1.692	Vogelbusch. (2015)
2005	Dry mill	Corn	1.2885	20	1.57	Shapouri, H. and Salassi, M. (2006)
	dm other table			20		
2005	Wet mill	Corn		20	1.57	Shapouri, H. and Salassi, M. (2006)
2005	Cane sugar (US)	Sugar from cane		20		Shapouri, H. and Salassi, M. (2006)
2005	Raw sugar (US)	Raw sugar		20		
2005	Refined sugar (US)	Refined sugar		20		
2005	Molasses (US)	Molasses		20		
2005	Beet sugar (US)	Sugar from beet		20		Shapouri, H. and Salassi, M. (2006)
2005	Cane sugar (Brazil)	Sugar from cane bz		20		Shapouri, H. and Salassi, M. (2006)
2005	Beet sugar (Europe)	Sugar from beet eu		20		
1982	Farmer corn	Small farmer corn		0.15	1.5467	Hunt, V.D. et al. (1982)

Note:

- Pages 227–230 provide entries beginning with a left-hand column of year followed by type of process followed by ethanol capacity, etc.
- Pages 231–233 provide the next set of columns to the right starting with Enzymes ($), Enzymes ($/gal ethanol), Yeast ($/gal), etc. each row aligned with the particular year and ethanol plant case started on page 228.
- Pages 234–236 provide the next set of columns to the right, beginning with Fuel Cost ($), Fuel Cost ($/gal ethanol), Fuel, etc. each row aligned with the particular year and ethanol plant started on page 228.
- Pages 237–239 provide the next set of columns to the right, beginning with Labor, Labor Cost ($), Labor Cost ($/gal), etc. each row aligned with the particular year and ethanol plant started on page 228.
- Pages 240–242 repeat the year, type of ethanol plant and feed stock, etc. returning again to the far left and starting with the top row again to list summary cost numbers for each facility examined plus showing a reference (already listed in the reference list).

REFERENCES

Aden, A., Ruth, M., Ibsen, K., Jechura, J., Neeves, K., Sheehan, J., Wallace, B. of NREL; Montaque, L., Slayton, A., and Lukas, J. of the Harris Group. 2002. Lignocellulose biomass to ethanol process design and economics utilizing co-current dilute acid prehydrolysis and enzymatic hydrolysis for corn stover. National Renewable Energy Laboratory. NREL /TP-510-32438.

Aries, R.S. and Newton, R.D. 1955. *Chemical Engineering Cost Estimation, McGraw-Hill Series in Chemical Engineering*, McGraw-Hill Book Company, New York, 15.

Bryan, T. 2015. *Ethanol Producer Magazine*. BBI International. Vol. 21(12), p. 21. http://www.ethanolproducer.com.

Chemical Engineering Plant Cost Index. 2011. http://www.nt.ntnu.no/users/magnehi/cepci_2011_py.pdf.

Duff, B. 2013. Tribal Opportunities: Ethanol Production. BBI Biofuels Course. http://www.bbifuels.com.

DuPont. 2013. Making Cellulosic Ethanol a Reality: By the Numbers. http://www.dupont.com.

DuPont. 2015. DuPont Opens Commercial Scale Cellulosic Ethanol Plant in Iowa. http://www.duPont.com.

Durante, D.A. and Sneller, T.C. 2006. A Guide for Evaluating the Requirements of Ethanol Plants. Clean Fuels Development Coalition and the Nebraska Ethanol Board, in Cooperation with the USDA. http://www.ne-ethanol.org. http://www.cleanfuelsdc.org.

Felix, E., Cardona, C.A., and Quintero, J.A. 2007. Chapter 5. *Technical and Economic Viability of Biofuel Production Chains*. Bioenergy and Food Security, Tanzania. p. 111–138.

Goldstein, W.E. 1999. Economic considerations for food ingredients produced by plant cell and tissue culture, in *Plant Cell and Tissue Culture for the Production of Food Ingredients*. Tong-Jen Fu, Gurmeet Singh, and Wayne R. Curtis, Eds, Kluwer Academic/Plenum Publishers, New York, Chapter 17, p. 195–213.

Goldstein, W.E. ed. 2014. *Pharmaceutical Accumulation in the Environment: Prevention, Control, Health Effects, and Economic Impact*. Taylor & Francis Group. CRC Press, Boca Raton, FL, Chapter 9.

Goldstein, W.E., Lasure, L.L., and Ingle, M.B. 1980. Product cost analysis, in *Plant Cell Culture as a Source of Biochemicals*, E.J. Staba, Ed., CRC Press, Boca Raton, Chapter 9, p. 191–234.

Hunt, V.D., Houghtton-Alico, D., Winston, S.J. et al. 1982. *Fuel from Farms: A Guide to Small Scale Ethanol Production*. 2nd Edition. United States Department of Agriculture and Office of Alcohol Fuels-United States Department of Energy. SERI/SP-451-519R.

Jessen, H. 2014. Abengoa plant has multiple feedstock capabilities. *Ethanol Producer Magazine*. BBI international.

Kazi, F.K., Fortman, J.A., Anex, R.P., Hsu, D.O., Aden, A., Dutta, A., and Kathandaraman, G. 2010. Techno-economic comparison of process technologies for biochemical ethanol production from corn stover. *Journal of Fuel*. Vol. 89. p. S20–S28. Elsevier.

Kumar, D. and Murthy, G.S. 2011. Impact of pretreatment and downstream processing technologies on economics and energy in cellulosic ethanol production. *Biotechnology for Biofuels*. Vol. 427. p. 1–19. http://www.biotechnologyforbiotuels.com/content/4/1/27.

Kwiatkowski, J.R., McAloon, A.J., Taylor, F., and Johnston, D.B. 2006. Modeling the process and costs of fuel ethanol production by the corn dry-grind process. *Industrial Crops and Products*. Elsevier. Vol. 23, p. 288–296.

Landress, L. and Loo, F. 2014. ICIS pricing. Reed Business Information, Ltd.

Portz, T. 2016. *Pellet Mill Magazine*. BBI International. http://www.bbiinternational.com.

Shapouri, H. and Gallagher, P. 2005. USDA's 2002 Ethanol Cost-of-Production Survey. United States Department of Agriculture. Office of the Chief Economist. Office of Energy Policy and New Uses. Agricultural Economic Report 841.

Shapouri, H. and Salassi, M. 2006. *The Economic Feasibility of Ethanol Production from Sugar in the United States*. United States Department of Agriculture.

Ulrich. G.D. and Vasudevan, P.T. 2002. *Chemical Engineering Process Design and Economics—A Practical Guide*. 2nd Edition. http://www.ulrichvasudesign.com.

U.S. Bureau of Labor Statistics. 2015a. Employment Cost Index. Historical Index Information. September 1975–September 2015.

U.S. Bureau of Labor Statistics. 2015b. Historical Inflation Rates: 1914–2015. http://www.bls.gov.

U.S. Energy Information Administration. Annual Energy Review. 2013a. Table 8.10. Average retail prices of electricity 1960–2011. http://www.eia.gov.

U.S. Energy Information Administration. Annual Energy Review. 2013b. Natural Gas Prices. http://www.eia.gov.

Voegele, E. 2014a. Abengoa celebrates grand opening of cellulosic ethanol plant. *Ethanol Producer Magazine*.

Voegele, E. 2014b. ICM retrofits Indiana ethanol plant for new owners. *Ethanol Producer Magazine*.

Vogelbusch. 2015. Planning a Bioethanol Plant. http://www.vogelbusch.com.

Walton, B. 2014. Circle of Blue. Water Costs. Water Pricing for 30 Major U.S. Cities in 2014 and the Percent Change from 2013.

Whims, J. 2002. Corn Based Ethanol Costs and Margins Attachment 1. Department of Agricultural Economics, Agricultural Marketing Resource Center (AgMRC), Kansas State University and Sparks Company, Inc.

Wirt, A. 2014. Biomass Program Overview. Poet-DSM Advanced Biofuels. http://www.poetdsm.com.

Wooley, R., Ruth, M., Sheehan, J., Ibsen, K. (Biotechnology Center for Fuels and Chemicals); Majdeski, H. and Galvez, A. (Delta-T Corporation). 1999. Lignocellulose Biomass to Ethanol Process Design and Economics Utilizing Co-Current Dilute Acid Prehydrolysis and Enzymatic Hydrolysis Current and Future Scenarios. National Renewable Energy Laboratory (NREL). NREL/TP-580-26157.

Wright, M.M. and Brown, R.C. 2007. Comparable economics of biorefineries based on the biochemical and the thermochemical platforms. *Biofuels, Bioproducts, and Biorefineries*, Vol. 1, p. 40–56. Society of the Chemical Industry and John Wiley & Sons. http://www.interscience.wiley.com.

Yang, R.D., Grow, D.A., and Goldstein, W.E. August 23–28, 1981. Pilot Plant Studies of Ethanol Production from Whole Ground Corn, Corn Flour, and Starch, *182nd American Chemical Society National Meeting*, New York, NY.

8 Considerations for Minimization of Pollution

8.1 INTRODUCTION TO POLLUTION CONSIDERATIONS

We must minimize fouling of the environment. This involves ensuring the pureness of air, water, and soil. The reasons for this are to preserve health, protect the planet, and maintain our well-being. For example, pollutants in the air can cause health problems such as asthma. Pollutants (even pharmaceuticals and personal care products from manufacturing or farming) can end up in our water or soil. These can cause significant health issues (Goldstein, W.E. 2014). Pollutants in our air can be chemicals or even mold spores that can also cause breathing and other problems (Goldstein, W.E. 2011). The range of pollutants normally seen in the atmosphere is shown in Table 8.1.

Air pollutants can also potentially affect the climate. For example, carbon dioxide and methane influence heat entrapment in the atmosphere and global warming. Global warming is the loss of temperature control of the atmosphere and the earth. This is a consequence of the sun's electromagnetic thermal radiation that travels to our planet and heats it, coupled with how well our planet traps heat, or releases it back to the outer space. Slight changes in the energy reaching our planet (vagaries of the sun's emissions) are theoretically catastrophic.

We are fortunately positioned quite favorably from the sun by natural forces and divine intervention. We are neither too close nor too far from the sun. However, this is not the whole story. How we manage the content of our atmosphere affects our life and life-forms. As experts have pointed out, the sun is not the sole factor. It is clear that humans can influence the planet's temperature control by causing emissions that end up in our atmosphere, leading to a chronic problem that may worsen with time (U.S. EPA. 2015a,b,c). This is termed an anthropogenic effect.

For example, carbon dioxide produced in many ways, such as from burning of fossil fuels, reduction of plant growth, and production of ethanol, accumulates in the atmosphere, apparently at elevated levels and may concentrate there. Carbon dioxide accumulates and is not depleted as fast as other gasses as a function of time. Massive convection currents sweep the atmosphere gasses, perhaps, in the author's conjecture, forming a boundary layer at the upper atmosphere. Carbon dioxide at higher concentrations at lower elevation may then transport by convection and diffusion across this boundary layer to accumulate in the upper regions of the atmosphere. The driving force to do this may be simply concentration difference or a yet undefined binding phenomenon for carbon dioxide in this outer region. Perhaps, diffusion and convection processes relate to processes for gasses described in the literature (Smith, J.M. 1970; Badino, G. 2009). Carbon dioxide also accumulates in the oceans, where the dissolution of carbon dioxide, forming carbonic acid among other entities, seems to be associated with lower pH and rising acidity (Solomon, S. et al. 2007; Ekwurzel, B. 2015).

TABLE 8.1
Average Estimated Composition of the Atmosphere up to an Altitude of 25 km (±0.5%)[a]

Gas Name	Chemical Formula	Percent Volume (%)	ppm[b]
Nitrogen	N_2	78.084	780,840
Oxygen	O_2	20.95	209,500
Argon	Ar	0.93	9300
[a]Water	H_2O	2.5	25,000
[a]Carbon dioxide	CO_2	0.0397	399 (December 12, 2014 mean est.)
Neon	Ne	0.0018	18
Helium	He	0.000524	5.24
[a]Methane	CH_4	0.000179	1.79
Krypton	Kr	0.000114	1.14
Hydrogen	H_2	0.000055	0.55
[a]Nitrous oxide	N_2O	0.0000325	0.325
Carbon monoxide	CO	0.00001	0.1
Xenon	Xe	0.000009	0.09
[a]Ozone	O_3	0.000007	0.07
Nitrous dioxide	NO_2	0.000002	0.02
Iodine	I_2	0.000001	0.01
Ammonia	NH_3	Trace	Trace

[a] Considered relatively variable gasses.
[b] Parts per million.

The earth's atmosphere is not deep. Experts say the area where nitrogen, oxygen, and gasses such as carbon dioxide reside is often taken to be about 30 km, which is 19 miles, or near 100,000 feet. This compares to the earth's diameter being about 8000 miles or about 13,000 km. The atmosphere is a thin layer around the earth. This layer contains a relatively small volume compared to the earth or the oceans to contain gas components or to store heat that arrives from the sun. The storage of mass or heat is of course temporary since these are distributed around the planet in many ways.

Carbon dioxide absorbs electromagnetic energy at particular wavelengths just like water. This results in heat absorption and emission of thermal radiation (Thomas, L.C. 1992). Furthermore, as reported in the literature, carbon dioxide and other gasses may accumulate in the upper regions of the atmosphere (Solomon, S. et al. 2007; Ekwurzel, B. 2015). Carbon dioxide may linger longer than other gasses such as methane. Its effect on heat transfer and accumulation may be accentuated if it concentrates in thin layers (meaning the carbon dioxide concentration may grow more than expected in a particular region of the atmosphere). The layer may serve as a blanket and is thermally active due to its emissivity or ability to radiate heat. The effects of this on global warming may accumulate with time and perhaps may accelerate since these global effects are nonlinear. It makes sense, therefore, to do something about the accumulation of carbon dioxide that occurs due to the activities of humans.

The impact of dealing or not dealing with greenhouse gas (GHG) emissions is dealt with in the literature (U.S. EPA 2015a,b,c). If one accepts the U.S. Environmental Protection Agency's (EPA) projections, then inaction is very costly and quite dangerous to all of us. There appears to be a correlation of global warming with carbon dioxide accumulation as covered in these references. Theories only predict what is likely to occur. The predictions often sound like "doomsday" scenarios. Such dire predictions may be "right on the money."

The most common components in the atmosphere at 25,000 m elevation are listed in Table 8.1 (OSS. 2011). Carbon dioxide as indicated at this elevation is 399 ppm or about 0.04% of the composition of the atmosphere. This compares to 780,000 ppm for nitrogen and 209,000 ppm for oxygen. Water at this elevation is 25,000 ppm. Despite carbon dioxide's seemingly relatively low concentration, it is very effective at helping to trap and emit radiant heat as noted. Too much of this is a bad thing as this influences the temperature to rise. If there were none of this, the earth would be a frozen ice ball, and we would not exist. The sensitivity to this component increasing over a long period of time is expected to be considerable (given the sun does not overwhelm this component by bursting with sunspots and excessively emitting thermal radiation to adversely affect the earth). The sun has not done this, of course, and it is not expected to, since you and I are still here and alive. Furthermore, these are just the numbers at 25,000 m.

The concentration of pollutants over a facility or farm areas where sheep are grazing, expelling out large volume of gasses like methane, will be far more significant in a shorter period of time and can affect respiration. Witness what happened in London in the 1600s where smoky effluent was so injurious to health (DeAngelo, L. 2012). Also note similar occurrences throughout the ages as documented in the literature (Georgia Institute of Technology). Therefore, how one argues about components at high altitudes affecting the planet does not affect the common sense conclusion that we ought to control emissions at low elevations and smaller areas to preserve health and quality of life, and maybe even our existence if we are negligent enough to let this continue unabated. There is no way (yet) to experimentally (non-*in silico*) simulate the effect of carbon dioxide on global warming covering all complex facets. However, this does not mean we cannot simulate small set of components in a systematic manner to check key aspects of theories.

Sometimes, one cannot conduct complete experiments to try and prove or disprove a theory. One cannot carry out experiments easily on a global scale (and neither would one wish to) because of the dangers involved. However, one can scale down experiments to test predictions considering a select number of variables and properly designed apparatuses and experiments. This is done in industry in many fields. The author led and participated in such work many times (see http://www.goldconsul.com). Such efforts could help refine theories and help deal with skeptics on the many complex aspects in this matter. Such efforts are sorely lacking. However, one can try. Furthermore, it just makes sense to take precautions and do something productive and useful with carbon dioxide emissions that make sense commercially.

8.2 A SEQUENCE OF HISTORICAL EVENTS THAT HELPED TO POPULARIZE ETHANOL

The government recognized in 1970 that the effluent coming out of the tailpipes of cars needed to be abated so requirements were issued for catalytic conversion. The results were favorable. In 1980, these demands were increased. In this period, it was recognized that the lead in gasoline was poisoning catalytic converters that were helping destroy pollutants exiting cars, so tetraethyl lead was removed. Removal of such lead avoided poisoning (not recognized at that time) that was, of course, harmful to people.

MTBE (methyl tertiary butyl ether), an oxidant, was added to gasoline to make combustion more complete and to complete oxidation of some pollutants. It was discovered that MTBE, a suspected carcinogen, was fouling the water supply. It was removed and this helped lead to ethanol being incorporated as another source of chemically bound oxygen to assist gasoline in the combustion process.

8.3 POLLUTANTS IN ETHANOL MANUFACTURE AND COMBUSTION

When ethanol is manufactured, carbon dioxide is released as a consequence of the reaction to form ethanol from carbohydrate. As suggested in this book ethanol should be produced, and it is necessary to find out how to make this production more useful and economical.

Each molecule of ethanol produced results in an associated molecule of carbon dioxide. For each 46 metric tons of ethanol produced, 44 metric tons of carbon dioxide was produced. Clearly, we are not going to continually discharge this into the atmosphere since this is akin to a form of suicide. Means to capture and sell, or use the carbon dioxide as a by-product, will and are being incorporated so that the ethanol industry will continue to grow successfully while minimizing pollution of the atmosphere.

Rather than release the carbon dioxide, it can be captured and recycled to produce another entity (ADM. 2013; Committee on Geoengineering Climate. 2015). It can also be liquefied or frozen to be sold as a by-product. This capture and the recovery of costs, or leading to a net profit when one manages carbon dioxide capture as a side business, must be economically justified by analysis and supportive research.

8.4 IMPURITIES IN ETHANOL AND GASOLINE, AND IN THE PRODUCTS OF THEIR COMBUSTION

The impurities in ethanol prior to distillation/extraction of fermentor contents to remove water are those present in the feedstock to the fermentation. These are mainly salts that will not exist in the ethanol product following distillation/extraction. For all intents and purposes, the ethanol in the blend made with gasoline for use in fuel is essentially pure. However, essentially pure implies that impurities do exist in relatively small amounts that can contribute to pollution later. Also, the way the ethanol is produced, and the way its feedstock is provided, affects how it contributes to air pollution. Indeed, one analysis indicates that gasoline and ethanol from corn contribute about the same level of pollutants considering only production. However, this particular analysis indicates that the potential contribution of pollutants from production using lignocellulose to produce ethanol is far lower (Hill, J. et al. 2008).

When ethanol is burned, it ideally forms carbon dioxide and water. Carbon dioxide is a pollutant as noted. However, the burning is not ideal so other molecules are formed that are products of combustion. Furthermore, the impurities present in the air that contains the oxidant (oxygen) contribute to forming other entities that represent pollutants. Even the major component of the air (nitrogen) is considered an impurity for purposes of combustion and reacts with oxygen to form NO_X compounds (nitrogen dioxide and nitric oxide—U.S. Environmental Protection Agency. 1993). Ethanol is co-combusted with gasoline. The number of products of combustion (carbon-based molecules and other molecules) is enormous (Flagan, R.C. and Seinfeld, J. 1988. Chapters 2 and 3).

Gasoline is purified from petroleum using distillation-based processes. It is an effectively defined mixture consisting of a class of carbons called alkanes. The distillation process also purifies the alkanes so that they are in a target molecular weight range required for use as gasoline. Small impurities are carried over in the process since the separation is not ideal. The burning of gasoline produces a myriad of carbon and nitrogen-based products.

When ethanol–gasoline blends are burned, what are the amounts of the impurities and how are they (or can they) reduced? The reactions do not go to ideal completion (to carbon dioxide and water) and intermediates escape such as carbon monoxide and forms of carbon, oxygen, and hydrogen, including aldehydes (Krishna, M.V.S.M. et al. 2011). Any sulfur present (as a fuel impurity) is burned and so sulfur chemically associates with the other molecules. Also, other entities form due to various influences, one being ozone.

Workers in Brazil apparently found that in measurements of the atmosphere above Sao Paulo that burning pure ethanol raised ozone levels even if carbon dioxide, hydrocarbon, and nitrogen dioxide levels were relatively lowered compared to gasoline (Peplow, M. 2014). The researchers in Singapore and the United States postulate that in the case of gasoline, nitrogen dioxide reacts with the hydroxyl ion, lessening the presence of this latter reactant to form ozone (Salvo, A. and Geiger, F. M. 2014).

This area is complex so it is possible that such pollution reduction is very locale specific, meaning one might need a different policy for best pollution control in each particular city because of local habits and preferences. Workers have studied and reported the emissions from gasoline burning engines in comparison to use of other fuels (e.g., Elliott, M.A. et al. 1955).

Workers have also studied burning of blends of gasoline and alcohol in different proportions using a fleet of automobiles. The design and implementation of the experiments provided some very interesting results (Knoll, K. et al. 2009). The experimental findings in this case was equally interesting since it pointed out the complexities in this kind of work and the variabilities that can result. The workers in this case noted that increasing ethanol content in the gasoline–ethanol blend, compared to use of pure gasoline, resulted in reductions in nonmethane hydrocarbons and carbon monoxide and increases in ethanol emissions (unreacted ethanol perhaps) and aldehydes. Production of aldehydes when burning ethanol seems to be a common characteristic. Perhaps, further work in catalysis applied to the catalytic converters may reduce the aldehyde content since this entity is undesired from a health standpoint.

Pollutants produced by combustion in automobiles are lowered by catalytic converters as engine components are being improved over time (Carley, L. 1996; CatalyticConverter.Org. 2015). The elimination of lead (tetra ethyl lead) as an additive was undoubtedly an important factor since lead is a poison and fouls catalyst in the catalytic converter. Catalytic converters are now quite stable as long as they are not physically damaged and typically last the life of the automobiles (Parker, A. 2016).

Catalytic converters break down pollutants to simplest least toxic forms (Nizkorodov, S. 2014). Gasses pass through reaction chambers where the catalyst is typically a noble metal such as platinum with the objective to maximize catalytic activity (ability to facilitate elevated rates of reaction per unit time) and stability (retention of activity with minimal loss over time). Patents in the literature cover this including polyfunctional rare metal catalysts imbedded in precise ways to manage elimination of pollutants such as unreacted hydrocarbons, carbon monoxide, and nitrogen oxides (Brunelle, J.-P. et al. 1983; Nice, K. and Bryant, C.W. 2015).

From the literature, several observations can be made. The fuel streams (be they ethanol–gasoline blends, or only gasoline) are relatively pure and modestly contaminated. Impurities are also provided from the air. Air quality varies locally and is maybe the major variable at times. The fuels in the blend may produce specific pollutants. For example, ethanol has a propensity to produce acetaldehyde (Stepanek, J. et al. 2010). The conversion of ethanol to products of combustion can be manipulated by varying catalysts (Idriss, H. 2004).

This means that reactions that proceed may result in combustion products and they are often quite similar. The levels are not that dissimilar unless one is comparing an engine running at winter versus one running at summer temperatures. A study in Europe produced some good and interesting data comparing driving conditions and two kinds of driving cycles with different blends (Suarez-Bertoa, R. et al. 2014).

The catalytic converters are now very stable in performance, and last a long time without losing their catalytic activity. Exhaust systems are advanced and do a good job purifying gaseous effluent from the engine before it reaches the atmosphere. Their improvements allow capture of pollutants to a greater degree with time, such as nitrogen and sulfur compounds. Since undesirable pollutants are still discharged, further progress in catalysis is desired.

If there are differences, they may be due to the native environment and the quality of the air that is used. In the end, with such advances, with all the vagaries that can occur with how well the car is maintained and the locale, there may (subject to statistical variation) in the final analysis be little practical difference in pollutants produced by ethanol–gasoline blends and those produced by gasoline alone (see Suarez-Bertoa R. et al. 2014).

That is actually a positive result since it means that both types of fuels are progressing to optimal pollutant minimization. Capturing the carbon produced is obviously a good thing so perhaps that can be cleverly done in future cars *in situ* with new car engine designs while maintaining elegance in style and aesthetics that are attractive to a consumer.

Furthermore, since the pollutants produced are a function of air quality, then newer and advanced designed cars and engines might offer ways to provide purified and enriched air maybe even approaching a safe supply of pure oxygen. This might be done using superadsorbents that release and bind oxygen rapidly coupled with cryogenic oxygen concepts.

Supply of pure oxygen would take care of dealing with the nitrogen in the air, which is a source of NO_X or nitrogen dioxide and oxide. Until such advances occur, then pollution reduction will be optimized as it has been, and differences between ethanol–gasoline blends and gasoline may become more of a minor variable in terms of pollutants produced.

In the end, the factors are expected to be such that ethanol–gasoline blends and pure gasoline will be essentially of equivalent value in the area of pollution and other areas as well. With advances noted in this book, both ethanol blends and gasoline are expected to be equally favorable offerings in many of the technical and scientific aspects. In the political area, ethanol use can provide nonwar options that the United States and others can use to minimalize blackmail tactics of nations that act like our enemies.

8.5 SPECIFIC ACTIONS ON POLLUTION AND JUSTIFICATION

Beyond what has been said, undesirable pollutants are still coming from tail pipes. In the case of ethanol, aldehydes are increased over gasoline use. The aldehydes are undesirable and are toxic. Through advanced work in chemical catalysis to improve the catalytic converter, the aldehydes can perhaps be removed by improved reactions to form carbon dioxide and water. Pathways and recommendations shown in the literature can be improved to either bypass the aldehydes or degrade them to a greater extent (Nizkorodov, S. 2014). Sulfur dioxide, which appears as a sulfur impurity in gasoline, should be eliminated using capture perhaps as a salt of sulfate, which can be a disposable chamber in the exhaust line perhaps just after catalytic converter. Since the amount of sulfur is small, perhaps this can be an inexpensive snap-in/out cartridge.

Minimization of discharge of carbon dioxide is very desirable as noted. Papers in the literature note how this entity in the upper atmosphere can actually force more heat input into the atmosphere due to absorption properties that cause heat to stay in the atmosphere in excess due to spectrally related phenomena (Bell, B. 2013; Harde, H. 2013). Heat incoming from the sun at shorter wavelengths passes to the surface while reflected heat at longer wavelengths is absorbed by entities such as carbon dioxide. In effect, a greater amount of net heat is forced into the atmosphere as the former author interprets.

Perhaps the carbon dioxide can be recycled and catalytically converted to fuel or perhaps it can be captured by binding or phase change to liquid or solid in a future car design. This is important to consider even if it is challenging since otherwise the emphasis on formation of carbon dioxide will be shifted to facilities such as those generating electricity where carbon dioxide is conveniently collected in mass amounts in one place. For many reasons, part of the effort in carbon dioxide removal should be shared by those involved in fuel preparation, and in design and production of automobile internal combustion engines.

REFERENCES

ADM. 2013. *A Climate of Innovation: Investigating Geologic Carbon Sequestration's Potential as an Environmental Solution.* Archer Daniels Midland Company. Decatur, Illinois. http://www2.adm.com/en-US/responsibility/Documents/Carbon-Sequestration-Brochure.pdf

Badino, G. 2009. The legend of carbon dioxide heaviness. *Journal of Cave and Karst Studies*, 71(1), p. 100–107.

Bell, B. 2013. Carbon dioxide and the fundamentals of heat transfer. *POWER Magazine.* http://www.powermag.com/carbon-dioxide-and-the-fundamentals-of-heat-transfer.

Brunelle, J.-P., Prigent, M., and Raynal, B. March 29, 1983. U.S. 4,378,307. Pollution Control Catalyst for Internal Combustion Engines Exhaust System/Catalytic Converter.

Carley, L. December 1996. Tech article: Catalytic converters. *Tire Review. Z Club of San Diego*, http://www.zcsd.org/tech-articles/catalytic-converters.php.

CatalyticConverter.Org. 2015. http://www.catalyticconverter.org/index.htm.

Committee on Geoengineering Climate. 2015. *Climate Intervention: Carbon Dioxide Removal and Related Sequestration.* ISBN 978-0-309-30529-7. http://NAP.EDU/18805

DeAngelo, L. July 3, 2012. London smog disaster. *The Encyclopedia of Earth*, England.

Ekwurzel, B. 2015. Global Warming. Union of Concerned Scientists. http://www.ucsusa.org/global_warming.

Elliott, M.A., Nebel, G.J., and Rounds, F.G. 1955. The composition of exhaust gasses from diesel, gasoline, and propane powered motor coaches. *Journal of the Air Pollution Control Association*, 5(2), p. 103–108. http://dx.doi.org/10.1080/00966665.1955.10467686

Flagan, R.C. and Seinfeld, J. 1988. *Fundamentals of Air Pollution Engineering.* Prentice-Hall, Inc., Englewood Cliffs, NJ.

Georgia Institute of Technology. Air Pollution. Lecture. History. http://www.aerosols.eas.gatech.edu/EAS%20Air%20Pollution%20Phys%20Chem/Intro1%20AP%20History.pdf.

Goldstein, W.E. 2011. *Sick Building Syndrome and Related Illness: Prevention and Remediation of Mold Contamination.* CRC Press, Boca Raton, FL. Taylor & Francis Group. ISBN 978-1-4398-0144-4.

Goldstein, W.E. 2014. *Pharmaceutical Accumulation in the Environment: Prevention, Control, Health Effects, and Economic Impact.* CRC Press, Boca Raton, FL. Taylor & Francis Group. ISBN 978-1-4665-1745-5.

Harde, H. 2013. Radiation and heat transfer in the atmosphere: A comprehensive approach on a molecular basis. *International Journal of Atmospheric Sciences.* Hindawi Publishing Corporation. Volume 2013, Article ID 503727, p. 1–26. http://dx.doi.org/10.1155/2013/503727.

Hill, J., Polasky, S., Nelson, E., Tilman, D., Huo, H., Ludwig, L., Neumann, J., Zheng, H., and Bonha, D. 2008. Climate change and health costs of air emissions from biofuels and gasoline. *Proceedings of the National Academy of Sciences*, 16(6), p. 2077–2082. http://www.pnas.org/cgl/doi/10.1073/pnas. 0812835106.

Idriss, H. 2004. Ethanol reactions over the surfaces of noble metal/cerum oxide catalysts. *Platinum Metals Review*, 48(3), p. 105–115.

Knoll, K., West, B., Huff, S., Thomas, J., Orban, J., and Cooper, C. 2009. Effects of mid-level ethanol blends on conventional vehicle emissions. SAE Powertrain, Fuels, and Lubricants Meeting. 2009-01-2723. NREL/CP-540-46570.

Krishna, M.V.S.M., Kumarz , S.N., Murthy, P.V.K., and Kishor, K. 2011. Control of Aldehyde Emissions from Copper Coated Spark Ignition Engine Fueled with Alcohol Blended Gasoline. ISSN: 2248-9622. 1(2), p. 337–340. http://www.ijera.com.

Nice, K. and Bryant, C.W. 2015. How Catalytic Converters Work. http://auto.howstuffworks.com/catalytic-converters1.htm

Nizkorodov, S. 2014. Ability of Catalytic Converters to Reduce Air Pollution: Measurement of Select Air Pollutants in an Exhaust. University of California at Irvine. Chemistry Lab Experiment. http://aerosol.chem.uci.edu/AirUCI_summer/PDFs/labs/2014_NOx_Lab.pdf.

OSS. July 9, 2011. Atmospheric Composition. Open Source System, Science, Solutions. http://www.ndsc.ncep.noaa.gov/.

Parker, A. 2016. How Long Does a Catalytic Converter Last? How Stuff Works. http://www.howstuffworks.org.

Peplow, M. April 30, 2014. Ethanol fuels ozone pollution. *Nature Magazine*. http://www.nature.com.

Salvo, A. and Geiger, F.M. April 28, 2014 online and June 2014 in print. Reduction in local ozone levels in urban Sao Paulo due to a shift from ethanol to gasoline use. doi: 10.10381/NGE2014. *Nature Geoscience*. Vol. 7. http://www.nature.com/naturegeoscience.

Smith, J.M. 1970. *Chemical Engineering Kinetics*. 2nd Edition, McGraw-Hill Book Company, New York, NY. p. 404.

Solomon, S., Qin, D., Manning, M., Marquis, M., Averyt, K., Tignor, M.M.B., and Miller Jr., H.L. (Eds). 2007. *Climate Change: The Physical Science Basis. Working Group I. Contribution to the Fourth Assessment. Report of the Intergovernmental Panel on Climate Change*. Cambridge University, Press, UK. ISBN 978-0521-70596-7.

Stepanek, J., Koci, P., Marek, M., Maunula, T. et al. 2010. Effects of biofuel blends on performance of exhaust gas catalyst: Ethanol and acetaldehyde reactions. *SAE International Journal of Fuels and Lubricants*, 3(1), p. 523–536. doi: 10.4271/2010-01-0894.

Suarez-Bertoa, R., Zardini, A.A., Keuken, H., and Astorga, C. 2014. Impact of Ethanol Containing Gasoline Blends on Emissions from a Flex-Fuel Vehicle Tested over the Worldwide Harmonized Light Duty Test Cycle (WLTC). *Fuel*. http://dx.doi.org/10.1016/j.fuel.2014.10.076. http://www.elsevier.com/locate/fuel.

Thomas, L.C. 1992. *Heat Transfer*. Prentice-Hall, Englewood Cliffs, NJ, p. 282. ISBN 0-13-384942-2.

U.S. Environmental Protection Agency. August, 1993. Air Quality Criteria for Oxides of Nitrogen, Report No. EPA-600/8-82-026 (1982a).

U.S. EPA. 2015a. Climate Change in the United States: Benefits of Global Action.

U.S. EPA. 2015b. Causes of Climate Change. http://www3.epa.gov.

U.S. EPA. 2015c. Office of Atmospheric Programs. EPA 430-R-15-001.

9 Ethanol Coproducts and Derived Products

9.1 INTRODUCTION TO COPRODUCTS AND DERIVED PRODUCTS

Distiller's dry grains including solubles (DDGS) was previously presented as an important by-product of the fermentation process to produce ethanol. Carbon dioxide is an important by-product as well of the fermentation process—that involves capturing the gas (to avoid pollution) where this by-product is preferably sold commercially. There are numerous technical and business possibilities that should be pursued to gain revenue from development of the carbon dioxide business.

Carbon dioxide is a pollutant produced by automobiles and other vehicles using internal combustion engines. These engines currently operate using fuel blends of ethanol and gasoline according to the renewable fuel standard federal statute. Research should be conducted to see how this gas emission can be lowered to minimize atmospheric pollution.

Ethanol's primary application is as a component of the ethanol–gasoline blend for fuels. It can be the basis for producing products such as ethylene and ensuing polymers as an alternative to production from petroleum-based sources to take advantage of what can be advantageous technology and business opportunities.

9.2 CARBON DIOXIDE PRODUCTS AND APPLICATIONS

Carbon dioxide production and sales is an established business, currently valued at almost $600 million in the United States (IBIS World. 2014). It uses encompass medical applications and oil drilling as examples. Twenty percent of carbon dioxide is applied to food processing and beverage production.

This gas is derived from steam reforming of natural gas to produce hydrogen as well as from ethanol production by fermentation (HIS. 2015). It is used in metal fabrication. Major markets for the product include China, North America, and Western Europe. The product should be preferably marketed with recommendations and associated technology for containment in use so that escape into the atmosphere is minimized.

Carbon dioxide is used in enhanced oil recovery and coal bed methane recovery. This is based on use as a compressed gas. The compressed gas can be derived on site by applying heat to solid or liquid to produce the high pressure gas. There should be opportunities in developing technology to deliver and contain the gas in use and prevent environmental release. With this information, the opportunities to grow this business while containing the product should be substantial (Adamson, K.-A. 2012).

In another development, a Canadian pilot project is under way to sorb carbon dioxide from the air into using a giant fan to blow air from the Pacific Ocean over lye-soaked plastic sheets, which bind the carbon dioxide for processing and purification (Dawson, C. 2014). The lye will facilitate forming a carbonate from the carbon dioxide. Capturing huge amounts of the gas this way will require substantial surface area. However, this sort of endeavor should be encouraged as long as it is economically feasible. There are patents in this area for removing carbon dioxide in an analogous fashion (Christiansen, D.E. et al. 1979).

This is ignoring the financial market that may develop that could include trading, future options, and arbitrage based on the carbon tax and trading based on emission controls. This is not selling the product, but seen by some as facilitating controlling emissions and the environment. Basically a party could operate their plant under the law and perhaps develop the ability to acquire credits in doing so that can be sold to another party needing them to stay within the law. Careful development will be required if this happens to avoid this becoming a basis for fraud or scams, and an uncontrollable and bureaucratic mess.

9.3 ETHANOL-DERIVED PRODUCTS AND APPLICATIONS

Ethanol is, of course, the major product. However, in cases of overproduction or development of new markets, ethanol properly purified for different applications can be a feedstock for other products. For example, it can be converted into ethylene based on dehydration (Chematur Engineering Group). This would be competitive to ethylene derived from petroleum refineries. Furthermore, the reaction is reversible under elevated temperature conditions so it may be advantageous to design the process to produce ethylene from ethanol or else ethanol from ethylene if the market need is there (Chematur Engineering Group). The technology is based on using a catalyst to convert ethanol to ethylene according to Equation 9.1 (the forward and back arrows indicate the potential for the reversible reaction).

$$C_2H_5OH \leftrightarrow C_2H_4 + H_2O \tag{9.1}$$

After purification, the ethylene is separated out by distillation. This, of course, provides entry to the polyethylene market. With technology advances, entry to other polymer markets is possible as well. Also, steps to expanding the business to other alcohols and products are certainly possible with a good technology and marketing base. Presumably, ethylene can be a feedstock for other applications based on catalysis and separations. The governing factor is, of course, economics. This is only a useful strategy if a special unique market is identified for such products. This fits into the extended discussion on the application of biotechnology to the development of chemicals (Committee on Industrialization of Biology. 2015). The ethanol and ethylene can be a basis for this if this avenue is desired in addition. This has to be on the minds of companies that have major chemicals and fermentation technology as a strategic opportunity. There are many historical sources of this effort. The writer participated on one such National Academy of Sciences panel in the mid-1980s (Tsao, G.T. et al. 1986).

REFERENCES

Adamson, K-A. Sept. 14, 2012. *New Markets Emerge for Carbon Dioxide*. Forbes.

Chematur Engineering Group. *Ethylene from Ethanol*, Karlsskoga, Sweden. http://www.chematur.se.

Christiansen, D.E., Cleski, W.J., Cortessis, P.G., Sih, P.H. Dec. 12, 1979. Process for the manufacture of sodium bicarbonate. European Patent Application 793009671. Publication No. 00005981. Applicant Stauffer Chemical Company. Westport, CT.

Committee on Industrialization of Biology. 2015. Industrialization of Biology: A Roadmap to Accelerate the Advanced Manufacturing of Chemicals. Board on Chemical Sciences and Board on Chemical Sciences and Technology. Technology. Board on Life Sciences. Division on Earth and Life Studies. National Research Council. National Academy of Sciences. ISBN 978-0-309-31652-1.

Dawson, C. March 3, 2014. Venture seeks to cut emissions. *The Wall Street Journal*. http://www.wsjd.com.

HIS. 2015. *Chemical Economics Handbook*. https://www.ihs.com/products/carbon-dioxide-chemical-economics-handbook.html.

IBIS World. 2014. Carbon Dioxide Production in the U.S.: Market Research Report.

Tsao, G.T. et al. 1986. *Bioprocessing for the Energy-Efficient Production of Chemicals, National Materials Advisory Board, National Research Council, National Academy of Science*. Publication NMAB-428. National Academy Press, Washington.

10 Petrochemicals as a Source of Ethanol

This book is focused on preparing ethanol from agricultural sources. It is of course clear that ethanol has been made from petrochemical sources. The primary route is to produce ethylene from petroleum and then oxidize the ethylene to ethanol in the presence of steam catalytically. This only makes sense if one has an advantageous position to use this method, since the focus is on renewable resources such as corn and lignocellulose used as feedstock for the fermentation process. In the end, the fermentation process must be less expensive to be sustainable from cost and environmental standpoints. It was mentioned previously that ethylene can be converted to ethanol under conditions that promote this reaction using chemical catalysis (Chematur Engineering Group).

One might plan for the possibility of using ethylene as a source in case the use of agricultural sources becomes a problem. However, investing to prepare for this eventuality is probably best postponed. A recent ethylene price was quoted at $0.625/lb or $1.4/kg (Yarns and Fibers 2016). Corn at $4/bushel amounts to $0.07/lb corn or $0.10/lb starch. With processing costs added on, and lignocellulose promising to be less expensive (once demonstrated), there is little reason to consider investing in processes to use ethylene as a hedge strategy. If one has such processes already available, contiguous to a starch process/ethanol plant, then by all means, it makes more sense to prepare for the contingency of using ethylene as more of an emergency contingency.

The conversion of ethylene to ethanol via catalysis may pass first through formation of ethylene oxide via oxidation followed by reduction (adding two hydrogen atoms) to the ethanol as shown in Equation 10.1 (Chemistry Assignment 2012–2013).

$$CH_2 = CH_2 \rightarrow CH_2OCH_2 \rightarrow C_2H_5OH \tag{10.1}$$

REFERENCES

Chematur Engineering Group. *Ethylene from Ethanol*, Karlsskoga, Sweden. http://www.chematur.se.

Chemistry Assignment. 2012–2013. Sources and Uses of Petrochemicals. http://www.chemistry-assignment.com/51-9-sources-and-uses-of-petrochemicals.

Yarns and Fibers. February 2016. Ethylene Pricing Trend. http://www.yarnsandfibers.com/news/textile-pricewatch/ethylene-price-trends-reports.

11 Conclusions, Recommendations, Research, and Business Directions

The Science of Ethanol presents an approach based on analyses and literature information that can result in success for the ethanol industry. This is based on the industry applying agricultural, therefore, renewable, resources. The suggestions leading to success are based on the writer's personal experiences. The success is not yet assured due to the presence of several obstacles.

After reading an article on Howard Buffett, the writer concludes that areas of the world that need improvements in agriculture would benefit tremendously by location of one or more ethanol production facilities within their territories (Munk, N. 2016). Ethanol makes money for American farmers. The same cash flow can result in developing areas. Profitable facilities can be established in central regions where crops are grown, if the right infrastructure is in place as described in Chapter 5. This cash flow should be an incentive to improve growing techniques in impoverished areas to both provide feedstock for the ethanol production facility and food based on agricultural sources. The facilities can be established by funds from charitable foundations, such as the one Mr. Buffett runs, where profits from the operations are used to improve agricultural production to supply feedstock needs for the facility, satisfy capital requirements, and also provide the basis for food crops to satisfy unmet needs.

The Science of Ethanol is targeted, chapter-by-chapter, to meet challenges to allow the industry to grow. Each chapter has a purpose in that direction. The pitfalls (maybe better termed obstacles, since those in the area may be aware of the challenges) are pointed out where appropriate.

- Chapter 2 on yeast biology provides essential recommendations since the yeast organism is the basis to produce ethanol from feedstock. One needs to understand the yeast selected for their facility, how it uses nutrients, and determine what limits yeast productivity by laboratory and then pilot plant studies scaled into manufacturing. Failure to do this properly will hamper progress.
- Genetic engineering can play a role in this if the genetically engineered yeast is tailored to solve practical challenges to productivity and promote scientific and engineering discovery. As covered in Chapters 2, 4, and 6, better understanding is needed for one's own yeast, how the yeast may be dispersed, form useful flocs, and retain stability over many fermentation cycles.

- The stability characteristic must be separate from any indication that the yeast, if released into the environment, can be hazardous to health through inhalation of small particulate matter that may cause pulmonary immunological reactions, any tendency of the yeast to mutate and lose its desirable characteristics, and any indication that a form of yeast (recombinant or not) can be harmfully infectious. Therefore, the yeast in use must have a good safety profile.
- Analytical characterization of the yeast from a production and health aspect is essential and recommended. Normal strains of *S. cerevisiae* are said to be safe for most people and have been used since ancient times in brewing and baking. However, some people, for example, those with irritable or inflammable bowel disorder (IBS), may react differently and in a negative manner to yeast consumed, so caution is recommended (Medline Plus. 2016; Wikipedia. 2016). Caution simply means adequate work in the analytical area to demonstrate efficacy for the process and human safety. Treating the safety issue lightly is most unwise (for obvious reasons related to regulatory oversight and potential litigation).
- Chapter 3 covers the main incentive for using ethanol in the modern world, to fuel internal combustion engines in automobiles. The main emphasis is to find ways to use ethanol more efficiently in blends with gasoline, and basically overcome its disadvantage in regard to lower energy density compared to straight gasoline. Ethanol is at a disadvantage to gasoline in this way as it is already partially oxygenated, and therefore provides less energy by combustion per unit weight.
- It is suggested that advantages for ethanol in friction reduction may be discovered due to ethanol in conjunction with study of materials used in engines, combustion formulation, and catalysis.
- Research should be directed to friction reduction in piston-cylinder operations where ethanol can possibly be a key component that can totally change the value of ethanol as a fuel. This friction reduction may also synergistically affect interconnecting areas of the automobile since more frictionless piston operation may result in ease of motion in downstream automobile parts.
- If nothing else, work in this area will spur other opportunities in engine design and improvement so that ethanol (even if not specifically responsible for the improvement) still results in less fuel used for engine operation. From a business standpoint, parties supporting such work may reap benefits from saleable or licensable technology in terms of patents and other intellectual property. Those parties may be able to take advantage of such improvements for captive use, perhaps through horizontal expansion of their business (carving out new areas that are profitable).
- Work on friction reduction should not be considered to be pursuing a digression or a tangent to the objective of improving prospects for ethanol. Such indirect work is a strategic move and essential since it is necessary to investigate and promote research that will help demonstrate ethanol's technical and business value. Otherwise, parties critically evaluating ethanol

may eventually see its use as an unnecessary cost. The product needs to be justified and this is certainly one promising avenue.

- Chapter 4 emphasizes process control that is best implemented through use of computer-based technology. However, process control is more basically established by controlling key variables allowing ethanol to be produced at minimum cost. Such control methodology can (and should) be developed independent of sophisticated computer process control since computer process-based control may not be possible at early stages such as in a developing nation. Proper process control starts by understanding how to optimize the mix of many variables important to the fermentation of ethanol. Use of computer process control, when installed, will be instrumental (in addition) in reducing fermentation costs as the writer and his associates (and others as noted in Chapter 4) have demonstrated.
- Chapter 5 focuses on the relative value of different feedstock sources and their availability throughout the world, country by country. It is clear that every country in the world can have successful projects developing and commercializing ethanol. The important steps and advice are covered in this chapter. Within each area of feedstock selection and use, there are challenges. The wise party will seek the help of experts rather than approach this alone.
- Chapter 6 points out how to study and understand yeast in terms of how it may interact in the system as part of an organized floc. This concept (introduced through use of a mathematical model) provides a way to optimize the yeast as a biocatalyst by using the model for optimal advantage. Use of models to understand and advance work is a technique that is practiced by those who are successful.
- Many models can be developed such as this one tailored to a facility. One can use work described in other chapters to determine how to produce yeast with longevity that is reusable, whether or not one chooses to operate a continuous process to make ethanol. A continuous process should be the goal, as it has the promise for sustaining the lowest cost operating commercial system.
- Chapter 7 covers a way to chart the industry and determine the key factors related to cost. As indicated in Chapter 7, the data can be examined and inferences made as to what should be emphasized in efforts to reduce costs and improve revenue. This work, in a good measure, hinges on technology development in research, pilot plant operations and in manufacturing. The analysis points out what those planning to use lignocellulose sources have to do in order to make their operation competitive with those using corn. The methods apply to use of any suitable global source. This follows the theme in the Science of Ethanol that any party anywhere can be successful if they take the right steps germane to their region and culture. A key step is identifying the right people to help one be successful.
- Chapter 8 deals with the important area of pollution. Steps need to be taken in this area to protect the ethanol business so it develops plus to be good global citizens. Pollution is a problem everywhere and fuel combustion contributes to this.

- Steps need to be taken in regard to pollutants (such as carbon dioxide) in the undesirable possibility that what some people perceive to be catastrophic comes to pass. One cannot "roll the dice" and take the chance that the extreme environmentalists are wrong. All the data are not in to judge the outcome. A lot of what is said in regard to carbon dioxide accumulation in the atmosphere affecting weather disturbances, and so on, is by correlation without absolute experimental proof. The fact that the dire consequences cannot be strictly proven at present is irrelevant.
- One must address matters through proper research to protect the development of ethanol and the global community in which we live. There is concern that there is the possibility of a nonlinearity occurring where, for example, carbon dioxide reaches a level affecting atmospheric events that become decidedly nonlinear and less controllable by climate change steps.
- This occurred to the writer when reading about carbon dioxide levels accumulating in the atmosphere maybe in a narrow region, thus becoming an extreme "thermal blanket" (see Chapter 8). In this case, weather-related events can dramatically (and not just gradually) worsen unless globally, action is taken to reverse matters for the better. This is not just an American problem. Asia, for example, is producing increasing amounts of carbon dioxide that surpass what is occurring elsewhere in the world (Zolfagharifard, C. 2016). Right now, China is the major polluter (it has the largest population). It is best to ignore the sensationalism of this reference, as people are rightly emotional about this topic and increasingly fearful. They want solutions but no one has yet provided anything that seems like it will solve essential problems.
- The solution to protect the development of ethanol (and our way of life) is to capture the carbon dioxide and find ways to use or recycle it (CO_2 is produced almost on a kg-for-kg basis with ethanol based on stoichiometry). As discussed in Chapter 8, it can be captured right at the source of production. Carbon dioxide (along with Distillers Dry Grains [DDG]) is a product of the process just like ethanol. So steps should be taken to do something useful with it as a process coproduct and not simply capture it and discard it.
- This idea has occurred to others as well. This just takes some imagination and hard work. For example, the writer just read about a party that has developed an experimental catalyst system to convert carbon dioxide to ethylene (Ondrey, G. 2016). Since ethylene can be a product of ethanol as noted in Chapter 9 and with prices that are above the price of ethanol currently ($1.40/gal or $0.212/lb for ethanol and about $0.45/lb for ethylene), then this is a possible by-product avenue (Platts. 2016). The economics based on technology development for this research finding have to, of course, work out and be convincing. This is just one example.
- Work should be directed into catalytic converter advances that can include recycling and converting carbon dioxide into something useful *in situ* as part of the automobile. Recycling carbon dioxide will require an input of energy (just as plants use solar energy to utilize carbon dioxide to for essential

carbohydrates, etc.). As an example, there is interest in examining photoelectric reduction of carbon dioxide using solar energy (Wikipedia. 2011). Also, this reaction can be facilitated by the right catalyst facilitating electroreduction (Choi, S.Y. et al. 2016).

- Take an example, burning 10 gallons (38 L) of fuel that is nine gallons (34 L gasoline and 1 gallon or 4 L of ethanol). By theoretical stoichiometry for combustion, ignoring other intermediates, which are preferably eliminated by modern catalytic converters, approximately 175 lbs or 80 kg of CO_2 is produced in burning 38 L of such a fuel.
- Assume the carbon dioxide can by bound by forming an adduct with a source of binding (Budd P. 2016). Perhaps, the bound or adsorbed carbon dioxide adduct could be formed from the carbon dioxide as the car is running. In this way, the carbon dioxide can be stored based on binding to an entity that is part of the automobile, with the entity then removed every time gas is replenished (or conveniently at home when the trip is over). That would be a way to capture and contain the carbon dioxide and then conveniently empty it or swap the reaction chamber for a new one. Since this is a gas with impurities, this is not a simple engineering challenge but it should be addressed. This is a research area that could be presented to the right parties with interest in protecting the environment while also protecting the businesses. Another idea is to draw carbon dioxide out of the air and sorb it to lye-soaked sheets (Dawson, C. 2016). Capturing carbon dioxide this way is not a new concept and also challenging, but certainly worth exploration (even if, upfront, the concept seems too hard to justify, as mentioned by a reference in the article).
- Chapter 9 deals with products of ethanol. This should be extended to by-products of the ethanol process, such as the just noted carbon dioxide. The by-products of the process (a DDG equivalent) will be unique to particular raw material resource, analogous to what is done in regard to use of corn producing DDG for animal feed.
- The technical program and markets for the by-products need to be developed since by-products are an important aspect to helping ethanol be profitable. For example, as noted in Chapter 5, sugarcane has several possibilities that may not have been explored as yet. Suggestions in that chapter in regard to cassava and rice, for example, point to several others. The by-products can be profit-producing feed sources for animals, or even human food sources, if sufficient creative work is carried out.
- Opportunities resulting in additional cash flow remain and need to be developed for animal feed and then human feeds linked to incorporation of beneficial microbial cultures to improve their ability to aid digestion and protect the animal against pathogens. Eventually human health will benefit from this work based in part on what is learned with animal experiments. There are products in this area. However, the products are not good enough and need to be better based on new research (DDG Users Handbook. 2012).
- The opportunity to create flavor systems within such feeds and foods is a particular enhanced opportunity. Animals, like humans, have taste preferences.

- An objective should be to maintain animal health without using antibiotics. Perhaps these culture systems can be synergistic with foods to be antimicrobial, as well.
- Certainly for human foods, no food or probiotics are sufficiently able to counter gastrointestinal infections, sometimes requiring use of antibiotics. This is a research opportunity that has great spin-off possibilities. It could rival ethanol itself as a business opportunity.
- The DDGs market is growing well beyond that of standard animal feed from corn as pointed out in Chapter 5. Microbial contamination in cattle meat is causing ailments. Antibiotics are given to promote growth and feed efficiency. However, since microbial resistance occurs, antibiotics may not helpful in preventing animal illness that is linked to proper animal growth and health.
- One source of the contamination is attributed to the grains fed to the cattle, and maybe the DDG is implicated as well (depending on the situation) in this contamination. This infection can presumably get in the meat that is eaten by humans and animals, thereby transferring the infection.
- For those with IBS, such infections are particularly woeful as the infection tends to linger resulting in gastrointestinal distress that does not go away easily. If meat is to be consumed, the bacterial load fed to the cattle must be kept down. It is necessary to find out if this is contamination coming in as part of feed, and if the animal's immune system is not performing as well as it should and is, therefore, compromised. It is suggested that this problem can be turned into an opportunity and help enhance the corn/ethanol industries' business.
- One important area related to Distillers Dry Grain Solubles (DDGS) potential microbial contamination and correction using beneficial organisms is that the presence of antibiotics in DDGS is a current controversy (DDG Users Handbook. 2012). The antibiotics pass into the DDGS (albeit at a low level perhaps) since they are added to the ethanol fermentation to cut down on microbial formation competitive to yeast. The problem is that these antibiotics may get into meat. Since the levels of these antibiotics in meat appear low, it is not a problem. The problem is that the low level of antibiotics could be a factor that facilitates contaminant organisms of a more health-harmful variety such as *E. coli* rather than spoilage organisms in the lactic acid bacteria family. If antibiotics are eliminated in fermentation of ethanol maybe there are questions about eliminating them in feed. Again this means that beneficial bacteria should be developed and used in feed including in DDGS. Until that can be done, heat treatment practices are recommended so that antibiotic use can be phased out.

REFERENCES

Budd, P. 2016. *Advanced Molecular Sieves Could be Used for Carbon Capture.* Cambridge University. UK. http://phys.org/news/2014-09-advanced-molecular-sieves-carbon-capture.html.

Choi, S.Y., Jeong, S.K., Kim, H.J. Bael, I-H., Park, K.T. 2016. Electrochemical reduction of carbon dioxide to formate on tin-lead alloys. *ACS Sustainable Chemical Engineering.* Vol 4. No. 3. p. 1311–1318.

Dawson, C. March 3, 2016. Venture seeks to cut emissions. *The Wall Street Journal.* B8.
DDG Users Handbook. 2012. *A Guide to Distiller's Dried Grains with Solubles (DDGS)*, U.S. Grains Council. 3rd Edition. http://www.grains.org/buyingselling/ddgs/ddgs-user-handbook.
Medline Plus. 2016. *Saccharomyces boulardii.* National Library of Medicine.
Munk, N. May 2016. How Warren Buffett's son would feed the world. *The Atlantic.* Vol. 117. No. 4. p. 70.
Ondrey, G. May 2016. *CO_2 to Ethylene.* Chemical Engineering. P. 8. http://www.chemengonline.com.
Platts. 2016. *Platts Global Ethylene Price Index.* http://www.platts.com.
Wikipedia. 2011. *Photochemical Carbon Dioxide Reduction.* Wikimedia Foundation, Inc. https://en.wikipedia.org/wiki/Photochemical_Carbon_Dioxide_Reduction. Last modified on September 21, 2016.
Wikipedia. 2016. *Saccharomyces cerevisiae.* Wikimedia Foundation, Inc. https://en.wikipedia.org/wiki/ Saccharomyces_cerevisiae. Last modified on September 29, 2016.
Zolfagharifard, C. 2016. Carbon emissions reach 40 billion tons high: World faces "dangerous climate change"—and China, the United States, and India are the worst offenders. http://www.dailymail.co.uk/sciencetech/article-2764323/China-US-India-push-world-carbon-emissions-up.html.

Index

D

E

F

G

H

I

L

M

N

O

P